MADE IN
☆ U S A ☆

MADE IN
U☆S☆A

THE SECRET HISTORIES
OF THE THINGS
THAT MADE AMERICA

Phil Patton

GROVE WEIDENFELD
New York

Published by Grove Weidenfeld
A division of Grove Press, Inc.
841 Broadway
New York, NY 10003–4793

Published in Canada by General Publishing Company, Ltd.

Library of Congress Cataloging-in-Publication Data

Patton, Phil.
Made in U.S.A.: the secret histories of the things that made
America / Phil Patton.—1st ed.
p.   cm.
ISBN 0–8021–1276–5
1. Technology—United States—History.   2. Industrial arts—United
States—History.   I. Title.   II. Title: Made in USA.
T21.P38   1992
609.73—dc20   91–19961
                                                                   CIP

Manufactured in the United States of America

Printed on acid-free paper

Designed by Irving Perkins Associates

First Edition 1992

1   3   5   7   9   10   8   6   4   2

Strange and hard that paradox true I give,
Objects gross and the unseen soul are one.

—Walt Whitman, "A Song for
Occupations"

Objects in mirror are closer than they
appear.

—Inscription required by law on right-
hand rearview mirror of automobiles
sold in the United States

☆ **Contents** ☆

vii

MADE IN
⭐ U·S·A ⭐

CHAPTER ☆ ONE

The World in a Package

THOMAS JEFFERSON WROTE the Declaration of Independence on a portable, folding desk of his own design. "It claims no merit of particular beauty," he wrote of the "writing box." "It is plain, neat, convenient, and, taking no more room on the table than a moderate quarto volume, it yet displays itself sufficiently for any writing."

The portable desk designed by Thomas Jefferson in 1775, on which he wrote the Declaration of Independence. (Smithsonian Institution)

Jefferson's traveling desk neatly packed his world up into a box. A tilted stand rose to support paper and pen, and a tiny drawer, cleverly opening to the side, held writing materials. It was a do-it-yourself project: Jefferson designed it himself and had it put together of mahogany by the cabinetmaker in whose house in Philadelphia he rented rooms while attending the Continental Congress.

Jefferson's desk was a model of what American things were to become: light, portable, flexible, and tailored to their owner. Doing-it-yourself required such devices; independence required the pursuit of invention. To assemble, denote, abbreviate one's world was to assert your control over it—and to take it with you. The American obsession with mobility—in space, in time, in class, in identity—so deep that it seemed to extend into the individual bones, lay behind the shape of all sorts of American things.

☆   ☆   ☆

WHAT HAVE WE made well? How and why have we seen ourselves as often reflected in things as in institutions or places or practices? Why have we sought individuality in the things we hold in common, invested ordinary things with extraordinary meaning, so often sought the spiritual in the material?

The common conviction around the world is that Americans are materialists; true, but in what way? How did devices become the vehicle of our desires—the desire at once for individuality and the desire to belong, the desires for individual power and individual distinction, so easily manipulated and apotheosized by advertising and marketing?

Our economy was built on the industrial techniques developed in the textile and clock and gun factories of New England. Today, those long disused brick buildings, punched out with great processions of brick-arched windows, house microelectronics enterprises—or museums. They have been bypassed, overgrown by condo developments or "inns" with gilded letters routed into thick plank signs—not unlike the stone walls of long untilled farms one comes across in the shade of second-growth forests.

Today, when we lament the decline of American manufacturing, decry the merely paper products of leveraged buyouts, we hear again and again that "Americans don't make anything any more."

But what is it that we used to build and make that inspired us, that we found beautiful—beautiful in shape and in engineering? What are these beauties, the beauty of the well-designed, well-thought-out, well-made thing, occupying that territory where engineering overlaps styling?

Every country has its representative objects. They are the stuff of national caricature and cliché: the Japanese transistor radio or the shoji screen. The German Mercedes or Braun appliance. The Italian espresso maker or Vespa motorcycle. Napoleon sneered at the Swiss as the people who invented the cuckoo clock, but he would surely have issued his army Swiss army knives—an equally persuasive national symbol—had they existed.

Nowhere have such symbols been as numerous, as valued, and as sought after as in the United States. We have shown an almost desperate hunger for such objects of self-definition, objects to sum up our character, epitomize our industry, mark our history. Presenting his original folding desk to his granddaughter's husband, in 1825, Jefferson wrote: "If then things acquire a superstitious value because of their connection with particular persons, surely a connection with the great Charter of our Independence may give a value to what has been associated with that. . . . Its imaginary value will increase with the years." He went on to envision the desk being "carried in the procession of our nation's birthday, as the relics of the saints are in those of the church."

But something more than superstition or imagination lies behind the national inclination to see larger meanings in ordinary things. Only in America do we, as a matter of convention, call our favorite objects "icons." The term has been expanded to cover trademarks, national monuments such as the Statue of Liberty, movie stars, and political figures. In its original meaning, of course, icon already suggested a physical representation replacing a spiritual truth. The iconoclasts of eighth-century Byzantium destroyed icons through the traditional proscription of Judaism—and Islam as well—against graven images. To its enemies, the icon represented the graven image warned against in the Commandments.

To deny the power of the icon, whether under the guise of "anti-materialism" or the Marxist critique of the "alienated object," is a latter-day form of iconoclasm and part of a certain hard Puritanism that lives on in American intellectual life—a relative of twentieth-century liberalism and a descendant of the anti-materialism of early New England. With our strong resistance to abstract ideas, sentiment and conviction have often grown solid around our attachment to objects, condensing on them like dew on grass.

Our icons tend over time to recede into gauze of celebration and nostalgia, like old dogs, buildings, and politicians. They need to be periodically reexamined—disassembled into their contradictions and complexities hidden beneath the sleek shells of mythology.

☆   ☆   ☆

TWO GREAT IDEALS shaped American design: that of the perfect model and that of the kit of parts. Objects such as Jefferson's desk have themselves been declarations of independence. American design has always sought to enable the individual to distinguish and improve himself. Even when mass-produced and universally owned, American objects were aimed at magnifying the power of the individual. The creation of standard, "democratic" models of products mediated the relation of individual to the society.

The melting pot may be the crucible of American design, but it not so much melded and molded foreign shapes into a common American one as jumbled the shapes into an eclectic stew and sought the common denominator. American design borrowed frequently and forgot to footnote its sources, while wrapping them in the mystique of the frontier. The log cabin, a Swedish and Finnish innovation, was seized upon by the larger British and German populations as the practical housing solution. It was not, however, until the 1930s that the myth of Jamestown and Plymouth colonists living in log cabins was demolished.

Our most prized objects have often been general types that could be individually customized. Because American objects tended to be aimed at large markets and produced in standard forms by machine, they appeared democratic from the standpoint of the large group but, crucially, not from that of the individual.

It was individualism we sought, far more than democracy. Characteristic American objects—the shapes and forms that have become iconic—declare the independence of owner or user more than they express participation in any social or political unity. They rarely represent claims of membership in a mass or common people. Even when they are "plain" or "basic," they are focused on the simple virtue of the individual. Even when they are mass-produced, they entice the single customer with their powers.

The story of our artifacts is the story of the long and never resolved struggle between the basic model and the customized object, between the unadorned object and the appliqué kits of decoration, between the engineer and the salesman, the sunflower and the gilded lily. National infatuation with Henry Ford's uniform Model T was succeeded by General Motors' success in selling variety—creating thousands of little infatuations out of the great infatuation with choice. Other companies continue to copy its example; "marginal product differentiation," they call it. The press release of a contemporary linen company boasts that its patterns of sheets, bedcovers, and other fabrics could be combined in so many different ways as "to make individuality a virtual mathematical certainty."

The personal computer has been wonderfully described by one of its leading developers as a "fantasy amplifier." So are cars and houses and guns and clothes; they are about power and imagination and social status. They do for our mental categories what simple machines do for physical power: they are levers of expression.

To heighten the power of the individual is a very different thing from "democratizing" the object. "Democratizing" is the means to the end: distinguishing the individual from the uniformity of the demos and making him himself wealthier or happier or more stylish than his fellows, or at least to be first—first on the draw, first to stake his claim, first on his block with a dishwasher or color TV. To keep up with the Joneses implies Mr. Jones's initial effort to stand out.

The irony of the American object is that it asserts the power and potential of the individual while possessing the generality of the type. That these efforts have in general only tended to make people more alike—that this common American drive has smoothed away the non-American differences with which many of them began, melting the ethnic and even the regional into the national market—is just one of the paradoxes of the process. The others, we know, are the mixture of efficiency with crassness, depth with superficiality, kit with kitsch. Americans have always possessed a remarkable ability to magnify their own sense of distinctiveness, despite overwhelming evidence of their sameness. There has always been a conflict between the pursuit of individual happiness and the equalizing factors of democracy. The equalization makes the magnification of the individual possible, but the pursuit of individual aggrandizement threatens democracy—that is the antiphony played out over American history.

☆   ☆   ☆

JEFFERSON KEPT HIS desk for fifty years, took it to France with him, to the shifting capitals of the United States, and finally to the White House. Geography—the extent of the country and its primitive transportation system—helped generate a desk like Jefferson's. Transportation has always been a dominant need and determinant of American life. Ships and boats, railroads and automobiles, aircraft and spacecraft, have been the means through which the shape of the land itself has shaped American technology and culture. The mobile life inspired a host of inventions and designs. What more characteristically American innovations are there than the railroad bogie and the cowcatcher? The bogie, the work of the early railroad developer Ross Winans, was the combination of twin axles on a swiveling base to help keep cars on the uneven and twisting rails of American tracks. The cowcatcher, the projecting plow at the front of the locomotive, testifies to the ingenuity of Isaac Dripps, a New Jersey mechanic of the 1840s. It did not catch the cows, of course, but pushed them and other obstacles off the tracks in front of the locomotive. In England, the track would have been fenced, but in the United States there was no money for that, and barely enough to build railways at all over great distances. Animals were often unfenced in those days, except in the more permanently settled areas of the country. Later, in the West, the railroad finally was enclosed— by the new machine-made fence, barbed wire, an American invention for which the railroads were the largest customers.

Distance demands transportation, which demands portability, which demands a special sort of object. Many American things unfold, from Jefferson's desk to the chuck box that Texas rancher Charles Goodnight stuck on a wagon in the late 1870s to feed his cowhands, to George Pullman's train berths, to the latest portable laptop computers—the electronic successors to Jefferson's desk.

The plank berth of the canal boat, unfolding like Jefferson's desk-top, was the size, Charles Dickens wrote in an 1842 letter, "of this paper—I know. I measured it." It was an unfolding bed even nar-rower than Jefferson's "moderate quarto volume."

That berth was the ancestor of George Pullman's bed, despite the legend that the folding beds in Pullman's railroad cars were in-spired by those in the miners' cabins of the Colorado Silver Rush of the 1850s. By then, Pullman had already lost a fortune touting a previous design and other inventors had worked on the pulldown bunk as well. But the story is appropriate nonetheless. The key was not invention but publicity and marketing. His first fancy railcar, the *Pioneer*, was made famous when it carried Lincoln's body home from Washington to Springfield.

American objects move, pack up, are portable. As Reyner Ban-ham puts it: "The man who changed the face of America had a gizmo, a gadget, a gimmick—in his hand, in his back pocket, across the saddle, on his hip, in the trailer, round his neck, on his head, deep in a hardened silo. The typical American way of improving the human situation has been by means of crafty and usually com-pact little packages."

☆   ☆   ☆

THE DRAWERS OF Jefferson's desk might have held sheets of his beloved graph paper. This paper was a French development—as much as the Cartesian system of coordinates or the political ideas of Montesquieu—but Jefferson's applications of it were distinctly American: generating vast schemes to survey and settle the conti-nent as well as serving his own spare-time architectural drawing. At the desk he helped lay the foundations for the land grid of the 1797 Land Ordinance, a self-generating system for selling land and cre-ating local political units, a sort of mechanical quilt on the land. It was the first of the great national grids that were to reticulate the continent: water, rail, highway, electricity, telephone, television.

Many of Jefferson's gadgets were acquired, like the polygraph, in Europe. The Duc de la Rochefoucauld-Liancourt, the French linen magnate and a cosmopolitan traveler, visiting Monticello in 1796, wrote of Jefferson that "his travels in Europe have supplied him with models." The writing box probably had a French model.

On his traveling desk, too, Jefferson studied the Old World for clues to making the New one work better. His idea for an improved plow came to him in Europe, where he was constantly noting small practical details of life in his notebooks—a gate in Brittany, a Dutch method for shaping the beams of houses. They are strange jottings, as if by a visitor from another planet, to whom all earthly novelties seem equal in importance.

But all the European models were modified to local construction and local needs; they were domesticated. Among Jefferson's most important popularizations was his Francophilia. After one of his trips to the Continent, he added to Monticello skylights of the sort he had seen in Europe, ripping out a whole room above his bedroom to make way for them. A plaster cast of the Maison Carré, the Roman temple in Nîmes, was shipped home to serve as the prototype for his design for the Virginia state capitol.

Over his lifetime, Jefferson designed and had several more portable desks built. They each contained a copying machine—first a copy press, which used a sort of crude carbon paper to reproduce a written page; then a polygraph, a mechanical copying apparatus that, through an elaborate series of joints and hinges, drove a second pen over paper to duplicate the writing of the first. They were desks turned into machines. His obsession with such devices led him to write more than a hundred letters—some of them while president—to Charles Willson Peale on the subject of copying, which was a fascination shared by James Watt, the inventor of steam engines, and Benjamin Latrobe, the engineer and architect.

Jefferson's desks were of a piece with his other gadgets: a rotating bookstand held open several volumes at once—"He would read one then t'other," reported his slave Isaac Jefferson. His swivel chair, a modified Windsor, enabled the omni-talented, omni-interested, and omni-occupied man to swing quickly from one task to another. His dumbwaiter and turning cupboards shielded guests from encountering the "servants," as the slaves were called. His bed, poised between bedroom and study, was watched over relentlessly by a clock and could be lifted like a Pullman berth to the ceiling. His chaise longue, equipped with a standing desk, allowed him to recline and work at the same time. All were devices for the magnification of the individual, machines for expanding the effect of one man's thoughts and intellect, and aimed at the great American ideal, convenience. In this they presaged a whole line of American devices.

The gadget is always a mechanical shortcut for reducing labor, which has always been the great pursuit of the American mechanic, at the factory scale or the scale of the individual in his den. It may seem odd for a man as little given to rest or recreation as Jefferson to invent for convenience, but it is convenience that frees time and energy—just as all the modern conveniences "liberate" the housewife.

Jefferson saw the universe as basically mechanical. He was a man with a modern sense of time. In the age of the rooster and church bell, it may have been shocking to wake up, as Jefferson did, so that the first thing you see is a clock. But today most of us wake up to the sound as well as the luminescent image of an alarm clock or radio.

Jefferson's love of gadgets, as well as his determination to record his letters, was so obsessive that it was almost pathological. He noted seemingly every penny he spent, but when he left the White House he did not realize that he was thousands of dollars in debt. He was the kind of guy who would have read *Popular Mechanics*. If his goal—not just an American one—was to create order in a disorderly world, as the most common explanation has it, there was also a flight from reality in the process. Some of the gadgetry—the dumbwaiter and the revolving serving facility, for instance—served to hide the fact of slavery from his consciousness, even while much of this furniture and equipment was fabricated by Jefferson's slave Thomas Hemming. Jefferson's chairs and desks were the forerunners of the patent chair in Melville's *Confidence Man*, designed to assuage not only the aching body but the troubled conscience.

Jefferson was a model, both conscious and unconscious, for future creators of American objects. His chaise longue inspired the designer Niels Diffrient's "Jefferson chair," with its seeming dozens of adjustable surfaces, all "ergonomically" shaped, and its swiveling arm for a computer monitor. But even the La-Z-Boy lounger in which Archie Bunker watched television is in some ways a bastard descendant of Jefferson's chaise.

Mark Twain was fascinated by the multitude of mechanical devices he found at Alexandroffsky, the Baltimore mansion of Thomas De Kay Winans, son of Ross Winans, the railroad pioneer. Visiting Winans in 1877, Twain wrote a thirty-two-page letter to his wife reveling in the gadgetry, which included a revolving dining table, a water-powered pipe organ, and an aquarium for raising brook trout from egg to table. This impulse later drove Twain into the disaster of the Paige typesetter, a mechanical wonder that drained his finances and optimism. Twain's fascination with the Paige machine, fabricated at the Colt arms works, was Jefferson's dream of the practical polygraph turned by the application of capital into a nightmare.

Even as Jefferson contemplated it from his desk, the process of mechanization was already at work in the mills of New England and at Paterson, New Jersey, where Alexander Hamilton and Pierre L'Enfant laid the basis for future American industry. L'Enfant designed raceways (an industrial approximation of the boulevards he laid out for Washington, D.C.) to provide water, liquid motive power, from the Great Falls of the Passaic, so that in years to come emigrants could sweat in the hot factories making silk dresses, so Samuel Colt could have a place to manufacture his first revolvers, and so the first locomotives in the United States could be built.

Such fascination with the mechanical, we now know well, destroyed the independent agrarian life that Jefferson idealized.

One of the opening titles of Chaplin's *Modern Times* blazons "a story of independence, of individual enterprise, humanity crusading, in the pursuit of happiness." Then Chaplin cuts to the famous montage of factory workers and sheep, token of an industrial system in which the drive for independence has been turned into dependence and the pursuit of happiness diverted to the satisfaction of invented needs.

Such was the landscape H. L. Mencken saw through a train window, one day in the twenties. Long, dreary rows of workers' houses outside Pittsburgh, punctuated with hideous churches and VFW halls, convinced him that the American possessed "a positive libido for the ugly." But he sounds oddly like early American travelers who saw in the scruffy cabins and stumps of frontier farms a characteristic national love of squalor. Each reaction is born from disgust with the haste and greed of industry—both individual drive and machine production.

☆   ☆   ☆

JEFFERSON'S VISION OF a nation of farmers unspoiled by what he believed to be the inevitable corruption of cities and industries was at war with his love of gadgets. He was fascinated by steam engines, which would shatter any semblance of the quiet agricultural country he imagined. The notion of a bucolic life made more secure by the introduction of the proper machines was refuted by the cotton gin, which buttressed slavery in the South. The yeoman became the blue-collar laborer, subjected to the squalor of "industries" Jefferson feared.

For all his gadgets, Jefferson remained suspicious of most inventors who went beyond the do-it-yourself. As the first Secretary of State he had initial responsibility for the U.S. Patent Office, and he may have signed some of the first American patents on his writing box. Despite his duty to sign them, Jefferson disdained patents as cheap commercial devices. He shared with Benjamin Franklin the belief that inventions were like gifts from heaven, bestowed for the benefit not of a single technician or capitalist but for everyone. But American patent laws offered little protection to inventors in any case; they guaranteed that the list of frustrated, impoverished, and swindled American innovators would grow far longer than the popularly taught register of successful inventive heroes. Eli Whitney initiated some sixty suits in defense of his cotton gin patent, but still made very little money from it. Americans ultimately decided to let the marketplace adjudicate patents—which the sewing machine combine did, often to the dismay of the small inventor.

So little respect did Jefferson, the inveterate borrower, have for patents, that he was sued by one early inventor: Oliver Evans. Jefferson claimed Evans's automatic grist mill, a remarkable assemblage of belts, pulleys, and hoppers that took whole grain straight out of the wagon, ran it through the stones, and poured flour out the bottom, was simply a combination of previous inventions. In a sense, this was true: Archimedean screws were used to transport the grain, but it was the first time they had been applied to solids instead of liquids. The question touched on what was really innovative in any design.

On the other hand, from Jefferson's desk came letters recommending the federal government's award in 1798 of a gunmaking contract to Eli Whitney, who built his famous muskets from interchangeable parts. (Jefferson had suggested such a scheme earlier, after hearing of the similar work of a Frenchman named LeBlanc.) Jefferson was thus a midwife to the birth of the "American system of manufacture."

☆　☆　☆

THE SYSTEM ASSOCIATED with Whitney's name became known as the "armory system," or ultimately the "American system." Behind the innovation lay the lack of skilled craftsmen and the abundance of inexpensive land for agriculture, which drained off potential workers. In Europe, over and over again, artisans resisted the adoption of technology that would displace them. LeBlanc's method was not adopted in France. In the United States, however, it was essential, as Whitney wrote, to resort to a system designed "to substitute correct and effective operations of machinery for that skill of an artist which is acquired only by long practice and experience."

So at first interchangeability was a means to fill a gap, not an end in itself. While the British form of the Industrial Revolution featured the application of power of coal to textiles and mining and later other processes, American technology depended on two key principles of industrialization: the production of identical, interchangeable parts (the "American system"); and, later, the movement of parts during assembly.

The key to the so-called "American system," explains David Hounshell, one of its finest historians, was the use of "a rational jig, fixture and gauging system." As Hounshell points out:

> The System was "rational" because it was based on a model, which in one sense can be interpreted as a kind of Platonic model in that armsmakers viewed the model weapon as an ideal form. All production arms were but imperfect imitations of this ideal (but real) model.
>
> Where dimensions and fits were critical, gauges were made based on the model, or ideal form. With such designed gauges and fixtures, parts produced in machine tools approximated comparable parts of the model.

The matching parts of each example of a product based on the model constituted a kit of parts, which in the classic assembly-line method of Henry Ford are brought together by transportation systems—the conveyor belt and overhead rail.

☆   ☆   ☆

THE AMERICAN SYSTEM also changed our sense of the object itself. At least as surely as the work of art is different in the age of mechanical reproduction—in the ways Walter Benjamin asserted in his famous essay—so the ordinary object is different in the age of its mechanical reproduction. The thing that was once a handmade product, unique in detail if not in general form, is transformed by machine and repetition into a *specimen* reflecting, a Marxist like Benjamin would argue, the alienation of the laborer in the alienation of the fruit of his labor.

In this system, the worker was as much an interchangeable part as the pieces of the things he made and, later, bought. The more the individual was seen as a democratic atom, an abstract worker, an abstract consumer, the more he turned to buying products that would distinguish him from others and display his individual character and differences. The lack of "particular beauty" in the product was soon remedied by the application of differentiating decoration.

A larger measure of social equality produces a proportionally enlarged compulsion for the individual to distinguish himself. Social mobility had as powerful an effect on the shapes of American objects as physical mobility. Joseph Whitworth, the English industrial genius who visited the United States in the 1850s, was to note a "tendency toward display" in American designs—by which he meant gaudy decoration. Edgar Allan Poe came to a similar conclusion in his essay "The Philosophy of Furniture," and offered an explanation with which de Tocqueville could have been satisfied:

> We have no aristocracy of blood, and having therefore as a natural, and indeed as an inevitable thing, fashioned for ourselves an aristocracy of dollars, the *display of wealth* has here to take the place and perform the office of the heraldic display of monarchial countries. By a transition readily understood, and which might have been as readily foreseen, we have been brought to merge in simple show our notions of taste itself.

Poe foreshadowed General Motors' Harley Earl, who was to argue that the tail fin gave the buyer "a visible premium for his money." No wonder "class" in America has been regarded as a quality to be assumed; such phrases as "he has class," or, "to class up the place" are more familiar and characteristic than any idea of belonging to a class, whether upper, middle, or working.

This upward-mobile side of the decorative tendency may seem to fall outside the Jeffersonian vision of America. But Jefferson, the man as opposed to the thinker, decorated his life with all sorts of

Europeanisms. In one sense, it is a consequence of industry, which could, and eventually was forced to, make new things constantly to replace old ones. Industry also made possible the application of all sorts of ornament: raised from cast iron, jigsawed or lathed into wood, stenciled on any material. Democracy makes obsolescence and decoration necessary: the individual has to struggle harder to distinguish himself, through better and brighter and newer models.

☆　　☆　　☆

WHAT IS STRIKING about all the various writing boxes Jefferson designed is their obsessively neat and specific compartments. Here is the list he drew up of the contents of one of them, carefully keyed to his drawing:

> wafers, sealing wax, seal, hair . . . box, shaving brush, ink phial, ink pot, dividers, drawpens, pencil, rubber, thermometer, magnetic needle, scissors, knife, whetstone, scale, razor strap, razors, toothbrushes, comb, night cap, ring dial, damping brush, scales, monetary steelyards, tape measure. . . .

It is a catalogue of personal trivial items—the kit of a life in a box—with overtones of nonsense: "Of shoes—and ships—and sealing wax— / Of cabbages—and kings." But for Jefferson, who mixed diplomacy to royal courts or worries about American naval strength with agriculture and domestic invention, it was a life in summary; more than an office away from home, it was Monticello in a box. Monticello, writes Jack McLaughlin, the most sensitive interpreter of Jefferson's home, is like Wallace Stevens's jar in Tennessee. The quintessential American object is like that jar, too: a work of artifice in the midst of nature, radiant of meaning and reference, alone but interactive.

> I placed a jar in Tennessee,
> And round it was, upon a hill.
> It made the slovenly wilderness
> Surround that hill.
>
> The wilderness rose up to it,
> And sprawled around, no longer wild.

Mason jar, 1858.

The jar was round upon the ground
And tall and of a port in air.

It took dominion everywhere.
The jar was gray and bare.
It did not give of bird or bush,
Like nothing else in Tennessee.

That jar is some decayed descendant of your good old Grecian urn—with its conflation of beauty and truth—but it also reminds me of the memorial urns of early nineteenth-century paintings, wrapped in sprays of weeping willow. It hovers in association somewhere between the Leyden jar in which Benjamin Franklin collected the clouds' kite-gathered electricity and the opalescent utility of Tupperware.

Our objects are models of ideas; they are kits of associations. American design is animated by the tension between the model—the single, universal appliance—and the kit—the set of parts from which one constitutes one's own version of the thing. There is a complex interplay to the relationship: a dialectic between the individual and the collective. When it enters the life of use, the universal object can become a personal one; collective designs can be used as a kit from which the individual can assemble his identity, becoming machines for self-improvement, self-expression, and self-invention.

Manufacturers, packagers, and industrial designers have traded on this tendency; advertisers have exploited it. The irony is that the evocation of power and possibility for the individual has been the most effective means of selling things in massive numbers. It may be true that there is no business in America *but* show business; most of the things we have made have been their own packages. Packages themselves have been a national obsession. We have been fascinated with the jar, the box, the barrel, the paper bag, the tin can (from its standard cylinder to the cam shape used for meat), the ice cream "bucket" later adapted for Chinese food and fish bait, the milk carton with its little roof, and of course the Coca-Cola bottle, that best known icon of the American way in a package. But what is also basic to packaging is the mysterious relationship of surface to solid, of the die-cut box to its assembled form, of the individual specimen to the mass—a series of relationships whose exploration was to become the stock in trade of Pop and other sixties artists.

The package is a dominating national metaphor; celebrities even hire firms to package themselves. In America, common things themselves have enjoyed a special status as packages in a different sense—containers for emotions, aspirations, and associations. De Tocqueville's theory of the American as a creator of social associations could be extended to our regard for the objects of daily life as well as the structures of social life. We defined ourselves by association with objects, and objects by their associations.

The associations of the object are about individualism—an idea and a word still so unfamiliar in the English of the time that the first translator of de Tocqueville felt the need to apologize for its use. Europeans early on noticed what they perceived as the materialistic side of American culture. But they had it backwards: it was a spiritualism of matter. And if, as Europe noted, we tended to regard simple packages as complete things, it was because we regarded things as mere packages. "Americans," declared Gertrude Stein, "are materialists of the abstract."

This was true by virtue of the search for individual identity and power—the insertion of the I between art and fact. We have made common objects models of more than simple practicality, even as we praise their practicality as the main thing. They have transcended not just physical reality but their own designs, to become part of our mental furniture, Gestalts far more deeply rooted in our thought than the pure shapes of geometries. But even the pagan iconology of our objects is leavened by a certain spirituality, a diffuse and precipitate religion—God is in every thing. "Godjets," Reyner Banham appropriately nicknamed our household devices, the little gods of the kitchen, mechanical and electronic lares and penates.

So many objects live an elegiac life in our memories and perceptions. Given the location, Stevens's gray jar might be crudely fired country pottery; but it is also easy to imagine it as a Mason jar, commonly used to hold fruits and vegetables, whose colors slightly heightened by pressure cookers suggest Kodacolor versions of themselves—as well as moonshine whiskey.

"Patented November 11, 1858," those jars used to say. Two billion have come out of the factory of the Ball Company in Muncie, Indiana, the model for "Middletown," and other companies, to cover the world. They were the invention of John Landis Mason, a tinsmith who made the first jar tops—the top was the key advance—at 257 Pearl Street in New York. (One rival called him a man "under medium size, dark complexioned, very nervous, a heavy drinker . . . a sort of a crank and very hard to get along with," according to a historian of glass jars in a text titled "Men Who Made Fruit-Jar History.") The jars were made by glass blowers who produced molds from Mason's patent specs and paid him a royalty. Clayton Parker, hired by a man named Samuel Crowly in southern New Jersey, supposedly blew the first one.

Mason's jar, which bore his name from the start, and the date of his patent, was a characteristic American device: a container, a package, capable of magnifying the power of the farmer and housewife, with a patented mechanical innovation (the screw top) for the do-it-yourselfer, a model and, with its standardized parts, a kit as well. If it was not the machine in the garden, it was a machine-made object for containing the garden, in which we saw ourselves darkly reflected and distorted, as in a Mannerist painting.

The Mason jar, whatever the brand name under which it was produced, was sold as a package in itself. Its successors, such as the milk bottle, with its characteristic silhouette, or the folding box used alternatively for goldfish and Chinese food, were objects as familiar as the Mason jar but still unassociated with a brand name.

The American propensity for packaging, which began with such literal innovations as the Mason jar, the paper sack, and the cardboard box, would eventually reach the abstract implications with which we are familiar: movie stars and candidates for office are "packaged," presidents raise public ire with new "tax packages," and on third down passing situations football coaches send in their "nickel and dime packages" of defensive backs.

Packing things in new ways became important in the United States because it was needed: the geographical dispersion of the population and the development of the mass market made it necessary. A whole range of unsung heroes of packaging emerged to create the shapes with which we were to become familiar.

In order to distribute beef from the American West in the days before refrigerated railcars, the tin-canning industry boomed, helped along by the Civil War's demand for transportable preserved foods. After the war, the corned beef can, with its tapering shape to allow for removal of the contents, became a familiar image, thanks to Libby, Armour, Swift, and other meat magnates, whose fortunes grew at the same exponential rates as those of the railroad kings whose systems distributed their products.

For the new department stores, the paper bag was a vital symbol of the fixed price, the neatly wrapped and standardized transaction that replaced the haggling of the small general store. In 1852, Francis Wolle patented the first paper bag–making machine; his Union Paper Bag Machine Company was to seize 90 percent of the market.

Luther Childs Crowell, a frustrated airplane designer fifty years ahead of his time, patented his own machine to make paper bags in 1867, then refined its design to produce the square-bottomed bag that became universal in department stores and later supermarkets. In the 1880s, Charles Stillwell developed the "self-opening" sack, with pleats or gussets, ready to snap open at the flick of the salesclerk's wrist.

Another paper bag maker, Robert Gair, watched his workers printing seed bags in 1879. A sharp raised rule in one plate cut the paper in addition to simply printing it and, the story goes, inspired Gair to develop the die for cutting a foldable cardboard box. He was hailed as "the father of the folding box," and by the end of the nineteenth century Gair City in Brooklyn was turning out millions of them.

In the way these warm brown packages turned almost miraculously from flat objects to solid ones, there was something deeply American, at once clever and false, like an instant Western town with false fronts and tents behind, or like the instant culture adopted by an uneducated industrialist grown rich, raising his European-inspired mansion like a circus tent.

With the invention of the machine-made container, the familiar printed packages of what, beginning in the 1880s, would become national brands followed: National Biscuit's Uneeda Biscuits, Kellogg's cereals, soap powders such as Sapolio and Gold Dust, and previously bulk products like Quaker Oats. With such commercial changes, packaging had superseded packing; the container and the contained entered into a new sort of relationship. Packaging changed what was inside, implied a special status for what might previously have been a generic product—"branded" it into something different.

Original prototype of Coca-Cola bottle, Earl Dean probable designer, 1915. The bottle was narrowed to fit existing bottling equipment and boxes, producing the familiar shape. (Coca-Cola Company)

The primary example of the package that became virtually indistinguishable from the product it contained was the Coca-Cola bottle, which achieved the fame previously enjoyed only by generic types of containers such as the Mason jar.

The Coca-Cola bottle demonstrated the power of sheer ubiquity to imprint an image and the sense of an object on the collective mind. It arrived when the drink itself was twenty years old. The beverage, its logotype, its diamond-shaped label, and its name were widely imitated by dozens of other colas. Coke's lawyers found that the variety of local bottle shapes and labels did not make their task of patenting easier. The impetus for creating a standard national container for Coke was to resist imitators.

Before the standardization of the Coca-Cola bottle, the beverage created by "Doc" Joe Pemberton (a patent medicine salesman who had previously pushed such compounds as Triplex Liver Pills) had been dispensed from soda fountains and distributed in a variety of bottles varying with the local bottler. At first, regional licensees such as Joseph Biedenharn of Vicksburg, Mississippi, bottled Coke in wire and stopper bottles. By 1892 the Crown cap, the familiar crimped metal seal, had appeared and the bottle narrowed its neck to accept it, producing something like the generic soda bottle of today.

One major bottler, Benjamin F. Thomas, argued in 1910 to fellow bottlers that "we need a new bottle—a distinctive package that will help us fight substitution . . . we need a bottle which a person will recognize as a Coca-Cola bottle even when he feels it in the dark."

In 1913, the company requested proposals for a distinctive, national bottle. In 1916, a convention of the loosely federated licensed Coca-Cola bottlers met in Atlanta and chose, from among several candidates, the design proposed by the Root Glass Works of Terre Haute, Indiana.

Root's Swedish plant manager, Alex Samuelson, headed up the creation of the design. But Earl Dean, a machinist and moldmaker, seems the more likely drafter of the shape itself, and an accountant named Clyde Edwards claimed to have shown Dean encyclopedia entries on the shapes of the coca and cola plants. The inspiration for the ridges is said to be the coca bean, but it is more likely the coca bean pod—which is not involved in the beverage at all. It may also have been inspired by the fashionable "hobble" skirts of the day.

The first proposal was much fuller than the final shape—imagine the familiar bottle crossbred with a Chianti bottle. But it did not fit the bottling machinery or crates of the day and so was slimmed down. To our eyes today, the revised version is a dramatic improvement. But had the bottle as originally designed fit the functional requirements, might it not have come to seem as familiar and beautiful as the version actually produced? Samuelson was credited on the patent, which was not issued until Christmas Day, 1923, but C. J. Root received a royalty of a nickel a gross, which was to make him the richest man in Indiana: in the next forty years, 6 billion bottles would be produced.

Even when appreciated by the eye rather than the hand, the tactile quality of the bottle—Benjamin Thomas's ideal of recognizability in the dark—was a major part of its appeal. Thomas Lamb, the designer known for his ergonomically shaped knife handles, praised its fit to the grasp. In addition to echoing, in reverse, the fluting of a classic column, the ribs had a practical virtue. Condensation drips down the grooved sides, allowing for a sure grip even on an August day in New Orleans. This practical feature, and the fact that the bottle's glass had been made thicker in areas on which it would strike when dropped, made the Coca-Cola bottle, it seemed to Raymond Loewy, "a masterpiece of scientific functional planning," as well as "a transcendental glass package." Loewy, who was to design soda fountain dispensers and other items for Coke, gave abundant evidence of wishing that he had designed the bottle itself. He had to settle for claiming authorship of the Fanta bottle, an elegant shape, but one that could not hope to compete.

Today, the fact that Andy Warhol and Robert Rauschenberg made use of the Coke bottle as a "found" object, framed in quotation, is testimony that it has achieved, like the famed Campbell's soup can, a status that in an earlier time would have been reserved for a natural or a generic-made object.*

A number of science fiction stories—and several advertising take-offs—turn on the idea of future archeologists unearthing a Coca-Cola bottle and being baffled by its shape and purpose, with the confusion of a contemporary student confronted by an amphora

---

* Andy Warhol, who made as effective use of the Coca-Cola bottle in his art as he did of the soup can, summed up the universality of the beverage this way: "A Coke is a Coke, and no amount of money can get you a better Coke than the one the bum on the corner is drinking. All the Cokes are the same, and all the Cokes are good. Liz Taylor knows it, the President knows it, the bum knows it, and you know it."

obviously unable to stand upright. The joke, of course, is that the
bottle is today so universally recognized that the opposite state of
affairs is not only humorous but almost unimaginable.

But there is no more dramatic use of the Coke bottle as a symbol
of Western civilization (or "Coca-Colonialism") than in the film *The
Gods Must Be Crazy*, whose whole plot spins out of the dropping of a
Coke bottle from a plane. What is it? Where did it come from? What
is it useful for? What does it mean? The landing of the bottle affects
the local bushpeople the way the landing of a flying saucer would
affect people in, say, Kansas. The Coke bottle in Africa is the jar in
Tennessee writ in global characters.

☆   ☆   ☆

FOR MANY YEARS, the location where each Coca-Cola bottle was
produced was molded into its base. The bottles were interchangeable among local bottlers and it became common practice for children to search for the most exotic origins. This, its particular green,
the chips and abrasions of its base, gave every bottle a sense of
history and mystery. You could look into the green the way you
looked into amber, seeking signs of the past. You would wonder
what lips had drunk from the bottle before yours. In the fifties, the
company estimated that each bottle during its lifetime would be
filled, emptied, and refilled an average of thirty-seven times.

In the old family house in North Carolina where I spent summers
as a child were assembled Coke bottles molded with especially distant locations. Beside them were ranked Mason jars full of vegetables
and jellies, their contents extracted from the vegetable garden strug-
gling against bird-filled blackberry bushes of the slovenly back lot.

That house was full, too, of older Mason jars and ancient whiskey
bottles, shaped like log cabins or beehives, of brown pottery churn-
ing jars and mountain baskets. It was also, however, full of objects
that were potential containers of family and historical meaning:
ancient tools, a flintlock rifle with powder horn, dozens of handmade ladderback chairs. And the strangest were the objects recently become old, such as the telephone, a Henry Dreyfuss design
of 1930 not so very different from the fifties models I knew, but
somehow a world away. The vacuum cleaner was an Electrolux that
looked like a lounge car on the *Twentieth Century Limited*. The refrig-

erator was a Raymond Loewy Coldspot, and it had run unfailingly
for a quarter of a century, deficient only in its capacity for ice cream.
My grandfather's huge La-Z-Boy sat in front of the gigantic brown
cube of an early television set—simulated wood grain, lacquered
and polished to a bright, impenetrable surface. Those objects were
already history to a child, but they did not yet have histories for me.
They were strange and fascinating containers, without contents.

☆　　☆　　☆

THE RULING MYTH about American designs is that they sprang
magically from the wilderness, the product of atavistic ingenuity in
face of an unprecedented landscape and experience, with few overseas sources. The "new man" makes "new things." Restored to contact with wilderness, the once again "natural man" makes "natural
things." The corollary is a myth of equally deep roots: that American design is naturally "plain" and "functional."

John Kouwenhoven's landmark book *Made in America* (1948) argued that what made American design special was that it was not
conditioned by ancient traditions of race or class. Kouwenhoven was
a major iconoclast of the high art prejudices of his time, but a tone
of condescension creeps into his accounts. He struck the note that
has dominated the discussion of American aesthetics ever since
when he described the "frequently crude but vigorous forms in
which the untutored creative instinct sought to pattern the new
environment."

Kouwenhoven saw conflict between the vernacular and the cultivated traditions. The best American designs constituted "a unique
kind of folk art," not a craft folk art but a technological one. He
described them as "the folk arts of the first people in history who,
disinherited of a great cultural tradition, found themselves living
under democratic institutions in an expanding machine economy."

It was natural, the argument went on, that this sort of aesthetic
should be unornamented. Vernacular, he wrote, was "likely to be
marked by constraint and simplicity . . . no room in such a tradition
for diffuseness, no resources to spare for the ornate. It was merely
sound sense to design a thing as economically as one could."

If Kouwenhoven's readers of 1948 had visited a department
store, they would have found a very different story. It would have

been almost impossible to find there inexpensive things without ornament—without decals, mock wood grain, chrome. All offered proof of how easy it was to decorate by machine: just as easy to cast, for instance, iron stoves with elaborate ornament as to cast plain ones. The same was even more true of plastics. There was no conflict between the application of ornament and the simplicity of mass production. Such design would have seemed far closer to Whitworth and Poe than to Kouwenhoven.

The same year that Kouwenhoven published *Made in America*, Siegfried Giedion, a refugee of the Bauhaus who came to Harvard thanks to his friend Walter Gropius, published his classic *Mechanization Takes Command*, designed in part by Herbert Bayer, who taught at the Bauhaus.

The gray bulk of Giedion's book is packed with illustrations of art: the wire models Frank and Lillian Gilbreth created to reflect the shape of workers' tasks are echoed in a Paul Klee reproduced nearby. A suspended bench of the type known as a "glider" is counterpointed by an Alexander Calder mobile. The implication is clear: these objects are found art—"anonymous history" of art, set beside high and identified art. For both Kouwenhoven and Giedion, American art was without known artists, the incidental result of the search for the efficient. But they persisted in presenting the vernacular as a parenthesis inside the main marching clause of art—art, that is, by artists.

Much of the evidence quoted by Kouwenhoven, Giedion, and others is commentary by European observers, marveling at the simplicity and directness of American equipment. Kouwenhoven and Giedion's argument that American design was proto-modern, innately functional and plain, was only part of the story. They depicted the American as the romantic natural man, the new man of the New World who, facing a frontier without precedents, invented shapes and symbols without reference to the past, with no time for indirectness or ornamentation. This is the myth of the New World as a Lockean tabula rasa, and the American as a recapitulation of the first man, a Robinson Crusoe. "In the beginning," John Locke had written, "all the world was America."

But if the new man was Robinson Crusoe, we should not forget how often he swam back to the wreck of his ship to salvage supplies. The American had Europe to cannibalize from. For the American, the Old World was not only a standard and a model—inspirational in some respects, cautionary in others—it ultimately became a kind of catalogue to be perused, a great kit of traditions and possibilities to be ravaged. With no direct tradition, all traditions became possible. And if man encountering a primitive situation again had no spare energy for decoration, as Kouwenhoven argued, it was equally true that he had no reason to expend energy to change anything of the Old World that worked in the New.

If the world was new, we should not forget how much more, in its mixture of the civilized and the primitive, it resembled the "uneven development" of the contemporary Third World, with its wild juxtapositions of primitive and advanced—the Jeeps and Jitneys painted up with tribal ornament, the mad prophets who rise to power through sermons on tape cassettes, the Motown songs heard on transistor radios in unelectrified huts, the designer jeans and running shoes worn in the bazaar.

As much as the machine in the garden, the characteristic American device may be the machine in the romanticized desert, the telephone booth miles from anywhere, the gas station, with its machine shapes, by the side of the road, the new cars helicopter-lifted to the top of Western mesas for ad shots, the experimental airplane, sleek and alien in the Mojave Desert, the photographer's wagon in Death Valley, the early car at the edge of the Grand Canyon, the Airstream nestled in the rhododendron cove or on the beach, the Model T in the mud, the locomotive on the prairie, the lunar lander on the Sea of Tranquillity. It is the power of that lone machine that masters the desert and, perhaps, turns it into garden: the great mantis-shaped irrigation machines of mechanized agriculture, sluicing their flow from dams that fill storied canyons with their lakes, or even the sprinkler in the hard-won front yard of a Tucson home.

This was the basis of European appreciation of things American—of things not artistic in themselves, but made artistic by being viewed in a European frame. Even Horatio Greenough, the American sculptor who was later rediscovered as a prophet of functionalism, had to go to Europe (in search of the true classic models) before learning to appreciate the "natural" beauties of the sailing ship at sea and the fast sulky.

It was the same basis on which Le Corbusier and other European modernists appreciated American grain elevators and factories. In 1882, long before Le Corbusier, Oscar Wilde arrived on the dock in

New York. He was European aestheticism embodied, and a crowd of reporters met his ship. The lily, symbol of the aesthetic movement, beloved of the Pre-Raphaelites—was his favorite flower and personal symbol. He told the reporters who met him that he sought beauty everywhere. "Could that have aesthetic value?" a reporter asked, pointing to a grain elevator across the harbor. "Might beauty then be in both the lily and Hoboken?"

Before his tour was through, Wilde had forgone wearing his lily boutonniere. Instead, on his tours through America he sported the sunflower. Wilde chose it shrewdly. It served him as a symbol of the new sort of beauty he was seeing—a simple, almost cartoon flower, of practical use, grown in front of sod houses on the prairie for oil and animal food—a flower whose beauty was unexpected and incidental, an icon of American beauty.

Europeans such as Wilde saw the particular American beauty in the plain and the machine made: American beauty was *incidental*. This is the other side of the notion of native intelligence: just relax and be your common selves and you will create the beautiful. Something of the Rousseauian Romantic and of the Pre-Raphaelite commingled here. Americans saw beauty as the province of the European—a kit of decorations from which they could select and apply to the plain thing. That was why Americans welcomed the Wildes and later Le Corbusier, Duchamp, and others: to provide them with the latest news on the progress of art in Europe. The visitors, for their part, wanted to see the new natural beauties of America. It was a paradox that is still with us. To apply the old romantic consideration of the work of art to an object never intended as art has been the century's longest-running experiment in irony—and one born of a European standpoint.

At the waterworks at Chicago, Wilde praised "the wonders of machinery; the rise and fall of the steel rods, the symmetrical motion of great wheels is the most beautifully rhythmic thing I have ever seen." Later, he noted that such works showed the American aesthetic at its best: when it was unintentional. The building that housed the machinery—the landmark Gothic Water Tower that would survive the Great Fire to become the centerpiece of a shopping mall—he found an abomination. Intended to match the beauties of the past, the building struck Wilde as "a castellated monstrosity with pepper boxes stuck all over it." Conscious efforts at art in America, he found, were almost always failures. About beauty, it was important for Americans not to be too earnest.

We might be surprised to hear the dandy invoking the functional; the champion of the lily bearing the sunflower instead. But Wilde was far from naive in indulging in its possibilities. "I have always wished to believe," was the way he put it, "that the line of strength and the line of beauty are one. That wish was realized when I contemplated American machinery."

Later, the logic advanced to just follow the demands of the equations and beauty will result, through the machine; the "line of strength and the line of beauty" would coincide. Here was the truth and beauty of the Keatsian jar, updated to the mechanical age and the mechanical land.

More often, however, the productions of American commerce were regarded like the famous piano player Wilde saw in Leadville, Colorado, performing under the sign: "Don't shoot the pianist; he is doing the best he can."

☆   ☆   ☆

THE BEST THAT could be done was a question of temporal values: the Western towns Wilde toured were false-fronted and grandiose because they were temporary—provisional stopgaps between frontier and future. The New World replaced the traditions of the past with the tradition of the future. A continent lacking both financial and cultural capital was working out do-it-yourself, ad hoc solutions, borrowing against the bright and inevitable future—for some new technology, new invention, new idea would surely come along before the old one was paid for. De Tocqueville talked to a man who explained to him the intentionally weak construction of steamboats: within two or three years, he was told, better machines would already have been designed, and they would be obsolete anyway. Mississippi riverboats were created for a life of three or four years.

In a hurry, Americans built to deadline. Despite the skills of craftsman and mechanic, it is a quality of American products to be no better than they have to be to get the job done. Planned obsolescence is not simply a creation of the twentieth century and the need to sell more products; it is built on a long tradition of the temporary

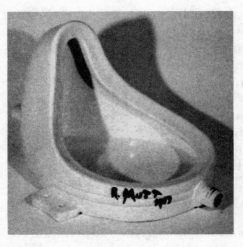

Fountain by R. Mutt (Marcel Duchamp), 1917.

and provisional, in turn built on a belief in progress that accompanies faith in the improvability of the individual.

☆   ☆   ☆

THERE WERE DANGERS in the European appreciation of the incidental beauties of the American object—dangers to art that were intentional. They were made manifest in Marcel Duchamp's idea of the "readymade"—an Americanism applied to off-the-rack clothing and other products that he used only after coming to the United States.

When Duchamp tried to display the urinal he titled *Fountain* at the Independents Gallery of 1917, even his fellow avant-gardists hid it behind a screen. As equipment for the male rest room, the urinal was a specific offense to tender womanhood. And the title, *Fountain*, plus the fact that it was displayed upside down, suggested a reversal of the process of urination itself: Duchamp was pissing on his audience. He was attacking "craft-based" art, high art, the whole European tradition of art.

Duchamp defended himself in the Dadaist publication *The Blind Man* in an anonymous essay entitled "The Richard Mutt Case," in which he showed the object to be a strange variant on the jar in Tennessee. "Whether Mr. Mutt with his own hands made the fountain or not has no importance. He *CHOSE* it. He took an ordinary article of life, placed it so that its useful significance disappeared under the new title and point of view—created a new thought for that object." When Duchamp placed that urinal upside down in a gallery—or attempted to—he had removed it not only from the literal plumbing but from the mental plumbing of functional context, commercial context, social context, and psychological context.

Duchamp's act of selection was a parody of the act of buying, out of a catalogue, not by active direct creation—craft if you will—but by selection. The urinal had been purchased from the Mott Street Iron Works, whose factory a few years before had been moved from the Bronx to Hoboken. Mott was one of the most successful New York ironworks, going back to the great iron age, when Daniel Badger and James Bogardus were creating whole buildings with iron pipes in the shape of Corinthian and Doric columns—kits, based on the shapes of the Old World. J. L. Mott, who began his career as a magnate of the iron stove, followed the course of civilization into the production of iron bathroom fixtures, covered with ceramic.

You could find urinals like the one Duchamp displayed in Mott's catalogue, or walk uptown from the gallery a few blocks to Fifth Avenue and 17th Street and find, if not a urinal, a toilet on display in J. L. Mott's own showroom, just as you could find a sofa or a piano. (Although recent research suggests that Duchamp's original was not a pedigreed Mott at all, but a urinal by a lesser maker—a mutt instead of a Mott.) If such items could be displayed as commodities without offending, why not as art?

But *Fountain* was an implicit attack on art, or a certain kind of art, the craft-oriented art that Duchamp condemned as merely "retinal." It was also a paean—just how ironic one could not be sure—to the beauties of the undesigned. Looking at a propeller on display at a show of industrial products, Duchamp had turned to a fellow artist and said that, in the face of such beauty, it was time for artists to give up.

Duchamp chose his readymades carefully, he said, looking at each of them a long time until he was sure that all vestiges of aesthetic appeal were absent. "My choice was always based on visual indifference and a total absence of taste, either bad or good." "R. Mutt" was possibly a pun on the German *Armut*, misery, possibly a reference to the Mutt character in the "Mutt and Jeff" comic strip. The urinal was something not just found but made—already made, before being transformed in context. Before being found, the piece had been designed, produced, and sold as part of a commercial and social process. The shape was not even "inherently functional," as was proven later, when bathrooms turned into showplaces for designers.

The bathroom was the first room to be modernized. Chippendale, remarked the designer Paul Frankl, "never designed a bathroom." It was a setting for "clean"—the word takes on new meanings for modernist design—shapes that James Cain wonderfully describes in *Mildred Pierce* as "a utile jewel." Henry Dreyfuss was designing toilets and other fixtures for Standard by the thirties, and Buckminster Fuller was proposing mass-produced, single-unit Dymaxion bathrooms. Within a decade, colored toilets, sinks, and bathtubs would become available. When Italian design began to emerge as a powerful force after World War II, one of the most noted designs, right up there with the Cisitalia or the Olivetti Lettera typewriter was a toilet designed by Gio Ponti, the father figure of modern Italian design and editor of the influential magazine *Domus*. Today, we have Warren Platner contouring fixtures—the ads claim—to those of the human form, and bathroom expert Alexander Kira examining the ergonomics of the urinal, toilet, and sink. In *The Bathroom*, Kira addresses this shaping to accommodate the contours and dimensions of the body. "The particular combination of size and shape may vary over a considerable range so long as certain criteria are met," he writes of the urinal. "When we consider that the stream assumes the form of a warped conical solid with a shifting base, it is obvious that the size and shape of the necessary container or enclosure are directly related to its distance from the point of origin in order to contain the stream completely." The urinal was no inevitable form, but the product of design, just as surely as a chair or table. It was also the result of mass production. There were, at least in theory, thousands of specimens of Duchamp's urinal.

Mass-produced objects such as the urinal, the shovel, or the bicycle wheel were identical because they were made from patterns, models. The physical item was a mere model of the ideal one. Duchamp understood this. For him a replica was as good as original, as he proved when the snow shovel he had picked up at a Columbus Avenue hardware store was mistakenly appropriated by a museum janitor in Minnesota for practical purposes and had to be replaced.

☆   ☆   ☆

IT WAS IN this spirit—as the experienced, even jaded European appreciating the simpler American with a dose of condescension—that the American plain style received its definition, even by Americans. What was aesthetic in America was—because found, incidental, therefore somehow *natural*—as natural as "the natural man." Machine-made shapes, oddly enough, were the work of this natural man. But how could the machine also be natural? Because its shapes appeared to proceed from the basic Platonic geometries.

"Machines are, visually speaking, practical applications of geometry," wrote Alfred Barr in the catalogue to the Museum of Modern Art's "Machine Art" show of 1934.

In that exhibit, curators Barr and Philip Johnson treated design according to an aesthetic of the found object. The Machine Art show accomplished something like Duchamp had: it took ordinary manufactured objects and presented them as aesthetic. Among the objects featured in the show was a Crane Company flush valve; it would have attached very nicely to the urinal Duchamp had displayed as *Fountain*. Some of the objects were literally designed by "function": the propellers, the ball bearings. They had to be shaped that way to work, and no other. They were geometric because physics is geometric and they were objects of motion.

But the arrangement of the objects, the way they played off each other, also was a critical part of the show's effect. The harmony of line and rhythm of shape praised by the show's organizers was supplied in part by their arrangement together—propeller lines echoing the curves of metal bowls, and so on. The assemblage was to modernism what the intricately spiraled pattern in which edge tools and Colt revolvers were arranged at world's fairs was to high Victorian decoration.

The Museum of Modern Art show provided the kernel for the Museum's design department: the permanent acceptance of machine-made objects as a form of art. So one could find at the Museum such items as radios and television sets and a GRiD laptop computer that recalled Jefferson's desk. And always foremost among the display were propellers of exactly the sort in which Duchamp had found the first challenge to high art: lovely, sensuous blades compounded, like Jefferson's ideal plow, of mathematics—equations turned into metal.

CHAPTER ☆ TWO

Blades

WHEN OLIVETTI, THE Italian corporation renowned for fine industrial design, decided to sponsor a project commemorating the
American Bicentennial, it chose to issue a portfolio of photographs
of old American tools. They were common hand tools, some easily
recognizable, some of mysterious silhouette and function. They
were photographed as art objects, hovering in space, by Hans
Namuth, who more frequently photographed artists themselves.
They were American tools seen as found art: blades as heroic as
Cycladic heads.

Walker Evans had taken strikingly similar photographs in the
mid-fifties, when he documented "The Beauties of the Common
Tool" for *Fortune*. And the sense of his appreciation was similar; the
hardware store was a museum, the tools themselves accidental art:

> Among low-priced, factory-produced goods, none is so appealing to
> the sense as the ordinary hand tool, hence, a hardware store is a kind
> of off-beat museum show for the man who responds to good, clear
> "undesigned" forms . . . each of these tools lures the eye to follow its
> curves and angles, and invites the hand to test its balance. . . . In fact,
> almost all the basic small tools stand, aesthetically speaking, for ele-
> gance, candor and purity.

A tool is the extension of a man's hand, the builder's medium. An
early American hayfork shows the shape of its user: two arms,
extended straight, above two curved tines or legs. The head attaches
to the handle. Like a mandrake or ginseng root, it is shaped vaguely
like a man.

But in the United States, even personal tools quickly became

34

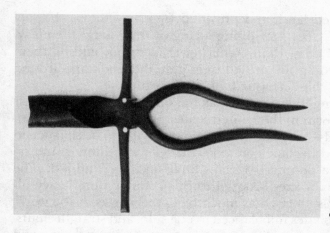

Hayfork,
nineteenth
century.

factory-made products; the extensions of a man's hands were made by other hands, using other extensions—machines. These machines helped make blades—axes, shovels, plows, saws, and the like—more or less uniform. And that uniformity, offering consistency and reliability, is the most basic form of branding. The earliest American trademarks are for tools: the famed Collins ax featured a crown sprouting a workman's strong arm and hammer, worthy of the baking soda box which it no doubt inspired, and the legend "*Legitimus*."

James Billington, in *The Icon and the Axe*, his classic work on Russian culture, opposes the ax—the medieval weapon and clearer of wilderness, the means and symbol of czarist power and implement of Raskolnikov's crime—to the painted religious icon. The two objects symbolize two poles of the Russian character. In America, however, the ax often was the icon, a tool viewed as an object of divine destiny, of human power and predation at once. It was an heroic object: the letter A stood for ax, not apple, in the McGuffey *Readers*. Walt Whitman sang the glories of the American ax, steelheaded and ash-handled, in "Song of the Broadaxe":

> *Weapon shapely, naked, wan*
> *Gray-blue leaf by red-heat grown,*
> *Helve produced from a little seed sown . . .*

No tool was more important to settling the frontier than the ax. The rifle and the knife brought the pathfinders across the Appalachians, but only with the ax could their stays become permanent. It built shelter, cleared land, and provided fuel. It was also the tool used in "girdling" trees—stripping a ring of their bark, then leaving them to die. The method was faster than simply cutting them down and the gaunt, gray, girdled trees gave the landscape around new settlements a characteristic look of desolation.

It is not surprising, then, that the ax was one of the first basic tools to take on a new form in the United States. Well before the Revolution, the twin head—with the advantage of having to be sharpened half as often—had become common. Blades were thinner, and on single-edge axes a counterbalancing poll had been added. The English and German axes brought to this country were heavy, almost medieval tools, derived as much from the war ax as the working ax, with wide blades imbalanced to one side, and straight shafts. They were perfect counterparts to the post-and-beam barns and houses they helped construct.

An ax in the right hands can become a fine instrument. Swung and balanced just right, it no longer seems the crude and heavy implement it does at rest. With the correct curve to handle and blade, it becomes light and quick. Benjamin Latrobe wrote that the American ax "worked more from its wedge-like form, its own weight, and skillful swing, which gives it impetus, than from any great exertion of strength on the part of the woodsman." Along with "maize and steam," he said, that ax had conquered the New World. In 1828, James Fenimore Cooper wrote appreciatively of the "neatness and precision of weight" of the American ax. Beside it, the English one still seemed crude, with its rounded eye and long heavy blade.

American axes became increasingly specialized as well. Some, like stoves and rifles and later reapers, took on the names of the states where they were most favored: Collins made "Southern Kentucky," "Michigan," "Michigan Pattern Double Bit," and "Michigan Phantom Bevel" models. Most famous of the general types was the broadax, sharpened on one side to shape logs for cabins, beams for barns, or puncheons for floors.

Another way of making tools light is to combine two of them into one—as in the double-bitted ax, which saved sharpening, far from home, or the peavey, named after John Peavey, a Penobscot lumberman who combined the tilting "cant hook" with the traditional logger's spike to join the ax, saw, and pickaroon in the lumberjack's repertory of tools. Like the culinary "spork" or the Army trenching shovel, a short-handled spade whose head tilted to form a sort of pickhoe, it joined two tools in one.

Before factories made axes, a man would have the village smithy hammer out an axhead, which he would then fit with a handle appropriate to his build. He kept the pattern for his own handle—a stencil, a template—hanging on the barn wall, most likely, and it was a set of specifications as personal as that for a professional's baseball bat. Even later, the head alone was made at the factory. The handle remained the province of its owner, and the line between individuality of the item and its universality was as clear as the delineation of wood and steel.

☆   ☆   ☆

As THE COUNTRY expanded west, the demand for axes increased. Guides for immigrants and settlers enjoined them to take only an ax with them into the lands of the West. Where they ended up staking claim, there was rarely a smithy. Still, it was not the manufacturing process that shaped the axes, but the selling one. Buyers came to trust the reliability of tools produced under known names and trademarks such as Collins. Men who once would have designed a tool themselves in collaboration with a blacksmith now found specific shapes aimed at specific tasks, trades, and geographies.

When George Sellers, the great Philadelphia machine maker, visited Great Britain, he found the English in Sheffield making older, heavier tools for their own market, but lighter ones to American design for the American market. American conditions had demanded a new ax shape, which individual blacksmiths had supplied. Now, however, it was impossible for the smiths to keep up with the demand; that was what Samuel and David Collins, two hard-goods dealers, realized, why they went into business, and why they applied the best machines and best machinists they could find to the job. Their Collins ax was one of the first nationally known, named, and uniform products.

They bought an old grist mill and converted its wheel to drive trip-hammers and furnace bellows. Massive grindstones, six feet in diameter, were brought in from Nova Scotia by water, sent up the Connecticut River, then hauled to the factory by six ox teams, to be used to shape the blades. In place of the old bellows, air for the forges was carried by a system of hollow chestnut pipes from a

massive wooden pump driven by a water wheel. By 1829 the
brothers were using coal instead of charcoal to fire their furnaces,
an indication of the steady high temperatures their work required.
They made their own steel from iron ore imported from Sweden.
The bit or edge piece of the ax was of steel, sandwiched between an
iron strip bent back upon itself to form an eye for the handle. At the
time Cooper wrote, a Collins workman could temper and forge
eight axes a day. He used the water-driven trip hammer, not simply a sledge; but like the blacksmith, he fabricated the entire ax
himself.

In 1831, in the midst of an economic boom, the company received an order for 1,500 dozen from a New York wholesaler (the
axes then sold for $20 a dozen). Before long, the Collins Brothers
employed six hundred men, turning out more than three thousand
axes each day, along with many other hand tools and a hundred
plows. This rate was made possible by the efforts of Elisha K. Root,
one of the great heroes of the machine-tool industry, who invented
and refined the machines to stamp and sharpen the axes.

Collins axes quickly gained first a national, then an international
reputation, as the company added machetes, manioc hoes, and
other items to its line. The British cutlers of Birmingham, well
known as knockoff artists, began imitating the Collins ax, right
down to the arm and crown seal and motto. In an 1858 court case,
Collins won damages from a number of firms involved in the practice. It was as important a moral victory for United States industry
as the victory of the yacht *America* in 1851. The key to the marketing
success the company enjoyed, the faith in the manufacturer evinced
by a buyer whose life might depend on the product, was suggested
by the court's conclusion: "Reputation is capital." For underfi-
nanced American companies, the statement was often literally true.

☆   ☆   ☆

AFTER THE AX, the plow was the next critical tool needed to settle
the land. The plow was the key instrument for cutting into the
tabula rasa of America. Jefferson, of course, was famous for working out scientifically the proper geometry of a plowshare. He used
mathematics to seek the right shape, streamlining the plowshare, in
effect, for easier passage through the earth and for improved ability
to shed or "scour" the soil, seeking the one right shape.

Jefferson was not the only American statesman to be fascinated by plows. In the tradition of Cincinnatus, called from the plow to serve his country, any properly democratic public servant should at least express the desire to return to that plow. Daniel Webster was proud of his design for a special deep plow, which cut a furrow two feet deep and two feet wide—just the thing for rocky New England. Webster confessed himself far happier behind his plow than addressing the Senate.

Jefferson's "Mouldboard of Least Resistance" received an award from the American Philosophical Society. Like so many of his ideas, this one had come to him in France, when he was serving as U.S. Minister there. But this streamlined terradynamic plow, mathematically designed for the American soil, was difficult for the average blacksmith to fabricate. It never became a practical influence in American agriculture; far more important were the more empirically derived innovations of less educated tinkerers.

Like axes, American plows took on new forms. First, however, a longstanding prejudice against iron plows had to be overcome—the

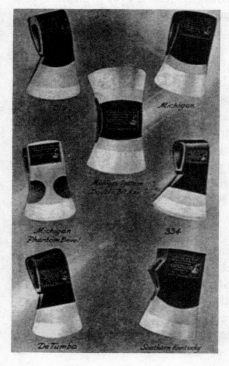

Catalogue page,
Collins Axe Company.

husbandry tradition of Europe held that they "spawned rocks," and British innovators had been limited by the superstition. Slowly, farmers in the United States began to accept the more durable and efficient iron in place of wooden plowshares.

As the frontier moved west, other problems arose. The soil of the prairies, for one thing, was sticky and tended to adhere to the iron plow. Sheet steel, first worked into a plow surface by John Deere in 1837—he hammered a share out of an old circular saw blade— turned the sticky soil but tended to wear through too quickly.

In 1860, an Illinois man named F. F. Smith wrote to Collins Brothers about a method he had developed for casting sheet-steel plows. In a singular instance of openness and imagination, the company seized on the notion and invited Smith to Collinsville, their company town near Hartford. Soon Collins had become a dominant factor in the plow market—internationally as well as in the United States. An English cast-iron plow of the time weighed nearly 250 pounds; an American steel plow just 40 pounds.

The tinkerers who developed new plows were only working out the ideals Jefferson had laid out from his desk. Plows assumed their own place alongside the ax in American iconology.

☆   ☆   ☆

IF THE PLOW broke the prairie, the shovel built the railroads that crossed it, dug out the basements for sod houses, and stoked the furnaces of the steam engines that dominated it. In the heraldry and poster art of the National Grange of the Patrons of Husbandry, the great farmers' self-help organization founded in 1867, the implement that looms largest is the shovel. The farmer portrayed as the Granger ideal is a far cry from the rough pioneer. He is a quiet man of bemused expression, with a gentle fringe of beard and a soft Yankee hat, looking thoughtfully off into space (meditating, perhaps, on the virtues of the soil), standing with his hands lapped over the handle of his shovel and his foot stepped up upon its blade.

This bucolic figure was the good American yeoman of legend, the man of the earth who was to fight it out with the oppressions of rate-fixing railroads and, later, with the violence of ranchers over the closing in of the open range. For while the ax cleared, the shovel built and nurtured. The shovel stood between ax and plow.

Display case, Ames Shovel Works, from 1876 Philadelphia Centennial Exposition. (Arnold B. Tofias Industrial Archives, Stonehill College, North Easton, Mass.)

The most famous shovel in America has been the Ames shovel. Ames shovels not only served the farmer but helped dig the successful Erie Canal and the ill-fated Hoosac Tunnel. They turned sod for houses on the Nebraska prairie and ferried gold ore from vein to pan near Sutter's Mill. They entrenched Gettysburg and Petersburg. They helped create the railbeds of the Union Pacific—the first bit of earth turned for the first commercial American railroad, the Baltimore & Ohio, was lifted in 1828 by an Ames shovel in the hands of Charles Carroll of Carrollton, then the last surviving signer of the Declaration of Independence. When Henry Ford decided to assemble American history as seen in tools, machines, and buildings at Greenfield Village, he and Tom Edison again broke ground with an Ames shovel.

The Ames shovel was also responsible for the creation of one of the country's earliest industrial dynasties, the model industrial town of North Easton, Massachusetts, site of the largest concentration of buildings by H. H. Richardson, an 1854 Harvard classmate, along with Henry Adams, Oliver's son F. L. Ames, and the family's chosen architect.

The classic Ames model was called the Old Colony shovel. Its curved steel pan was three to four pounds lighter than English competitors'. This factory-made item looked like its handmade cousin, but in fact was very different. The top model, the "Antrim patent," featured a steel plate with iron straps welded to it. While the first American shovels were hammered out by blacksmiths, by 1840 the Ames shovel was fired at more than 2,000 degrees in increasingly sophisticated furnaces.

The handle, too, changed. The handmade models featured a U-shaped handhold at the end of the shaft, carved out of a solid piece of Michigan or Maine ash. In the factory version, the ash was split, bent in a V, and closed with an additional piece at the top.

Beginning with the basic Old Colony shovel, the American shovel took on specialized forms. Before the invention of barbed wire, the Ames brothers produced a special shovel for planting the thorny Osage orange in cattleproof hedges. The Ames Company produced brass shovels for loading sugar and grain, military shovels, even toy shovels.

The shovel works were founded in direct response to the Revolutionary War by one John Ames, a blacksmith with ambitions, a classic Yankee. He was also a close descendant of the English Puritan theologian William Ames. With the beginning of the war, the British restrictions on colonial production of metal products became moot, and demand increased. Ames may have begun production as early as 1776, in West Bridgewater, Massachusetts. His son Oliver founded his own branch of the business in 1803, locating near the falls of the Taunton River in North Easton. By 1844, when the plant was turning out some twenty thousand dozen shovels a year, he had turned the business over to his sons, Oliver and Oakes.

It was the sons who made the family famous. Oakes Ames was popularly known as the King of Spades and had supplied virtually the entire shovel and tool stock of the Union Army. Just a few days before Abraham Lincoln was assassinated, he asked Ames to take over direction of the floundering project to build a transcontinental railroad. Among other qualifications, Ames had valuable connections with Boston capital. "Ames, you're the man to take this thing in hand," Lincoln told him. "It will make you the most famous man of your generation." The Ames brothers put up a million dollars of their own money and convinced Boston financiers to add another million and a half. While Oakes became president of the railroad, Oliver was elected to Congress and served as pointman for the railroad's lobbying there. He handed out generous portions of stock in the railroad's financing arm, the infamous Credit Mobilier. By 1869, the Ames shovel works was producing 120,000 dozen shovels a year.

☆ ☆ ☆

NOT LONG AFTER the last spike of the railroad was driven at Promontory Point, Utah, the Credit Mobilier scandal broke. Congressmen had been bought off with the stock; as much as a third of the money the federal government had advanced the railroad had been misappropriated. And Oakes Ames became a national villain.

Oakes was convinced that everything he had done was completely proper in light of the overriding goal of completing the railroad. In fact, he had unintentionally made the scandal public as a consequence of an effort to root out some of the more notorious profiteers, such as railroad vice president Thomas Durant.

Nonetheless, Oakes was made a scapegoat, censured by Congress, and died a few months later. The family commissioned their house architect, H. H. Richardson, to design a grand monument to both Oakes and Oliver Ames near the Transcontinental Railroad. It is a pyramid of rough stone, faced with bas-relief profiles of each man by Augustus St. Gaudens, standing rough and lonely on the plain as a shovel stuck into the soil by a workman who has quit for the day.

The Ames family was not entirely beloved even in North Easton, the company town they had made. They had Richardson design a free public library—the characteristic philanthropy of the day—but their offering of a town hall, also by Richardson, faced with a park designed by his friend Frederick Law Olmsted, was rejected, apparently as an unseemingly intrusion on the pretense of independent government. Instead of North Easton Town Hall, the rock perched structure is known simply as the Ames Memorial Hall.

The company, however, flourished until the 1920s, when the supply of largely Irish labor the Ameses relied on began to grow restless. A second plant was established in Parkersburg, West

Virginia, where wages were lower and jobs scarce enough to earn the company a more genial reception. But here, too, the company faced a classic labor confrontation, when the Congress of Industrial Organizations began a membership drive in the thirties. Violent skirmishes and lasting bitterness resulted. The family sold the business in the fifties, and the original shovel shops—rangy, low brick buildings—later became offices and labs for the Wang Computer Corporation.

☆   ☆   ☆

THE SPECIALIZATION OF shovels was a particularly American fascination. Frederick Taylor, the "father of scientific management," did extensive time and motion studies of stokers and other shovel users at U.S. Steel. He determined that the optimal load for shoveling was exactly 21 pounds and proposed redesigning shovels so that the worker used the strongest muscle in his body, the back, instead of his arms to do the lifting. For U.S. Steel, Taylor recommended fifteen different types of shovel, each for handling a different material or task. Ames and its competitors went on to produce a variety of such types. The common American shovel today, in the suburban garage, on the local construction job, or opening up irrigation ditches on the farm, remains in many cases an Ames product.

☆   ☆   ☆

BEFORE THE REVOLVER, the frontier sidearm was the Bowie knife, a cartoon Excalibur for everyman, a short sword for the land of the Tall Tale, the Southwestern frontier where it was born. Its social role corresponds to the agricultural role of ax and shovel. The Bowie knife stands in the same relation to a physical object as the legends of Jim Bowie and Davy Crockett stand to the lives of real individuals. And it aimed to give its wielder the same enlargement into legend that it gave its supposed inventor, Jim Bowie. The tipoff to what the Bowie really stood for is that, among the many various forms of the knife, there exist many examples whose handles are carved in the shape of the half horse, half alligator that was the mascot of Davy Crockett, a sign of his boast and braggadocio.

Jim Bowie's legend has been conflated with those of Daniel

Boone, who killed a bear with a knife, and Davy Crockett, with whom he died at the Alamo. Crockett, thanks to the legend makers, came as close to mythical embodiment as is possible for a living human being. Celebrated in spurious autobiographies and almanacs for years after his death, Crockett, like a Caesar, entered into folkloric divinity. Bowie followed him in a similar direction, but not quite so far. Both shared martyrdom at the Alamo, where one false claim had it that more Mexicans were killed by Bowies than by firearms; but Bowie himself betrayed his mortality by spending the fight in a sickbed.

The weapon's nickname, "Arkansas toothpick," reflects the same spirit of exaggeration. So does the familiar frontier saying, "Use a Bowie and save ammunition." Frontier toughs would lean back, consider the blade of a good Bowie, and say things like "You can tickle a fellow's ribs for a long time with a knife like that before you'd make him laugh."

There was little innovative in the design or construction of the knife, except for the evil upturn of its tip, a scooped top blade, great weight and length—as much as 18 inches. It was equipped with a wide guard to protect the hand and allow the user to pull it out quickly after each thrust. The steel was tempered, but the blade as often made from an old file. Black walnut was favored for the early handles.

The shape of the tip could have been taken from the upward-bending tusk or tush of the razorback hog, a bastardized creature sprung from the interbreeding of domestic swine with the wild—and a fit symbol for the frontier degeneration into the primitive. But

Bowie knife.

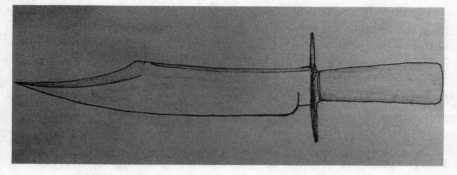

it also carried a touch of the exotic: of the scimitar, or at least the Arab-influenced dagger of Spain, brought in turn to New Spain. Before the Bowie, the frontiersman used a simple butcher knife, and possibly a penknife and a special small knife for cutting the patches of cloth that were critical to his rifle.

Most early accounts of the Bowie refer to its utility—for hunting, skinning, and so forth. One claims that Rezin Bowie, Jim's brother, thought up the knife after he cut his hand using an unguarded knife to kill a wild cow. But it was certainly of minimal use in skinning. The trappers, who knew of these things, used the Green River knife, made by J. Russell of Greenfield, Massachusetts, for that purpose, sharpening it on one side only. Of course, this knife also had its fighting utility. The phrase for stabbing a man was to "give it to him up to Green River"—plunging it to the stamped trademark. Later, "up Green River" became applied to any job done thoroughly.

But the Bowie knife was not a utilitarian blade; its design and purpose were neither for working nor hunting but for killing human beings. It was the objectification of an ethic of personal value and force. The cult of the individual, reduced to his lowest terms: personal honor, the gambler's honor, and the gambler's chances. It was a tool for fighting.

Born in Arkansas, the Bowie came into its own on the east bank of the Mississippi, in the camps and "stands" along the Natchez Trace or the seedy riverside dives of sin city Natchez or Memphis—towns built above the river and, it was punned of both, "built on a bluff." To call a man a liar on the frontier was the critical insult and challenge to his honor. It was calling the entire bluff of his character. The bluff called, in a dirty card game or dueling insults, the Bowie was pulled from its belt scabbard, or more conveniently, from behind the head of the owner. The resulting fight showed who was "the better man"—there was precious little left of the old chivalric premise of divine intervention in such battles. It was character, somehow, rather than the degree of divine favor, that was invoked and exposed by such encounters.

☆   ☆   ☆

BOWIE GAVE HIS original to an Arkansas blacksmith in 1830, to have a copy made. He was one Thomas Black, born in Hackensack,

New Jersey. The legends assign varying degrees of responsibility to Black for the shape of the knife, but he is credited with the strength of its blade. It was tempered steel, and some claimed that Black had "rediscovered the Damascene secret"—the lost method of tempering Dasmascus steel so sought by the Crusaders. According to another legend, he was on his deathbed and called someone in to take down the secret of the process—only to find that he had forgotten it.

Bowie was not an attractive character. He had a terrible temper. He was not so patriotic and devoted an American that he did not accept Mexican citizenship in order to live in Texas, still one of its provinces. And Bowie had made a fortune not just in land speculation, common enough on the frontier, but in slave trading and smuggling. By one account, he bought slaves from Jean Lafitte, the pirate who helped out Andrew Jackson at the Battle of New Orleans, for a reputed dollar a pound.

Bowie had first become convinced of the need for a special fighting knife, legend has it, after a dastardly pistol attack on him while he was unarmed and had to borrow a knife from his brother that bore the famous shape. Its worth was later decisively proven on September 19, 1827, the date most commonly given for his famed fight on the Vidalia sandbar near Baton Rouge. There, he sliced up a gang of vicious opponents; shot twice and stabbed himself, Bowie killed one man and wounded another. The unevenness of the numbers is typical of Bowie stories—as is the suspicious lack of provocation on Bowie's part.

The Vidalia sandbar battle was the aftermath of a formal duel fought with pistols by two other men. Apparently, the traditional means had not exhausted honor's demands: the episode can stand for the transition from the old code of the duel to a new, frontier version. At the sandbar the knives came out when the single-shot pistols had been exhausted. Robert Crain, one of the combatants who faced Bowie and his friend Sam Cuney, reported what happened next:

> I wheeled and jumped six or eight steps across some little washes in the sand bar and faced Cuney. We fired at the same moment. His bullet cut the shirt and grazed the skin on my left arm. He fell. Jim Bowie was at that time within a few feet of me, with his big knife raised to lunge. I again wheeled and sprang a few steps, changed the

butt of the pistol, and as he rushed upon me, I wheeled and threw the pistol at him, which struck him on the left side of the forehead, which circumstance alone saved me from his savage fury and big knife.

Knife fighting was the poor man's version of the duels that Andrew Jackson, for one, and Burr and Hamilton fought—more or less old-style pistol duels. The pistols of the day misfired so often as to provide a merciful margin of chance for their handlers that was entirely absent from combat with the Bowie. The frontier code duello—part parody, part purification of the old code—offered some new twists. One version featured the opponents locked in a darkened room. Bowie supposedly invented the method of dueling in which one arm of each opponent is bound to the other's for a struggle, presumably to the death.

It was in this style of encounter, in 1829, that Bowie faced the notorious Natchez gambler John Sturdivant. Bowie won the fight—slicing the tendons in Sturdivant's knife arm and disabling him—but declined the *coup de grâce* and released the man. Rubbed wrong by this mercy, Sturdivant sent three hired desperadoes after Bowie. Bowie cut off the head of the first, disemboweled the second, and split the head of the third. The symmetry and detailing of the various stories of the result possess something of a Norse saga.

The Bowie became the weapon of choice for the duels that were becoming indistinguishable from street fights. What the Colt revolver and the justification by quick draw were to become for Dodge, the Bowie was for the Southern frontier. By the late 1830s, Alabama, Tennessee, and several other states had attempted to outlaw its use.

More and more, the individual causes served by the Bowie became caught up in ideological ones. Many came to bear such slogans as "Death to Abolitionists" or "Death to Slavery." On the floor of Congress, in the spring of 1860, Congressman John Fox Potter of Wisconsin took on Roger A. Pryor of Virginia with his fists. Pryor challenged him formally, but terms were never settled and the duel did not take place. Potter's offer of a Bowie-style closed-room encounter was turned down as barbaric—the Yankee had outfrontiered the Virginian.

The Civil War provided further proof that the Yankee had adopted the Bowie legend. Union soldiers packed Bowies almost as often as the Rebs, who also adapted the knife as a bayonet—what implement came closer to rendering in steel the chilling edge of the Rebel yell? The British forges at Sheffield turned out thousands of the knives for both sides, sometimes stamping them with appropriate slogans. John Wilkes Booth flourished a modified Bowie, with the catchy sales mark "Rio Grande Camp Knife," onstage at Ford's Theater as he cried: "Sic Semper Tyrannus!"

That appropriately theatrical flourish may have been the Bowie's last great role. After the Civil War it lost favor, finally supplanted by the Colt revolver. An early (1840s) Colt design, in fact, featured a Bowie knife attached to the revolver as a bayonet, providing reassurance to those who still didn't trust a pistol mechanism to do the job.

The Bowie's legend, however, persisted. An 1870 photograph of gunfighter Wild Bill Hickok shows him with a huge Bowie in his belt, unsheathed and approaching a foot in length. The Marine Kabar, carried ashore onto the beaches of Tarawa and Iwo Jima, was the Bowie's descendant, although more a macho symbol and impromptu machete than a fighting weapon.

As a staple in Hollywood's prop departments, the Bowie represented the American frontier fighting ethic as depicted in westerns. But its meanings have expanded: Rambo carries an outsized, Bowie-type knife, and so does Crocodile Dundee. A television show of the fifties about Bowie sought to capitalize on the Davy Crockett and coonskin cap craze. The emphasis here was on throwing the knife—a counterpart to Crockett's marksmanship. In the mid-sixties, when a would-be English rock star named David Jones was looking for a zippier name, he took that of the knife. It was the toughest, most American-sounding name he could find. But this was the Bowie as pure legend and prop. The real end of the Bowie might be marked on the day in the 1880s when Theodore Roosevelt, setting out for the West and ranching, decided he needed a Bowie for his equipage. He had it specially made and engraved—by Tiffany.

CHAPTER THREE

The Equalizer, or the Confidence Machine

"God created all men but Sam Colt made them equal."

—A BUMPER STICKER SEEN FREQUENTLY
IN THE AMERICAN WEST

*"Be not afraid of any man no
matter what his size;
When danger threatens, call on me,
and I will equalize."*

—INSCRIPTION ON A COLT REVOLVER

J. FRANK DOBIE, the Texas chronicler, recounts the story of a judge in the early days of the republic who confronted a Bowie knife–wielding frontiersman. The man stood up in the courtroom—probably a saloon, temporarily so designated—and said, "This is how we keep order in these parts."

"Not any more," the judge replied, "now this is the law," and gaveled the man down with the butt of his Colt revolver.

So the gun established itself as the heir to the blade. "When guns are outlawed, only outlaws will have guns," runs the gun lobby slogan. And when, as in the West, everyone is the law, everyone has to have one.

50

Colt revolvers, 1850s. (Springfield Armory National Historic Site)

American firearms embody the American cult of the individual driven to its most reckless, violent, and exaggerated form, but they also embody the essence of American innovation in manufacturing: interchangeable parts and a universal model. The gun can give anyone power, but when everyone has a gun, what power does anyone have? The paradox of equality was that each man asserted his individual superiority by carrying a gun virtually identical to that of all his fellows. In the gun, the American intersection of individualism and equality was made physical.

What was critical to the success of all the classic guns was the machines used to make them and their incorporation of machinery. The guns resembled the machinery that made them: such devices as the turret lathe corresponded to the design of the revolver itself. The final product reflected the means of production: American guns were themselves small factories of firepower, assembly lines of destruction. They were not only guns made by machines, they were guns made into machines. And the reason they were this way was to magnify the power of the individual, whether it was the cowboy with his six-shooter or the gangster with his Thompson submachine gun.

The first great American gun was the Pennsylvania, or Kentucky long rifle—the name, significantly, changed with the advancement of the frontier. It was a handmade weapon, with one exception: its barrel was rifled, or lined with a pair of spiraling tracks that imparted spin to the ball. The spin functioned as a crude gyroscope, so that the ball kept on course, boring itself through the air. And these tracks were traced into the barrel from a wooden pattern, a tracked wood or metal rod that served as a sort of primitive machine tool.

The technique of rifling had been developed in Central Europe, in mountainous areas of Germany, Austria, and Switzerland. Gunsmiths who came from those areas improved the design for American needs, lengthening the barrel to increase accuracy, and lightening it by shaving the sides of its cylinder down to an octagonal or hexagonal section. The other crucial innovation was the addition of a linen patch between charge and ball. The patch expanded into the grooves, serving as a kind of gasket to capture as much of the exploding gases of the charge as possible, and allowing the ball to firmly grip the grooved rifling.

The stock, carefully carved of cherry or maple, was turned down toward the shoulder, keeping the barrel on a neater line with the eye, and its end was cut in a deep half-moon curve—slipping onto the shoulder with intuitive ergonomics. Cut into the stock was the famed "patch box," a compartment for storing cartridges—homemade as they were—and patches. Some Pennsylvania rifles featured sighting tubes, without telescopic lenses, but offering a light shield that made for better sighting in the high-contrast dappling of the forest.

At least as early as 1720, the rifle was made with the use of a primitive machine tool—a shaft or dowel with the pattern of the rifling carried out in spiral grooves. The fallible eye of the craftsman was done away with in the use of this tool (which did, of course, have to be carefully shaped by eye).

The narrowing of the bore and lightening of the bullet, the reduction of the charge and the lengthening of the barrel to increase the efficacy of the rifling, were all critical to the hunter, roaming far from home, with the necessity of hitting on first shot. The increased accuracy and range of such a weapon had everything to do with survival on the frontier, where life had returned to dependence on the hunter. Another name for the rifle was "squirrel gun," and the affectionate female nicknames—"Ol' Bess"—suggest the domestic familiarity of the gun.

At the time of the Revolutionary War, the frontier was still western Pennsylvania. When frontiersmen from this area joined up with the Continental troops around Boston, the range and accuracy of their weapons astonished the Massachusetts militia. They pumped round after round into a seven-inch target 250 yards away. Another account of a demonstration of the Pennsylvania rifle cited how nineteen of twenty bullets were placed within an inch of a nail at sixty yards. The standard smoothbore musket had a range of barely half that of the long rifle, making redundant the order to hold fire until the whites of the enemy's eyes were visible.

The accuracy of the long rifle became legend. In "Fenimore Cooper's Offenses," Mark Twain makes hilarious fun of the shooting feats attributed to the Deerslayer, who in Cooper's novels talks to his rifle. But Cooper's tales of shooting prowess are only slightly exaggerated versions of those Revolutionary War stories. The range of the Pennsylvania rifle of course meant a departure from established norms of warfare both in tactic and in etiquette. It made sniping possible, and a long-distance version of the sort of warfare conducted with muskets from behind stone walls at Concord and Lexington. It was a weapon of guerrilla war, as shown in the decisive series of battles in the South that most histories underplay—at places such as Cowpens and Kings Mountain. To the British, it must have seemed a virtual terrorist weapon.

Roger Burlingame, the historian of technology, argues that although only a small fraction of the American soldiers were equipped with it, the use of the gun led to the British selection of Hessians (some of whom had German jaeger rifles, the long rifle's ancestors) as their mercenaries. The British command knew well the German sources of the weapon, and believed the Hessians to possess the same weapons and same skills. But the American gunmakers who turned out the Pennsylvania rifle were from a different part of Germany; and it was not until they were confronted by the frontier world where survival depended on hunting (a state of affairs long gone in European life) that the original jaeger rifle was refined into the new, American one.

George Washington was at first taken with the rifle, and planned to outfit as many of his men as possible in frontier buckskins to create the impression of more riflemen. But the rifle did not lend itself well to the traditional tactics of European warfare, which still saw the bayonet as supreme. The musket was to be fired once or twice, in this thinking, as a supporting actor for the bayonet. Washington decided to fight a relatively conventional war, not a wholly guerrilla combat, and after he brought in Baron von Steuben to train his troops in the methods of Continental warfare, the rifle was virtually eliminated from his army. When guerrilla fighting did take place, as in the South, the rifle proved its worth. But it did not win American independence all by itself.

The problem was not the rifle, it was the riflemen. They were frontiersmen, used to coming and going when they pleased, ill suited to military life of any sort, and especially to the formalities of conventional warfare. Washington's early affection for the frontiersmen soured when he saw how poorly they adapted to camp discipline and to the siege warfare around Boston during the early days of the war. Even Andrew Jackson's troops, bearing the long rifles now called "Kentucky," after the latest frontier, proved as difficult to discipline as they were effective at such battles as New Orleans. Before Jackson could get his men to New Orleans he had faced several mutinies and had recalcitrant soldiers shot to restore discipline. But it was there, and not during the Revolutionary War, that the first real encounter between an army of rifled frontiersmen and Continental-style regulars took place.

The long rifle never became a standard U.S. Army weapon. At the Battle of New Orleans, the long rifles used by Jackson's troops were those of the backwoods militia—personal arms, not standard federal issue. The tradition the Kentucky rifle created was to conflate the individual right to own guns with the right of the states to preserve their militias.

☆　　☆　　☆

THE GUN THAT did become the standard Army weapon was a copy of a French musket—produced, among others, by Eli Whitney, whose contract is traditionally cited as the charter of the American system of mass production, with division of labor and interchangeable parts. For generations, the story was told and taught to schoolchildren: how, after inventing the cotton gin, Eli Whitney went on to create the very foundations of American industry, developing the system of making things with interchangeable parts that was the veritable Declaration of Independence and Constitution rolled into one for American industrial methods, and, since it provided items for the democratic masses, a sort of economic bill of individual rights as well.

Here, in brief, is what happened. On May 1, 1798, Whitney proposed to Oliver Wolcott, the Secretary of the Treasury, to deliver 10,000 to 15,000 muskets to the United States government. War with Revolutionary France seemed likely. French privateers already roamed off the American coast. The government armories at Springfield, Massachusetts, established in 1794, and at Harpers Ferry, West Virginia, set up two years later, did not have the capacity to produce the arms needed—some 50,000 of them. Arms had previously been purchased from Europe; indeed, the 1763 French Charleville musket, supplied to Continental troops by the French ally at the end of the Revolution, still served as the model for American-made weapons. A sea war would threaten the overseas supply.

There was no question of innovation in the design of the gun itself. Whitney was instructed to replicate the Charleville musket, which was similar to the standard Brown Bess used by both the British and the Continental regulars. The innovation lay in Whitney's idea of how to produce this already antique design. "I am persuaded," Whitney wrote, "that Machinery moved by water adapted to this Business would greatly diminish the labor and facilitate the Manufacture of this Article. Machines for forging, rolling, floating, boreing, Grinding, Polishing etc. may all be made use of to advantage."

Whitney was a desperate man: he needed the advance on the government contract to keep his business afloat. For years, he had been seeking unsuccessfully to capitalize on his cotton gin. But Southern planters copied the invention without compensation and the legislatures of Southern states had so far failed to aid Whitney. Among others, his friend Thomas Jefferson had attempted to help him, but the patent law was inadequate to defend Whitney's claims.

At the time Whitney obtained his contract, twenty-six other

contracts were let, some of them to men working on similar ideas. In England, Marc Isambard Brunel (for ship's-tackle pulley blocks) and Joseph Bramah (for locks) had already demonstrated the practicality of the theory. Another contractor, Simeon North, who built pistols in Middletown, Connecticut, had created in his shop the filing jig—a device that directed the motion of the worker's file to produce a part of uniform dimensions.

Whitney's strength was in clearly stating the aims of a system of interchangeability and colorfully boasting about them. He shared the basic ideas of LeBlanc, the French gunmaker of whom Thomas Jefferson had learned while serving in France and whose similar methods he had urged on the new federal government. "The tools I contemplate," Whitney wrote, "are similar to engraving on copper plate from which may be taken a great number of impressions exactly alike. . . . One of my primary objects is to form the tools so the tools themselves shall fashion the work and give to every part its just proportion—which when once accomplished will give expedition, uniformity and exactness to the whole."

Only long practice, he argued, could teach an individual mechanic "the art of giving a particular uniformity of shape to particular substances." Whitney's aim was not cheapness, but the use of the machine as a replacement for the skill of the laborer. For the government, the attraction of interchangeability was utility in the field. One observer of Whitney's operation wrote that "quality improved . . . without increasing expense." The idea of reducing costs through economy of scale associated with more recent "mass" production methods was not the thrust of Whitney's plan.

Whitney found it hard going. It was one thing to use triphammers to make axes that were basically all the same shape, but could vary in size and shape within reason, and another to produce many small parts to close tolerance with machinery. First, there was the difficulty of creating the machines at the water-powered factory he set up in Mill River, site of an early grist mill, at Hamden, near New Haven, Connecticut. He also discovered the difficulty of producing parts of irregular shape. Replicating the "accidents" of shape was harder than Whitney thought it would be.

Then there was the problem of labor. Whitney demonstrated a fear of and contempt for workmen—still known as "artists," "artisans," or "mechanics." He deplored the lack of skilled "artists." "I have not only the Arms but the Armourers to make," Whitney wrote. He worried about training men who might hold him up for "a six pence" increase or defect en masse to a competitor. He talked, in 1799, after finding out just how hard the job ahead of him was, about the possibility of bribing workmen from the national armory at Springfield to move to his operation and bring with them knowledge of what machines the government was already using. A visitor to the Whitney armory noted that "any person of ordinary capacity would soon acquire sufficient dexterity to perform a branch of the work. Indeed, so easy did Mr. Whitney find it to instruct new and inexperienced workmen, that he uniformly preferred to do so, rather than to attempt to combat the prejudices of those who had learned the business under a different system."

It was a sentiment that would be echoed by virtually the entire Hall of Fame of great American manufacturers. When in the 1980s Roger Smith of General Motors committed his company to a plan of massive automation, he made the same mistake. Attempting to learn again from the Japanese what they had learned from observing the American system, Smith saw only the robots, the machines, and not the men. His effort failed, whereas more labor-intensive cooperative ventures between GM and Japanese companies succeeded.

This attitude toward workers was to become deeply seated in American life and business. The dominance of the system reduced the importance of labor. Early on, labor was expensive—land was cheap and farming was always an alternative for the laborer. The great immigrations lowered the cost of labor, but it was unskilled and recalcitrant. And a system that treated the laborer as simply another machine—or rather, as the residue of the non-mechanizable—was bound to cause both resentment in labor's ranks and disdain for it on the part of the capitalist. Within a few years, another gunmaker, Samuel Colt, was also touting the virtues of inexperience in workers—it made them easier to shape into parts of his system.

By 1801, the "Artist of his Nation" had completed only five hundred muskets. Fulfillment of the contract would eventually take Whitney eight years instead of the specified two. It was testimony to the skillfulness of his lobbying that he obtained the necessary extensions and advances on the job, and that he obtained a new contract when war again loomed in 1812. But Whitney's greatest success was in creating the impression that he had invented and brought to complete realization the system of manufacture with interchange-able parts.

The climax of the story came in the first week of 1801, when, with outgoing President John Adams and incoming one Thomas Jefferson in attendance, Whitney demonstrated his system of interchangeable parts, choosing, he said, parts at random and as-sembling them into a complete gun lock. Almost two centuries later, someone took the time to survey the parts used in the demonstra-tion. In 1960, Robert S. Woodbury reported that Whitney had rigged the test. Desperate to keep his contract, despite missing deadlines, he marked his cards, secretly inscribing the musket parts used in his demonstrations. No one in the room caught the decep-tion. In 1785, Jefferson had himself assembled, from random choice, locks made on similar principle in France, by the gunmaker LeBlanc. He knew it could be done, and apparently was not invited, nor did he ask, to try it now. By the following September, the first five hundred muskets of Whitney's order had been finished (by whatever means), inspected, and approved by the government.

Woodbury showed that Whitney must have known that the likeli-hood of his building the necessary tooling and producing the mus-kets on time was wildly optimistic. These parts were only marginally interchangeable, at least in the beginning. David Hounshell's fur-ther investigations of the way things were done at Colt and the Singer sewing machine plant revealed that the ideal of interchange-able parts had never been realized in those brick New England factories. (At Colt, for instance, Hounshell reports such telltale signs as the huge numbers of files in the company tool inventory—the smoking gun of the fudging of the ideal.)

If these revelations show anything, it is that the ideal of inter-changeable parts was so powerful that it allowed wish to overcome reality. Identity of parts was as compelling an ideal—and as imper-fectly realized—as such other American ideals as equality of social and economic opportunity.

☆   ☆   ☆

AT LEAST SINCE Woodbury's debunking of the Whitney legend, it has become clear that the national armories at Springfield and Harpers Ferry did more than Whitney in establishing a system of uniform parts. There, others such as Simeon North and John Hall, less publicized and less remembered, accomplished much of the work that realized the ideal Whitney made vivid and sold to the government.

John Hall of the Springfield Armory was issued a patent for his machinery for manufacturing a gun "so as to conform to a model" in 1811. A government compensation board, assembled to evaluate his claims of priority, found his system "unique"—a clear rebuttal of Whitney's claims.

The presence of two armories furthered the usefulness of interchangeability: a lot of Springfield-made guns should be mixable, for cannibalization of spare parts, with one from Harpers Ferry. By 1815, interchangeability was a military specification. At a demonstration in 1824, one hundred guns from different armories were disassembled, their parts mixed, and the weapons assembled from parts chosen at random.

These conflicting claims by the first of what we would today call defense procurement officials and defense contractors are less important than the main fact: that government sponsorship of work aimed at creating firearms with interchangeable parts was the mainspring in the development of American industry, and ultimately in the shaping of American design. It was the need for a native firearms industry that inspired the move to what became known first as the "armory system," and then, as it was applied to other industries, the "American system" of manufacture.

The firearms industry also was the first to point up one of the problems of machine-made goods: that it was far easier and cheaper to keep making an older product than to change the machines to incorporate improvements in design. Thus during the Civil War, the North failed to take advantage of its superiority in firearm manufacturing by rejecting breechloading and rapid-fire weapons, even though the occasionally fielded Henry rifle was feared among the Confederates, who called it "that damned Yankee rifle that can be loaded on Sunday and fired all week."

But Whitney stands neatly at the head of the family tree of the American system, if only as a symbol. The Whitney plant was eventually sold to Winchester in 1888 for production of its .22-caliber rifles. From Hamden, experts fanned out to new factories and machine shops. Horace Smith, a workman at Whitney's armory,

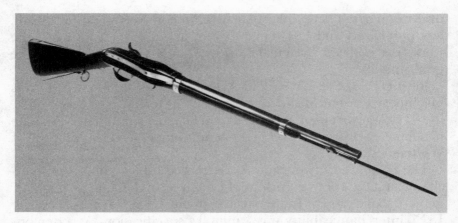

Musket manufactured from interchangeable parts, 1840s. (Springfield Armory National Historic Site)

went on to found the Smith & Wesson Company. Samuel Colt placed one of the first manufacturing contracts for his revolver with Whitney's heirs in 1847, and imitated their methods when he opened his own factory later. A. F. Cushman, Francis Pratt, and Amos Whitney moved from Colt to the Phoenix Ironworks, then opened their own soon-to-be-famous machine-tool shop after the Civil War. The resulting company would, of course, later produce aircraft engines as well.

Even when the degree of interchangeability was lower than at government armories, and even when the product being made was simple, the application of machine tools had dramatic effects on the shape of American objects. The Colt revolver was a success due to the machine-tool genius of Elisha Root, who was hired away from Collins Brothers, the ax- and plowmakers. The Winchester rifle that "won the West" was said to be as efficient at dispatching Indians as the railroad builders it protected were at laying rail. The Remington repeating rifle company moved naturally from the production of firearms to that of sewing machines and typewriters.

☆   ☆   ☆

WHILE THE MACHINES that Alfred Barr praised as "practical applications of geometry" in his catalogue to the MOMA Machine Art show may have been geometric, the products of those machines were not. One of the greatest difficulties in producing his musket, Whitney wrote, was that "the conformation of most of its parts corresponds with no regular geometric figure."

The only way to standardize was to make a set of gauges. From the viewpoint of the machine-tool designer, the shape of the gauges was incidental—it varied according to the product. It was in a sense a found object, a readymade.

In 1913 and 1914, Marcel Duchamp created his *Three Standard Stoppages* and then used them to lay out his *Network of Stoppages*, a series of lines that bears a vague resemblance to a map of a golf course. These bore roughly the same relation to craft art as machine-made products did to handmade ones.

The stoppages were made from three one-meter-long strings, randomly dropped onto a canvas a meter below. The strings were glued to the canvas in the patterns in which they fell. They were then each traced into a rulerlike piece of wood and cut out, forming something like the French curves of a draftsman's set. Duchamp played off the regular standard of the meter to create "templates of irregularity"—"standards of accident," in the words of the critic Joseph Masheck. Their shapes suggest the gauges—rulers for shape, not length—used in gun production. Chance has been made into model, just as the incidental shapes of function—products of metallurgy, of the chemistry of gunpowder explosion, of the state of the machine art, of the ergonomics of hand and shoulder, and of some residue of aesthetic volition—are made into a model in the machine tools of the armory system of manufacturing.

☆   ☆   ☆

SAMUEL COLT, WHO was to create the classic piece of personal fire-power that became known as the Peacemaker or Equalizer, got his start as a showman. He toured Ohio and Mississippi river towns as "Dr. Samuel Coult of London, Calcutta and New York," demonstrating the wonders of nitrous oxide, "laughing gas," to eager audiences.

On the Western rivers the figure of the frontiersman—the legends of Davy Crockett and Jim Bowie—turned into the protean one of the Confidence Man. Part showman, part salesman, part gambler, his rise is recorded in Melville's *Confidence Man* and his typology traced in Mark Twain. He was always a figure who transcended his profession.

Sam Colt was a twist on this type. His story was not one of the poor boy who wins fame and fortune due to his inventions. His mechanical skills were at least equaled by his ability to win friends and influence people in power. Born in 1814, he was given his first pistol, a rare and precious item at the time, at age seven, some accounts have it. His family connections helped him get his start. His 1836 revolver patent was made possible by the intervention of Henry Ellsworth, a patent commissioner who also happened to be a friend of Colt's father. As a teenager he went to sea, where, according to legend, he soon learned about class distinctions—his pistol was taken away; only officers were allowed to possess firearms. On shipboard, the tale goes, he was inspired by a ship's wheel to carve his first wooden model of a revolving pistol.

In fact, revolvers of various types had existed before Colt and he had dreamed of building one before he left shore. Colt's, or his company's, contribution was to match the new generation of machine tools, the percussion cap, and later cartridge ammunition with the new demands of the frontier—demands for firepower to kill Indians, Mexicans, and outlaws, and impose law and order.

Colt returned home and took up the showman's trade. Among his earliest public performances was the detonation of an electrically ignited submarine mine, staged in 1829, when he was just fifteen. The explosion drew quite a crowd, one of whom was an impressed young mechanic, Elisha K. Root; but the crowd's enthusiasm turned to displeasure when the mine doused them all with muddy water. Root helped Colt make a run for it, the story goes. Colt's brother John, who had slipped into the sort of Natchez-under-the-Hill life of dissipation where the Bowie knife reigned supreme, subsequently invited him to Cincinnati, where Colt and his laughing gas became an attraction at a local "museum." From there, he worked his way down the rivers to New Orleans.

His experience would help in the firearms business. In 1836, Colt set up his first company to manufacture the revolver in Paterson, New Jersey, where the Passaic River was penned into channels that drove the silk industry. But he went out of business in 1843. The cause was a familiar one: shortsighted military procurement policies. The Colt was judged unfit by a test panel of Army officers. Colt lobbied to arrange another test—by the Navy, in the historically well-founded belief that one branch of the service would be likely to disagree with another. But that trial, too, went against him. Four years later, Colt traveled to Florida, where the Army was attempting to subdue the Seminole Indians, to point out the superiority of his firearms against the guerrilla-ambush techniques of the Indians. He made the sale but lost the payment check when his boat capsized on the way home.

In 1846, a Captain Samuel Walker of the Texas Rangers handed Colt a salesman's dream: the proverbial unsolicited testimonial. "The Texans," Walker wrote Colt, had such confidence in Colt revolvers that "they are willing to engage four times their number. . . .

"In the Summer of 1844 Col. J. C. Hays with 15 men fought about 80 Camanche [*sic*] Indians, boldly attacking them upon their own ground, killing & wounding about half their number . . . without your Pistols we would not have had the confidence to have undertaken such daring adventures." (Walker was later killed in the Mexican War—by a lance.) Walker placed an order. Colt had no factory, so, neatly tying his revolver to the family tree of the armory system, he contracted much of the production out to Eli Whitney, Jr., who ran his own machine shop.

General Zachary Taylor made a well-publicized order of Colts for the Mexican War, and by 1859, Colt was distributing an eight-page list of officers who endorsed his guns—many of whose minds had been eased toward the free expression of their opinions by the engraved presentation models Colt gave them. Some of these were accompanied by lithographs by George Catlin, an artist whose vision of the perpetually wild West Colt shrewdly patronized and used for advertising. Others bore images from standard books of bank-note engravings.

Soon Colt headed an industrial giant. He was vocal in his belief that untrained workers, "without prejudice," were the best for his system. His chief interest in craftsmanship was for the ornament applied to the plain, factory-made, machine-tooled product. He had confidence in machines, but very little in men. David Hounshell shows that even Colt never achieved the ideal of perfectly interchangeable parts. Final fitting was done by hand. Men individually filed the parts to shape; men individually filled the gap between the ideal and the reality of interchangeability. Men, working by hand, were a necessary, residual irritation.

But Colt's disdain of previously trained personnel did not extend to technicians: to persuade Elisha Root to leave the Collins Axe Company and superintend his armory, he doubled Root's salary, making him, it was said, the best paid "mechanic" in New England. Root was to succeed Colt as head of the company.

Colt didn't mind selling his weapons to both sides of wars, as in the Crimea, or fulfilling orders from J. E. B. Stuart and the Southern state militias right up to the first shot at Fort Sumter, or even inspiring conflicts if he had a chance. With his profits, Colt built himself a mansion outside New Haven, named "Armsmere," executed in the same carnival-version Arabic architecture that P. T. Barnum chose for his "Iranistan" down the post road in Bridgeport. It was no accident that Armsmere resembled a huge landlocked riverboat, such as those where Western confidence men plied their trade. In selling his pistol, Colt was also selling confidence; his gun made the skinniest dude the fightin' equal of the biggest and toughest outlaw.

☆   ☆   ☆

RATHER THAN THE gun, it could more truly be said that the cartridge won the West—beginning in 1873, the ammunition for the Colt revolver and the Winchester rifle, the model 73, was interchangeable. The metal cartridge—developed before the Civil War, but used by Colt in its revolvers beginning only with the Peacemaker of 1873—was the most interchangeable part of all. Colt at first missed out on the metal cartridge, a blindspot comparable to Edison's failure to grasp the virtues of alternating current or Ford's failure to understand the need to improve the Model T.

In 1869, the Rollin White patent of 1855, covering the design of revolvers to accommodate cartridges, expired. White, a former Colt employee, had left the company, apparently after his idea failed to win favor there, for Smith & Wesson, which was more enthusiastic. A few years later he was in a position to ask Colt, whose engineers had by now been trying to design a cartridge revolver that would not violate White's patent, for $600,000 for his share of the rights to it. Along with the fee for Smith & Wesson's share of the rights, that meant that Colt would be forced to cough up more than a million dollars, an unthinkable sum. Fortunately for Colt, the late 1860s were the years of a postwar slump in the arms market, denying Colt's competitors their newly acquired patent advantages. The company's new revolver, finally developed on a separate design, was ready in 1873, a decade after Colt's death, and the Army enthusiastically adopted it, in plenty of time for Indian wars in the West and the booming of the cattle towns.

The six-shooter of gunfights was the Single Action Army model—the Equalizer or the Peacemaker—the weapon of choice in Dodge City and Abilene, Yuma and Tombstone. A man took pride in his revolvers—especially a lawman or gunfighter. In July 1885, for instance, Bat Masterson wrote Colt management from the Opera House saloon in Dodge City to order eight rubber-gripped Single Action Army model Colts.

The weapons of Colt and his competitors at Remington, Winchester, and Smith & Wesson became more tightly machined and perfectly interchangeable with each passing year. These firms added something to the weapons that Whitney's single-shot muskets—more primitive than the Pennsylvania rifles of the time—did not possess: their own internal machines, mechanisms for serial production of fire. Each of their weapons, at various times, was called "the gun that won the West." Their common principle was the aim of using equal parts to make equal men.

☆　　☆　　☆

WHEN A GUN makes everyone equal, no one can be without one. Mark Twain reported that while he never used his revolver, he would have felt almost nude in a mining town without one. Among the advantages of the revolver was that it could be carried almost anywhere and that it could be easily concealed. Hollywood was inaccurate, it seems, in creating the cult of "the draw." Fights rarely turned on simple speed in extracting the firearm from holster and pulling the trigger; face-to-face standoffs on Main Street were uncommon. But in creating the myth, Hollywood may have succeeded in extracting the dramatic essence of the real gunfights. According to

Roger McGrath, who studied contemporary accounts and statistics of several violent Western towns, most of the gunfights were matters vaguely related to "honor" in the sense of "who was the better man." The revolver was not terribly accurate—Mark Twain admitted he couldn't hit a thing with his—a fact that gave a roulettelike quality to the true Western gunfight. The gunfight converted the cavalier tradition of the duel into a contest basically of chance—appropriate in a world where overwhelming boredom was eased by gambling. In a kind of revolver roulette, deciding on "the better man" was left up to the mechanical equalizer.

But the revolver also put an end to the old-style fights of the Southern frontier. The six-gun gave one lawman—or the vigilante, for that matter—a shooting chance against a mob. The history of the cattle towns, as historian Robert Dykstra has shown, is the history of public guns gradually putting private guns under control. The shoot-out at the O.K. Corral was precipitated by lawmen attempting to disarm gunslingers.

☆   ☆   ☆

IF THE ARMORY system was born in what we would today call defense procurement, its decline in America may have been presaged the same way. The Springfield Armory produced the basic Army infantry weapons, including automatic and machine guns, the Browning Automatic Rifle, and the M-16, until 1967, when it was finally shut down and turned over to the National Park Service. Colt continued to produce weapons for the military; but in 1988 it lost the contract to produce the basic infantry weapon, the M-16, to a Belgian firm with a factory in South Carolina. The problems were with quality and labor relations—the same problems Eli Whitney struggled with. By 1989, the arms manufacturer was for sale by its parent conglomerate, which had borrowed its name and its trade image of the rampant and unbroken horse, and the state of Connecticut stepped in to buy partial control. But already, some years before, the Colt automatic had been replaced as the standard military sidearm by the Italian Beretta—a weapon whose name was shortly to be borrowed by General Motors as the name for a Chevrolet model.

Colt's civilian market for semi-automatic weapons remained active. As the instrument on which the "Syndicate" rode to power, the

Thompson submachine gun, the "Tommy," became a great symbol of individualism and free enterprise run amuck. The first examples were manufactured by Colt. Developed at the end of World War I, the Thompson just missed the war in France, but became famous in the wars of the streets of Chicago as the instrument of choice for the Valentine's Day massacre and very nearly as legendary as the Colt Navy or Winchester 73.

More recently, the semi-automatic "assault rifle" such as Colt's AR-15 and its cousins have assumed the place in media mythology once occupied by the Tommy gun. It is the enforcer for the drug dealer, of course, but defended by the National Rifle Association as part of what its television ads call "our heritage." In an ironic footnote to both the social and the technological history of the gun, just as the Colt gunmaking arm was being sold by Colt Industries and foreign-made models rushed in, the president of the United States was signing stern legislation regulating the sale of such weapons: in the future, no such guns were to be sold—unless they had been manufactured in the United States.

CHAPTER FOUR

Little Factories

ON THE MANTEL of my cousin's house in western North Carolina sits a family heirloom: a Seth Thomas clock, dating probably from the 1840s or 1850s. It is shaped like a small reliquary or shrine, its case peaked like a little chapel's. It was probably sold by a peddler, and it is inconceivable that the ancestor who bought it could have had any use for it—except as a status symbol, as the first machine in his home. He got up, worked, and went to bed by the sun, attended church by the bell in the steeple, had no stagecoach or railroad to meet on time. But owning a clock was a token of being part of the mechanical revolution, a symbol of being up to date. The armory system, applied to clocks, made them inexpensive enough for even cash-poor, land-rich farmers to buy.

The clock was the most prominent household example—followed shortly by the sewing machine—of the way in which the American system developed to manufacture firearms was quickly adapted to other products.

A clock was to furniture as jewelry was to clothing: a symbol and a decoration. A clock, after all, was a kind of model of the Newtonian world, like the orrery Thomas Jefferson admired. But as a practical matter, in those days, only the professional and above all someone who dealt with railroads had to be very specific about time. (It was not until the 1880s that the railroads, over rural resistance, forced the synchronizing of clocks and the creation of local time zones retreating hour by hour to the West.)

The clock was also the ruling device of the factory system, which demanded schedules and punctual arrival. Workers would check

Seth Thomas clock,
circa 1830. (Smithsonian
Institution)

into the factory on the redundantly named "time clock." David
Landes notes that a clock was often the prize offered to factory
workers with the best on-time records—a practice still echoed in the
proverbial gold watch presented on the retirement of a devoted
employee. To possess a clock was to own an object of the new
industrialization, which turned hours and minutes into interchangeable parts.

Eli Terry had begun making wooden clocks in the conventional,
handmade manner. Then around 1802 or 1803 he began to construct a thousand wooden clocks using machinery, in a water-
powered factory in Plymouth, Connecticut. By 1808, he had developed something akin to mass production of his shelf clocks; by
1816, he was using such tools as circular saws.

Water-driven machinery was being used in little central Connecticut towns in the 1820s to produce wooden clocks sold by peddlers.
For this trade, Terry—who began by selling the clocks door to door

and ended up renowned as the father of Terryville, Connecticut—developed the shelf clock, a mantel-sized spring clock that did not require the high case of the familiar grandfather clock, most of whose height held nothing but pendula. Clock prices fell to less than $10 with mechanization, and observers noted that even cabins too poor to possess chairs sometimes boasted mantel clocks.

Southerners grew more and more resentful of Yankee peddlers and the clock manufacturers behind them, in part because they drew rare hard currency out of the area, in part because they reminded the South of its own lack of modern industry. The peddler seemed to sum up the worst in the Yankee—his guile in placing clocks on credit, for instance, knowing their owners would never be able to part with them and would have to pay up.

A number of Southern states passed laws regulating peddlers and setting high license fees for their operation. Terry and others responded by establishing factories in Virginia and South Carolina where the clocks were assembled from kits of parts manufactured in Connecticut.

Two other major figures in the clock industry, Seth Thomas and Chauncey Jerome, got their starts in Eli Terry's factory. Jerome, son of a poor ironmonger, put to work in his childhood making nails, was a self-trained carpenter who went to work with Terry in 1816 making shelf clocks and later spun off into his own business. In 1837, Jerome produced a brass clock, smaller and more accurate because it was not susceptible to changes in humidity.

Before long, Jerome was even exporting his cheap clocks to England, where their price struck the proud British clock industry as prima facie evidence of shoddiness. But, notes David Landes, "the American clocks were good enough." They soon seized a significant portion of the British and then the European market. By the 1860s, the Germans imitated American methods. The noted firm of Junghans began making clocks "on the American principle," as it advertised.

Watches took longer to adapt to the American system. They demanded a more rigorous fit to the parts, a tighter tolerance: 1/5,000th of an inch. Before the Civil War, the United States did not possess even a crude watchmaking industry. There was, reports Landes, "simply no pool of cheap skilled labor to sustain a cottage watch industry like that of Europe." But the 1850s saw efforts to adapt to watches the system of interchangeable parts that had worked on clocks. With several partners, Aaron Dennison began what became the Waltham Watch Company. One of his first moves was hiring a skilled machinist from the Springfield Armory. Waltham was soon joined by Elgin, Ingersoll ("The watch that made the dollar famous"), and others.

The American watchmaking system stood in contrast to Switzerland's, with which fine watchmaking was virtually synonymous. There, in such towns as La Chaux-de-Fonds, the old craft style of manufacture continued, with some intrusions of the machine on a system whereby the work was put out to workers in their homes. The Swiss watch worker produced 40 watches per year; by 1900, Waltham was turning out 250 watches per worker per year.

One of the first designs by an art student in La Chaux-de-Fonds, Charles-Édouard Jeanneret, was for the case of such a watch. He was later to adopt the name Le Corbusier, preach to the world about a house that was a machine for living and a chair as a machine for sitting, and wax lyrical over American industrial products. When he spoke of a machine, it was the watch that inhabited his sense of the word, well before he fell in love with airplanes and automobiles.

The methods of Waltham inspired another young man, in the United States. Long before he ever got involved with automobiles, Henry Ford had the idea of making simple, cheap watches by mass production. In the 1890s, while maintaining generators for Thomas Edison's Detroit Electric Company, he worked out a plan by which he estimated he could turn out 600,000 watches a year. The question was how to sell them. Unable to provide an answer, Ford cast about for another type of product.

☆ ☆ ☆

PERHAPS EVEN MORE than the clock, the sewing machine figured as the very epitome of the mechanical intrusion into non-mechanical life—and particularly the industrial into the domestic. That is why it figures in the famous Surrealist phrase, joining "the umbrella on the operating table."

Part of the sell was ease of operation—"even a child can run it"— a familiar pitch, to be repeated for everything from the first lawnmowers to the first Xerox machines to the first personal computers.

The sewing machine sat on the box in which it arrived; later models hid inside pieces of furniture when not in use.

The basic shape came early and has not changed—that of a head, grazing or feeding, something almost gargoylelike—even though there were variants: the fox, the dolphin, the animate arabesque of the basic head and body arcing across a plane, were to undergo many transformations, often in a simple effort to subvert patent law.

The machine came enclosed in a black sheet-metal case, painted with flowered ornament whose every bright red or yellow brush stroke was identical to a petal. Even the model built in 1851 to secure a patent was decorated with painting. The famous Singer New Family machine of 1858 was advertised as "a machine decorated in the best style of art, so as to make a beautiful ornament in the parlor or boudoir."

The readymade clothing industry bought most of the early machines, but it remained a relatively small sector of the economy until the Civil War. The real market lay at home. It was tapped in the 1850s by Isaac Singer and his marketing man, Edward Clark. In 1856, Clark introduced an installment plan. The installment plan meant breaking up the payment into pieces—an economic kit and division of labor. Also viewed as a rent-and-buy scheme, it led to a jump in sales from 883 in 1855 to 2,564 in 1856.

After a patent pooling agreement of 1856 ended litigation over the rights to the various features that together constituted the machine, Singer also began to offer trade-in allowances for older machines in 1857. Singer accepted in trade not just its own but also the cheaper, less dependable machines of low-end competitors, which were simply destroyed. The company trumpeted the idea of "cleansing" the market and improving the impression of the public of the sewing machine. It also helped close the price difference between its own relatively high-priced machines and the cheaper ones such as the Willcox & Gibbs. In the 1850s, the price of the machines fell by half. The Singer machine was still among the more expensive—$125, to $50 for the Willcox & Gibbs and others.

In the same year, Singer opened a parlorlike salesroom in Manhattan to create a natural homey ambiance for the cool metal machines. The salesroom was housed in Singer's building on Mott Street, which was itself a kit of iron parts, the creation of Daniel Badger, pioneer of iron buildings. Upstairs from the showroom,

ILLUSTRATIONS

OF THE

GENUINE HOWE MACHINES.

Letter A Machine.

Letter B Machine.

Letter C Machine.

Letter D Machine.

Variety of sewing machine models. (Smithsonian Institution)

roughly similar but not yet interchangeable parts were fitted together in a mass production system, with hand adjustments as the
last part of the manufacturing process.

Singer's success came through such finely appointed salesrooms,
like great parlors in which any woman would be comfortable, and
through innovative selling terms. These helped compensate for the
machines' relatively high prices. Singer created a spiritual air
around his machines that spoke at once of progress, leisure in the
parlor, hominess, and prosperity. It was in this spirit that Louis
Godey, publisher of the women's magazine *Godey's Lady's Book*, called
the sewing machine, "next to the plow, Humanity's most blessed
instrument." Humanity's most blessed device, however, was chiefly
for American humanity. By 1880 there were still only about 15,000
machines being produced each year in Europe, while one moderate
size company alone turned out 50,000 in the United States.

Whereas Singer chose high quality and a high sheen of respectability rather than low price as its sales strategy, the cheaper Willcox & Gibbs machine was produced by more and more sophisticated
machine tools to increase interchangeability and quality. The master here was Henry Leland, who was to go on to pioneer the Cadillac, the first car with interchangeable parts, and the Lincoln.

Many women bought sewing machines to work on the putting-out
basis—sewing cut goods into partially or completely finished products as supplementary income. The machine brought more division
of labor and concentration of size in the clothing industry. It was
ready just in time for the Civil War, whose armies offered measur-
able bodies of men from whose dimensions standard sizes were
modeled.

The sewing machine marked a key departure: it was one of the
first machines whose savings in labor were converted not into higher
productivity but into more elaborate production of individual
items. Historian James Parton recorded that the sewing machine
had led to more elaborate decorative stitching: one stitch by hand
was now replaced by ten made by machine. The sewing machine
fancified garments as the bandsaw fancified houses. Similarly, its
cousin the typewriter was to increase the weight of business and
government correspondence.

☆   ☆   ☆

THE YANKEE PEDDLERS who brought clocks and eventually (as in the case of Faulkner's narrating salesman Ratliff) sewing machines to the backcountry also carried with them a generous supply of such Yankee notions as mechanical peelers, corers, and pitters. These kitchen tools were an offshoot of more serious toolmaking: one refugee from the Ames shovel operation ventured into production of his own apple peelers.

Few of the devices caught on—except for the eggbeater. Before the blender, before the food processor, the symbolic piece of kitchen technology without which no truly modern cook or housewife felt fully equipped was the eggbeater. While Europeans tend to continue to beat eggs with a whisk, the geared eggbeater has been standard in American kitchens at least since the 1850s. Of all the common "food-processing" devices, only the coffee grinder—a miniature mill in the kitchen—preceded it. It was a piece out of the fabric of Yankee ingenuity, as were the apple peelers and cherry pitters that accompanied it on the peddler's cart. But while all those peelers, corers, and pitters were plentiful in the files of the Patent Office, they never became universal appliances, with reduction in price accompanying expansion of production. Dover was the best-known name in eggbeaters, and not expensive; the 1897 Sears catalogue offered a Dover priced at 9 cents.

With all sorts of crimped gears, beaters of curious twist and bulge, and handles set straight up like an old-fashioned drill or tilted like a sword haft, old-time eggbeaters make up a miniature compendium of the mechanical engineering of the time. Like the clock above the mantel, the eggbeater in the kitchen was a symbolic bit of modern mechanism, a piece of the industrial age. They are, in fact, tiny turbines, models of the wheels and gears of larger factories.

In elementary science books, a comparison with the eggbeater is always used to explain the workings of the differential gear in an automobile, converting the drive shaft's power to the rear-axle power. Before that, it could have served as a model to explain the workings of a water turbine at a flour mill. The gears and works of the eggbeater were never covered, but proudly exposed. That is its visual appeal: the way the turn of the handle, via gearing, produces the dematerialized transparency of the spinning beaters, akin to a mechanical whirlpool or tornado.

But the successors of the eggbeater were different. When they appeared, early in the twentieth century, the guts of the first electric beaters were as nudely displayed as their mechanical forebear. But from the twenties on, the motor and gears of the kitchen mixer, the blender, and the food processor were quickly covered with vaguely streamlined, then later European, School of Ulm shapes.

The arrival of these machines depended on the development of very high-speed electric motors; from there, the technology matured quickly. Soon, they were sold like cars, on the basis of marginal stylistic differences and in some cases, mechanical ones: the Osterizer touted the fact that it was compatible with Mason jars; you could screw on, blend, cap, and "put up," all in one jar.

The Waring Blendor, with its association with the bandleader, was a marketing ploy. The machine, spelled "Blendor" to make a trademark possible, was invented by a slightly crackpot tinkerer named Fred Osius, who approached bandleader Fred Waring for backing. Waring was taken with the device. Six months and $25,000 later, in September 1937, it was introduced to the public. The Waring succeeded through its association with the current fad for blended drinks; sales were boosted by a marketing tie-in with Ron Rico rum.

The button wars of the 1970s saw manufacturers add more and more speeds to their devices—each speed only fractionally different from its neighbors. The wars ended with some eighteen speeds, the practical limit for variations and also the limit, it seems, to the imaginations of the people mandated to name the buttons—for example, "flash blend"—and identify functions for them.

☆   ☆   ☆

SUCH MACHINES BROUGHT an almost comically industrial seriousness to the home. But the earliest machines in the home did all they could to domesticate themselves. From its beginnings, the typewriter modeled itself after the sewing machine, and the first models hid under floral-painted cases like home sewing machines. Only when its maker discovered the office market did the new machine flourish and its look change to serious mechanism.

Typewriter retailers often display antique machines in their windows. They are resolutely mechanical objects, some lovely, some gawky and awkward. The business point is to suggest just how far the typewriter has come: sleek electronic models are set beside the pioneer devices. Inside their long, sleek, black and colored cases, the wide-carriage models big as El Dorados, the current models are fighting a rearguard action against the computer, bits of whose technology they increasingly incorporate. The early typewriters look complex and heavy, and as proud of those qualities as one of Admiral Dewey's battleships. The electrics and electronics, by contrast, hide inside their shells whatever mechanism they cannot replace with energy, whatever levers and gears cannot give way to plastic-encased silicon.

The typewriter was born in Milwaukee, a city that in the 1860s was a center for the mechanical ingenuities of German immigrants. In the machine shop of a mechanic named Charles Kleinsteuber, Aaron Latham Scholes, inventor of a machine for printing page numbers on blank books, met Carlos Glidden, creator of an improved plow called "the Digger," who was to become his promoter and backer. Both men, it turned out, were having Kleinsteuber fabricate the first models of their inventions. Glidden suggested the application of the principles of Scholes's paging machine to a more versatile writing machine, and became his partner. Together, they signed on Charles Densmore, a newspaper editor, and in the process acquired his capital investment—which amounted to all of $600.

By the 1870s, too, the Remington Company was suffering from a decline in its arms production business. Scholes and Glidden were able to persuade Remington to manufacture their machine. Despite demonstrations at the 1876 Centennial Exposition in Philadelphia—many a visitor sent home the first typewritten card or letter his fellows would have seen—the typewriter was slow to catch on. Remington was to spin off its typewriter division into a separate company, apparently just as the corner was about to be turned.

The typewriter took so long to become established in part because its backers attempted to present it as a device for the home and specifically for the study—for the "literary man," the clergyman or author—rather than as a business machine. Their model was the sewing machine, with its painted case, aimed to blend into the parlor. The Remington No. 1 appeared in a black case with colored floral ornaments.

"The Type-Writer in size and appearance resembles the family sewing machine. It is graceful and ornamental, making it a beautiful piece of furniture for any office, study, or parlor," ran a Densmore, Yost advertisement of 1876.

"It is to the pen what the sewing-machine is to the needle, and its use is destined to become as universal as culture and enlightenment . . . for the simple reason that it will save as much time and labor in our business houses as the sewing machine saves in our dwellings."

To Scholes and Glidden, the benefits of the typewriter seemed obvious—but they were not the benefits we would imagine today. The appeal of typing initially lay not in its speed, nor in the ability to make copies, nor in the clarity of its letters, but in the fact that it looked like type—that it wrote type—and turned any document into something resembling a printed page. Theirs was a prototypical version of the notion of computerized "desktop publishing" of our own era.

Mark Twain and his friend and fellow writer Artemus Ward stopped in front of a store one day in 1874 and saw the machine being demonstrated. Only after putting down his $125 for the machine did Twain realize that the woman using it had memorized the phrase she typed over and over—"The boy stood on the burning deck"—and that operating a typewriter was far more difficult than it looked. But according to legend he was to submit the first typewritten manuscript for publication: *Tom Sawyer*. Twain saw the typewriter in the same way he saw the Paige typesetter—into which he sank most of his fortune and enthusiasm for technology and ended up losing both. It represented another installment of the dream of the power of print; after all, as Marshall McLuhan argues, the first system of interchangeable parts had been invented by Gutenberg.

Between 1874 and 1878, only 4,000 or so typewriters were sold. By 1886, however, the number was 50,000. Early advertising touted the typewriter's usefulness in taking down telegraphic messages. But at the American Telegraph Company, the story goes, an employee named Thomas Edison personally rejected the invention. The typewriter key was the telegraph key multiplied, put on an assembly line—Scholes's crude prototype was that telegraph key set to print a single letter. In the finished machine, these keys were assembled, like so many workers, on a frame, each called to duty in sequence.

It took the development of touch typing to spur the typewriter's

Early Remington
typewriter, 1878.
(Smithsonian
Institution)

growth as an office instrument. The look of the machine now
changed, too, to match the office. By the time Remington offered
the No. 2 model, the case had been abandoned so that the works
stood naked on a frame, rather like a small loom. The story is well
known of the origins of the QWERTY keyboard. The arrangement
of letters in the QWERTY keyboard was a compromise between
ease of use—frequently used letters deployed beneath most agile
fingers—and intentional difficulty. The typist had to be slowed
down in order to assure that adjacent keys would not be struck in
quick succession and jam. But the idea of memorizing the keyboard
came only late. And, several recent studies suggest that even the
more "logical" layout of the Dvorak keyboard—created by a student
of industrial efficiency expert Frank Gilbreth—does not result
in significantly faster typing. The key to speed is memorizing
the keys.

This was all the more important in the early days of typing
because the line being typed was not visible to the typist. The first
typewriter whose letters were instantly visible was not produced
until 1890. As late as 1902, Sears was still advertising its "writing in
sight" feature as an innovation. For a while, some manufacturers
offered machines with complete, separate sets of keys for upper and
lower case. Not until 1878 did the first model appear with a shift
that offered both lower- and upper-case letters.

Indeed, the problems with the first typewriters, whose type

IBM Selectric I electric typewriter, 1961; design by Eliot Noyes.

resembled that of printing far more than does that of contemporary models, was that its products were often mistaken for printed materials. Personal letters were mistakenly discarded as printed "junk mail." And typewritten letters were dismissed as impersonal—a perception that is still quite alive today in the realm of personal correspondence. The notion of the signature as the personalizing touch came late; for twenty years, typed letters also ended with the name of the sender in typing.

For the expanding world of office work, the typewriter did more than speed things up, it also made multiple copying easier. It made use of the new carbon paper (patented 1869), which was to be employed in making masters for Thomas Edison's mimeograph machine. To be "compatible" with machines, as no less an observer than Le Corbusier noted, stationers standardized the size of paper and therefore the size of file folders and file cabinets.

Writers who followed Mark Twain to the keyboard found the typewriter changing their styles. For Henry James, the use of a secretary with typewriter changed his way of writing. The machine, he said, "fairly pulled the stuff out" of him. But he could not conceive of the machine without its operator—nor of himself as operator. Secretary and typewriter were one—they were the machine. T. S. Eliot, composing at the typewriter himself in 1916, found a very different result: "I find that I am sloughing off all my long sentences which I used to dote upon. Short, staccato, like modern French prose. The typewriter makes for lucidity, but I am not sure that it encourages subtlety."

The typewriter was a tool especially associated with the American writer. Isaac Bashevis Singer felt "really American" only when he bought his first typewriter—even though it was a Yiddish model. "In America," he said, "everyone had a typewriter." Hemingway's style seemed perfectly represented by the image of his hairy hands curved over the keys of his black portable—and that was the image that Scribner's kept on the back of his books for years. (Although in fact he tells of writing his stories and novels mostly with pencil, the typewriter stood as a symbol of his journalistic background and the topical material of his novels.) The work of Hemingway's imitators and successors, the kind of novels whose cover lines of "stark and powerful" Nabokov mocked, had to have been written on typewriters. So did Kerouac's *On the Road*, delivered to its publisher in one great typed scroll of telex paper and written in a style that elicited Truman Capote's *mot*: "That is not writing, it is typing."

The notion of coupling a typewriter with an electric motor to even the force of keystrokes was not realized until 1933, when manufacturers electrified it by adding the type of short-burst motor Charles Kettering originally developed for the cash register. With the shaping of modern typewriters, the look of office machines found a median between the home appliance and serious industrial equipment: it took the automobile as a de facto model. IBM and Remington electric typewriters of the early fifties, when the machines began to replace manuals, resembled the bulbous cars of the day. The first IBM Selectric, its curves crimped in no-nonsense edges, looked like, say, a Ford Falcon, the compact car Robert McNamara made a success at Ford around 1960, just as the Selectric was appearing.

The creator of the classic Selectric I was Eliot Noyes, who came to IBM from the office of Norman Bel Geddes and the Museum of Modern Art. The machine incorporated a technical advance: a "golf ball" of type replaced the arcuate arrangement of individual letters on their shafts. Noyes used this ball, which tilted and rotated rapidly, as an inspiration. Its movement seemed to radiate to the outside, to impress a rounded shape to the exterior shell Noyes created. The Selectric aimed at humanization of the typewriter, or "demechanisation," as Adrian Forty puts it, in an approach similar to that taken by Olivetti in Europe. Noyes said: "We tried to emphasize the singleness and simpleness of form by making the whole shape something like that of a stone so that you are aware of the continuity of the sides and under the machine and over the top."

The Selectric was created from clay models, molded by halves backed with mirrors to provide a full image, such as auto designers use. It was then replicated in wooden, and finally aluminum models, which Noyes himself filed to the required fineness of detail. The rounded case and face left nothing visible of the keys except their tops. The result looked almost organic—Noyes had written the catalogue for the Museum of Modern Art's 1940 "Organic Design in Home Furnishings" show, which focused on biomorphic and other natural shapes—such as a rock in a Japanese garden or a Noguchi sculpture.

It was the last step, perhaps, in the effort to present the typewriter not as something resolutely mechanical—almost Rube Goldberg—but as something natural to human ways of moving and thinking. The electronic typewriters and word processors that followed the typewriter were to achieve this goal almost perfectly.

☆   ☆   ☆

WHAT THE TYPEWRITER was to the secretary, the cash register was to the clerk, but the point of the retail machine was less to maximize speed than to minimize theft. It was famously "the machine that kept them honest." Each in its own way became an icon of efficiency. Both devices reduced human error—one the vagaries of handwriting, the other the vagaries of morality.

The first cash register, patented by James Ritty in 1878, looked like a clock, and like the clock it was aimed at regulating the work habits of employees—to keep them honest as well as prompt. "Ritty's Incorruptible Cashier," it was called, asserting the moralizing power of machinery.

Ritty had been the proprietor of a "cafe saloon" in Dayton and found himself plagued by petty embezzlement. On an ocean cruise, he is said to have observed a machine that recorded the rotations of the ship's propeller on a dial. The device he patented recorded transactions on a similar clock face.

The dominating characteristic of the company that sold these selling machines was sales technique. The formidable boss of National Cash Register (NCR), which turned Ritty's invention into a major product, was a man named John Patterson, a gruff, sloganeering booster who turned the company's Dayton, Ohio, headquarters into the high temple of high-pressure sales.

Patterson had bought one of Ritty's cash registers for a dry-goods store he ran to cut out petty theft. In 1884, Patterson bought the whole cash register business for $6,500. The company's chief asset was a storefront factory; it had sold only a few dozen machines. The next day Patterson thought better of it and tried to have the deal voided, in vain offering $2,000 to be released. It was the last doubt he would show.

To make NCR pay, he set up the first sales school and turned sales education into a sort of religious revival meeting, full of slogans. He demanded that his salesmen dress well. He invented the sales quota and the guaranteed territory assigned to salesmen. He used statistics to develop the sales quota—employing exactly the sort of statistics that the cash register could compile for his customers. Patterson added features to the machine: merchants could now total their sales by categories. Customers were provided with printed receipts to facilitate exchanges.

The new system also printed "Thank you call again" on each receipt. It was perhaps the first instance of packaged politesse, modular hospitality, canned kindness—the grandfather of such banalities as the smile button and the digital "Have a good day." It was a gesture in keeping with Patterson's basic notion: that he was selling a service, that he had to keep in touch with the customer. It was of a piece with his saturation advertising, addressing specific needs of potential customers, with the memorized 450-word sales talks he taught his salesmen, and with the flip-card talks he delivered, employing five-point lists.

The cash register was a machine that demonstrated a conviction in the corruptibility of man, and his redemption—or at least the removal of the temptation that corrupted—through machinery. Under Patterson, NCR would sell 1.2 million cash registers by 1913.

Cash register, 1879; clock face for indicating prices. (Smithsonian Institution)

One of Patterson's top salesmen was Thomas J. Watson. When he was hired in 1895, Watson was twenty-one years old and a former bookkeeper from Painted Post in upstate New York, with a background selling pianos, organs, and sewing machines for Wheeler & Willcox—the sewing machine concern formed when Wheeler bought Willcox & Gibbs. Most recently, Watson had failed at turning a Buffalo butcher shop into a chain on the model of Woolworth's.

Within a few years, however, Watson helped NCR achieve a near monopoly of the business by setting up used cash register sales businesses to undercut rivals, and falsely promising the imminent arrival of new machines that would match the improvements boasted by competitors and also be less expensive.

Watson was a star NCR man, and he became Patterson's favorite.

He, like Patterson, believed in the proper wardrobe: the white shirt that was to become legend at IBM. "Clothes don't make the man," he would admit, "but they go far toward making the businessman." He would also take with him one of Patterson's many slogans: "Think."

In 1910, the American Cash Register Company, headed by former NCR employees, some still smarting at having been fired by Patterson, sued NCR for antitrust violations. Patterson and Watson were convicted and sentenced to fines and a year each in jail. Both men won a new trial on appeal, but Patterson fired Watson and moved to rehabilitate his image with the assistance of a natural disaster. In 1913 a massive flood hit Dayton, and Patterson threw all his efforts into helping the citizenry. His resulting popularity prevented the case from returning to trial. Watson, meanwhile, joined the Computer-Tabulating-Recording Company and vowed to have his revenge by building it into a company as big as Patterson's. Patterson had a pattern of nurturing promising executives into protégés, then firing them. Cutting short all the saplings around him, he left the company leaderless after his death.

Another employee was a tinkerer and former telephone lineman named Charles Kettering. When NCR realized that its increasingly elaborate series of punch tapes and printed receipts required more than human motive power, Kettering and a team closeted themselves in a workshop named "Inventions 3" and developed a special electric motor for the job, one producing the powerful, repeated short bursts of power the machine needed. Kettering went on to develop a telephone credit-authorizing system, called the O.K. Telephone, the predecessor of credit-card authorization machines.

Later, disillusioned with Patterson's tough-minded regime, he began to moonlight for Charles Deeds, an NCR executive with a hand in the new world of motorcars. Several NCR alumni, fired by Patterson, had gone into the new automobile industry. One was Edward Jordan, founder of the motor company whose car bore his name. Working nights in a barn for Charlie Deeds, Kettering became more and more taken with the automobile, and in 1909 he left NCR to devote full time to it. Within a few years he would replace the crank on the automobile, as he had replaced the crank on the cash register, with a high-torque, short-burst electric motor—the self-starter.

When Patterson bought NCR, the wooden fronts of his first cash

National Cash Register models, 1890s and 1919. (Smithsonian Institution and NCR Inc.)

registers were decorated with painted flowers. But as time went on they evolved in design, growing more like strongboxes, but with decorations, polished to high glow, that spoke of gentility as well as respectable wealth.

The NCR register was a sort of physical version of the NCR salesman himself. NCR cash registers were turned out in as dapper fashion as Patterson demanded his salesmen be. NCR's body styles—sometimes lower, sometimes higher—varied their cast ornament in near lockstep with the bank notes and postage stamps of the day.

A cash register had to fit in with store décor and yet, since stores were many and various and the cash register's shape one, it could not be overly specific in style. Like the salesman, it had to get along with everyone. So by the thirties, the basic NCR model had become all shiny stainless steel in the hands of an industrial designer, Walter Dorwin Teague. Alfred Barr and Philip Johnson included a specimen in the Museum of Modern Art's Machine Art show. That glistening shift of highlights and shadows was the ornament of the machine age; in adopting it, NCR was doing nothing more than it did in the Gilded Age, when it donned stamped-brass arabesquerie. That they were durable machines is shown by the fact that they are still at work all around us. Well before old telephones, old cash registers, with their solid, chunking metallic sound, entranced us. That the object summed up the company—whose ideal was an object that sold itself—was shown in 1939, when the NCR pavilion at the New York World's Fair was opened. It was a building-sized version of one of those stainless-steel registers—designed by Walter Dorwin Teague—in whose numerical windows the Fair's daily attendance was toted up, a shining little temple to clerical virtue.

CHAPTER FIVE

Easy Chairs

WHAT COLLINS DID for the ax, and Ames for the shovel, Lambert Hitchcock did for the chair. He contracted with the state of Connecticut for prison inmate labor and applied water-driven machinery to produce a product that standardized common designs, stamped his name on it, and cut prices. His chairs traveled: they went in pieces or whole, by peddler's wagon and even muleback, from the town created around his factory, Hitchcocksville, Connecticut. From 1821 on, he sold parts—chair kits—to dealers in Charleston and other points of embarkation for the backwoods. His chairs were designed to move—to travel, but also to sell—by providing the potential buyer with an item of perceived elegance previously unthinkable for the less than wealthy.

Hitchcock's innovation, however, was not interchangeable parts but interchangeable decoration. He adapted the Yankee machine-tool school of thinking to decoration. Beginning around 1818, he pioneered the use of complex sets of stencils, which he called "templets," to create painted and gilded fruit and flower patterns adapting folk methods to the factory. His "templets" were analogous to the templates and dies of the machine tool. Crude, amateur stencils had been common, but Hitchcock marbleized, wood-grained, striped, and banded his furniture to approximate, in two dimensions, the three-dimensional ornament of traditional furniture, which he had neither the skilled workers nor the money to provide. He borrowed shapes from Duncan Phyfe—and the Windsor chair and common rockers. There was even a Hitchcock painted portable writing desk not unlike Thomas Jefferson's.

He began the use of decoration to sell to the middle classes: the

Basic American
ladderback chair.

graining and staining, the painting and stenciling are all ancestors
of the sort of machine-made ornament, in wood or iron or painted
tin, that was to be so detested later by the high culture. In retrospect, the meretricious shapes and decoration of Hitchcock furniture, which turned craft elements into machine ones, can be seen as
the ancestors of the toaster covered in simulated wood grain with
accents of floral decal, the "marbleized" plastic radio case, and the
automobile with painted pinstripes and simulated wood grain.

While Hitchcock's chairs were moving across the country, other
sorts of chairs that themselves moved were changing the idea of
seating. Just as Americans learned about new-fangled luxuries they
would never enjoy at home while staying at hotels, they have found
themselves better seated first on railcars, then in automobiles and
on airliners than at home. Today, the most comfortable chair owned
by the average American is in his car. "The automobile manufacturers," Walter Dorwin Teague, the champion of streamlining, argued in the thirties, "have made, in the past few years, a greater

contribution to the art of comfortable seating than chair builders had made in all preceding history."

Like the auto seat, American chairs move. Some chairs are modeled on those of transportation. Some are simple chairs that can themselves easily be moved, folded, or broken down. Some do their moving in place—like the classic rocking chair or its heirs, those elaborate mechanical thrones capable of an unlimited number of positions. But at the extremes—the basic model and the kit of possibilities—are two basic types: the simple ladderback chair (commonly called a "plain" chair) and the easy chair (the Barcalounger, the La-Z-Boy, and their like).

☆   ☆   ☆

THE MOST AMERICAN of chairs, all will agree, is the rocking chair—even though no one can prove it is originally American. Benjamin Franklin was not its inventor, despite legend. The American, ran the classic European observation, must be in motion even while sitting. The rocking chair, like chewing tobacco or gum or like whittling, summed up a characteristic American restlessness and need for activity. The clear inspiration for the rocker was the cradle. Rocking was thought to be soothing and ameliorative. In the crudest homes, it was a pleasure that loomed disproportionately large. Visitors were always offered the rocker.

But even the chairs without rockers were rocked. The characteristic American slat-back or ladderback chair was made for leaning back in. Even the Shakers, hardly the most casual of people but the ones who brought this chair to a state of stern and sleek refinement, understood this. They put various devices on the bottom of their ladderbacks to allow leaning. One was patented in 1852 by George O. Donnell. His patent, number 8,771, protected "a new and improved mode of preventing the wear and tear of carpets and the marring of floors, caused by the corners of the back posts of chairs as they take their natural motion of rocking backward and forward." Natural or not, the Shakers did not hold with it. They forbade rockers among themselves, except for the infirm, and sold most of their production to the world outside—the world that Donnell soon joined: the Shakers disapproved of patents also.

The Shaker ladderback was a rationalization of the type familiar

Steven Holl's Shaker-
inspired "Linear"
chair, 1980s. (Pace
Collection)

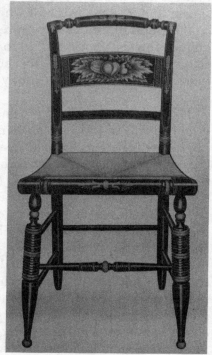

Hitchcock chair, 1840s. (Index of
American Design)

in Europe from medieval times—straight legs and backs made up of horizontal slats. Its ancestors were familiar in Holland and arrived in New Amsterdam; its Italian cousin, the chiavari chair, was updated by designer Gio Ponti in 1955 as the Superleggera. Steven Holl's 1980s "Linear" chair is another abstraction on the same theme: a meditation on the Shaker version of the ladderback.

You can easily follow the evolution and variation of the ladderback from the upright chairs of New England—straight and stiff, fit for a Brewster or Mather—through their intersection with lighter Dutch and German models, down through the mountains where they become as lean and bony as the Appalachian pioneer, their stretchers unlathed, but their backs bent back in relaxation now. Their finials first lost their ornamental forms and then flattened into "mule's ears," and the oak splint or rush-woven seats gave way in the South to cowhide as the river of cultural migration fanned out into a country-and-western sort of alluvium. Legs grew shorter and squat in the crudest backwoods version; in the plantation big houses, legs became higher and the chair taller. (Short legs, folk saying had it, were a sign of white trash.)

The finest of those chairs were hickory, their legs not a fraction of an inch thicker than they had to be, their slats like miniature barrel staves, their roundels no thicker than a first grader's pencil, their main members changing subtly from the bottom of the leg to the smooth upright of the back.

The ladderback was light enough to transport easily; it could even be broken down for movement, its roundels or stretchers slipping out of the legs, its seat easily replaceable. (A number of such chairs, it is reported, were taken from the rural South to Washington during the Poor People's March of 1968, borne along like spirituals and blues songs as reassuring artifacts.)

The Hitchcock factory created disassembled versions of more high-falutin' styles, such as Sheraton. And with the director's chair—a piece of military camp furniture with origins in Greece, used in the Civil War and made popular in the 1940s—and the suburban folding aluminum chair, the tradition continued as the sharecropper moved into the mobile home. The folding chair was always respected in this country. George Hunziger made folding chairs first, then fixed ones; his chairs rode with the eminent Bostonians privileged to ride the showcase train on the Transcontinental

Railroad, accompanying the adjustable mechanical chairs of the railcars.

Siegfried Giedion accomplished the pioneer investigation of the ways the railroad chair was adapted to barber chairs, dentist's chairs, invalid's chairs—there was a clinical quality to the type—and finally home recliners. To many, such chairs exemplified a passion for movement while seated that reflected an almost pathological restlessness.

By 1856, when Herman Melville wrote *The Confidence Man*, the adjustable chair was already mature enough to experience his derision. He saw it as a metaphor for the restlessness of the American soul, darkly perceived. He wrote of a "Protean easy chair" like so many patented in the 1840s and 1850s, that was "so all over be-jointed, behinged, and bepadded, everyway so elastic, springy, and docile to the airiest touch, that in some one of its endlessly-changeable accommodations of back, seat, footboard and arms, the most restless body, the body most racked, nay, I had almost added the most tormented conscience, must, somehow and somewhere, find rest."

This was a chair for an anxious man in an anxious age, and the idea that seating reflected a wider cultural outlook persisted into the twentieth century. The designer Paul Frankl argued that "people sit differently today than they did a hundred years ago, [in] more natural positions of repose." Siegfried Giedion argued something similar: that the cultural ideas of correct position for sitting had changed.

The development of railroad seating gave impetus to the study of human proportions—ergonomics. From the quasi- and pseudo-science of the nineteenth-century patent rocker, which reinvented the wooden rocker as a spring-supported mechanical one, ergonomics moved to a study of the actual body, in specific and in average. The impetus here was with commercial travel. In 1945, a Professor E. A. Hooten studied the proportions of 3,867 adults to improve the ergonomics of rail seats, and the process was continued by the auto industry. If the nineteenth century saw the railroad chair come home, the twentieth saw it conflated with the automobile and airplane seats in brand-name "loungers." From Barcalounger, based on the "scientifically articulated" chair of Dr. Anton Lorenz (formerly of the Kaiser Wilhelm Institute in Dortmund, the

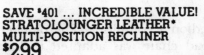
Advertisement for
Stratolounger
recliner, 1989.

company's literature boasted) to the La-Z-Boy, offspring of a folding bed and modeled on a folding camp-style chair, to the "organic"
forms of the Eames chair and Ottoman, the goal was the same—to
be as comfortable at home as on the train or plane. For such an icon
of the Moderne as Kem Weber's Airline chair, the frame was tapered and trimmed as if it, like a moving vehicle, had to reduce
wind resistance.

The ruling motive behind all these chairs was an ideal of ease that
would have been familiar to Jefferson, with his Windsor-based recliner and reading chairs. Jefferson had adapted the Windsor for a
writing desk and turned it into a revolving chair—an innovation the
Shakers were later to claim they were the first to make. Political
opponents derided this chair "in which you could turn your head
without rotating your rear end." If he got uncomfortable there, he
could slip over to a Windsor matched with a footstool and an
overhanging low table. The ideal of comfort transcended class. The
dominance of the articulated lounge chairs for the manly den ran
the social gamut, from the Eames lounger down to the Archie
Bunkerlounger—the Middle American mechanical easy chair,
padded now with cushions and springs, and skinned in tough fabric
or Naugahyde. The Eames chair took the approach of fixed

flexibility—the Bunkerloungers, of all brand names, featured everything adjustable, like car seats. There were, of course, so many
possibilities that no one could figure out the most comfortable
settings. It was a chair of inexhaustible possibilities.

These chairs combined the seat and Ottoman into one unit.
Charles Eames's chair kept them separate but still related—like a
house and garage. Less adjustable, attempting by shape alone to
offer the body the same variety of positions the other chairs offered
mechanically, the Eames chair has a beauty that comes from the
relations of its component shells in space, a beauty that is part of the
era of its development but also reminds you of concrete shell buildings of the heroic period of Expressionist engineering—of Nervi
and Saarinen—as well as certain contemporary cars. This was "organic" design at its best, and the chair took on a certain cachet. It
was purchased by those who looked down their noses at the recliners; it was listed at the top level in Russell Lynes's famous chart of
highbrow, middlebrow, and lowbrow items. And its price was about
twice that of the big padded chairs. But its sitting qualities were not
unlike those of the Barcalounger and La-Z-Boy. By 1966, Barcalounger was trying to pick up "design-conscious" customers by
commissioning Raymond Loewy to reshape its chairs.

Such chairs became symbols of the self-satisfied American,
watching television, and settling back into his prejudices like cushions. Easy in a Stratolounger—the name suggesting an airliner or
intercontinental bomber with equal readiness—the world was laid
out beneath him.

The sixties went to the other extreme: that of the hang-loose
beanbag and its cousins, of dripped or piled plastics, or in cute Pop
shapes—the giant baseball glove or the 1969 "Volkschair," based on
the body of the Volkswagen Bug and offered, of course, in green. In
such a chair, your every bone was allowed to do its own thing,
resulting in a general assault on your skeletal establishment.

☆　☆　☆

THE SEARCH FOR the "scientific" chair gained momentum as ergonomics developed, especially in the office of Henry Dreyfuss.
Part of the discipline of "human factors," ergonomics expanded on
its success on the rails by joining the war effort. The Pentagon had
made use of it designing tanks and aircraft during World War II

(the La-Z-Boy company built some of that seating). The military had been a source for such developments before: during the Civil War, Quartermaster General Montgomery Meigs, a brilliant engineer, was developing a sizing system for uniforms that would survive into civilian ready-to-wear clothing.

Henry Dreyfuss carried ergonomics into virtually every corner of civilian life. In his *Designing for People* he created two fictional everypeople, Joe and Josephine, bearing the proportions of the average American man and woman. Le Corbusier had used an ideal—the Greek Golden Section, the ratio of eight to thirteen—to devise his Modulor system of standard dimensions and a vision of the "typical" modern man as exactly six feet tall (a statistic, associates said, taken from his reading of English detective stories). Dreyfuss had the real number, the national average: Joe was five feet nine inches and a fraction.

The Dreyfuss office had been a consultant for Lockheed's Constellation and Electra airliners, and Niels Diffrient applied the dimensions when he worked on the interiors of those planes for American Airlines. With the real-world numbers of ergonomics, the designer could argue for the shape of his chairs, trays, and other equipment with more than subjective evidence. "We could," Diffrient told the designer historian Arthur Pulos, "at long last, present our concepts to engineers with more than an aesthetic rationale . . . we gained stature and a new power."

Diffrient, who began in Dreyfuss's office, refined this study of human dimensions in his *Human-scale* series of books, published beginning in 1974, and applied the results in his Advanced Operational Task Chair—a name worthy of the Pentagon—for the furniture firm Knoll.

Diffrient carried his fascination with ergonomics into the Jefferson chair, a bizarre combination of a recliner that resembled a dental chair and a swinging stand for computer monitor and keyboard. Diffrient's Jefferson chair is his ultimate statement of these unarguable numbers; but in appearance it is intimidatingly mechanical and slightly frightening, with suggestions of invalidism. It seems designed to minister to exactly the sort of spiritual uneasiness the Confidence Man addressed.

☆   ☆   ☆

JUST AS THE railroad car evolved from simple style and democratic arrangements to class divisions, the office chair "for everyone," to fit any body, has grown from the single mechanism of the late nineteenth century into a careful hierarchy of chairs (with headrests and heights of varying sizes) to indicate the progression from secretary to CEO, like the progression from Pawn to King.

The Alma brand Zucomat chair, for instance, boasts a scientifically designed back, control of seat height and back angle, and "comfort flex." But for all its science, the chair comes in a hierarchy of shapes, from "operator's chair" (low back and no arms) to workstation (higher back), to guest chair (higher back still), and on through manager's, executive's, and director's chairs. In these last three, the back has reached its maximum, and the distinction is made by the application of varied pleating of the cushions.

☆   ☆   ☆

THE REAL DESIGN triumphs came in the most common of chairs—resort and lawn chairs. The Adirondack chair, associated with resorts and camps, with an origin in "rustick" twig-and-barked-plank furniture, has proven a remarkably adaptable shape over the years. The Adirondack sums up a familiar American trait: a stylized attitude toward the natural. It was the product of the great camps and resort hotels of the Adirondacks, Berkshires, and eventually Maine, where the cult of the "rustick" reigned in post–Civil War America. This cult first took the form of uncomfortable chairs of twisting limbs with the bark left on—so-called "stick" furniture. But the search for the natural was even then in conflict with the appreciation of comfort, and the Adirondack struck a balance.

With its long, tilted back and seat angled against level arms, the Adirondack chair suggests a simple cabin or a dock on a lake. The Adirondack chair's angle is optimistic: it sits like a tail dragger airplane ready for flight. Its builders eschewed the lathe, that basic tool of most furniture makers, and with it gave up the curve. In its resolute use of straight lines and simple boards, in the way it assembled planes into a three-dimensional sculpture of uncommon comfort, the Adirondack chair was as visionary as the furniture of Gerrit Rietveld, the Dutch prophet of De Stijl. It is almost a cartoon of a larger, upholstered easy chair, a drawing of an easy chair broken down into planes and rebuilt in wood, like a Cubist sculpture.

For such a seemingly simple item, the Adirondack lends itself to an amazing variety of stylization, as witnessed by variants advertised in catalogues and magazines. The boards of its back can take on all sorts of variations beneath the blade of a jig saw: the chair comes in Gothicized versions, Deco ones, gently streamlined ones, and so on. In the Deco version, the five planks that make up the back have been rounded at the top to suggest a thirties skyscraper. The arms are similarly curved into little wings, and the beams of the seat are tapered like the stern of a sailboat or the rear of an airplane. An optional trim footrest or stool ramps up like a loading dock to the front of the seat. They are painted up in colors such as Chinese Red and Mariners Blue, whereas the traditional item came, as a rule, in green or white. They are cedar, while the original was pine or occasionally oak. Upscale garden catalogues are full of variants on the Adirondack, their shapes gently softened to imply the respectability of old summer places that have been in the family for generations.

American aluminum and nylon-webbing chair.

The next revival will no doubt be of the folding aluminum lawn chair, whose green and white webbing, once universal in American suburbia, is at once the realization and a parody of the Bauhaus ideal of inexpensive democratic seating based on modern industrial techniques.

Born of the need to employ the vast aluminum capacity created by World War II, it is composed of four simple angular C shapes of tubing and an abbreviated basketry of plastic webbing—as if the seat of a woven oak splint or textile tape chair had been blown up in scale. The arms tend to be stamped in a channel pattern whose abbreviated ornamentation would have to be called Moderne. California beach style brings it inside, to the living room. But even so positioned—as camp—the chair was valued for comfort, informality, and convenience.

Light enough for a four-year-old to fold and carry, its extruded tubular frame recapitulates the seats of the Chrysler Airflow or the Piper Cub. And it is more comfortable than 90 percent of the chromed tubular-steel chairs you will find at high-design outlets. In fact, it is too inexpensive to be produced any longer by American industry and has for some years come almost exclusively from the Orient. It has become something rare for furniture, a true commodity product.

The aluminum lawn chair has been underappreciated, largely because of the places it has been seen and the people who have been seen in it: it is associated with ugly backyards full of birdbaths and the ugly backsides of fat women in curlers. But with its deceptively simple combination of squared-off loops and the carefully calculated amount of give to the webbing, it is a small marvel of engineering. In economy of means, simplicity of concept, and low cost—in its possession of our favorite national virtues—the aluminum lawn chair is also true heir to the basic ladderback.

CHAPTER ☆ SIX

Kit Homes: Log Cabin and Balloon Frame

DURING THE CAMPAIGN parades for the election of 1840—the famed "Tippecanoe and Tyler too," or Log Cabin campaign—the centerpiece of torchlight parades was often a quarter-scale model of a cabin.

This log cabin is in our heads: the perfect cabin, replicated in bottle, box, and tin, or flattened to figure in quilt and coverlet. It is strangely abstract; the logs are dowel-like, generic, barkless. It is a prop, a model. And no such model has become a more subliminal part of American minds. The log cabin is not only the basic American icon of yeoman birth, but a model for the basic ideal American house that was to have spiritual descendants in houses ranging from bungalow to ranchhouse.

The amazing thing about the log cabin is how quickly it turned from the necessary to the picturesque. More than a living space, it became a container of national myth. Ancient in conception, European in origin, it nonetheless became a symbol of what was new about life in the New World. It has survived as a cliché symbol of naturalness and natural virtue, from Uncle Tom to Agee and Evans's sharecroppers. It crops up in all sorts of places, as a device of iconic rhetoric. A postcard showing an allegedly happy black Southern family in front of their cabin. A hamburger restaurant on Central Avenue in Los Angeles. A cabin-shaped car deodorizer, to dangle from the rear-view mirror.

100

Log cabin model used
in 1840 presidential
campaign.
(Smithsonian
Institution)

I remember very distinctly from childhood the day in the mid-
fifties that Vermont Maid maple syrup ceased to be packaged in
small, log cabin–shaped tins—with the chimney as pour spout—
and was henceforth available only in glass and later plastic bottles.
At the same age, I was playing with Lincoln Logs, the invention of
John Lloyd Wright, the son of Frank Lloyd Wright. According to the
accepted story, however, it was not any prairie construction that
inspired the younger Wright, but a scene he witnessed in Japan in
1916 when he was in his early twenties. Accompanying his father to
the site of the famed Imperial Hotel, he saw workers lifting great
timber beams into place.

Apparently a young man still in search of vocation, John Wright
produced the toy version of the logs two years later. (His father's
own rapturous memories of the aesthetic and intellectual benefits of
playing with Froebel blocks in his childhood may have been an
influence as well.) It was not until the mid-twenties that Lincoln
Logs caught on. They came in a box bearing a rustic framed picture
of Honest Abe, and the legend: "Interesting Play Things Typifying
'the Spirit of America.' All forms of log construction can be 'worked
out' with Lincoln Logs."

The packaging suggested a physical model of virtue and industry, as well as an educational kit teaching the physical facts of

construction. The toy fulfilled the former purpose better than the latter. Making windows and doors, I well remember, required supporting the long logs with little stubs of small logs, making the whole set obey a logic of construction that was its own, closer to Lego than the real thing. But as early as 1865, a man named Joel Ellis sold a toy called the Log Cabin Playhouse which anticipated Wright. It was marketed mainly on the strength of the contemporary apotheosis of the rail-splitter.

☆   ☆   ☆

THE WORD "CABIN"—earlier terms included the simple "log house," as well as "pen and crib"—came to signify the basic unit of habitation, an American version of the English cottage. How many times

Log Cabin Syrup
advertisement, 1919.

have you been told that someone has "a cabin up at the lake," only to arrive and discover a small transplanted suburban house?

From the log cabin to Levittown, Americans searched for the basic house, the universal housing appliance. This basic housing model has been based on a kit—a kit of materials, from logs to two-by-fours, and a kit of styles, from gingerbread Gothic to bungalow Empire, from prefabricated houses shipped west to Gold Rush California to Sears, Roebuck houses out of a catalogue, from the houses built to plans in builders' books to the five plans in the developer's office. The American has built a house—a commodity to be bought and sold—which must then be made into a home.

Every so often, when the country at large notices that there are not enough well-built houses and apartments, we hear again a refrain that started almost a hundred years ago. "Why is it," the question goes, "that we can't build houses in factories, the way we build other things—cars, refrigerators—so well?"

But we can, we do, we have: our houses have been kits.

The log cabin was the first of these kits, built of a repeated element whose thickness nature itself presented with remarkable consistency, and whose length was dictated by the strength of the average man who had to lift it. It was a revival of a Neolithic building form. When the drafty imitations of English wood-and-wattle houses built by the amateur builders failed at Jamestown and Plymouth, other pioneers turned to a long disdained method of building that had survived, as it survives today, in Scandinavia and Russia.

By 1876, when Custer died and the nation celebrated its centennial, the idea that the first settlers had stepped off the *Mayflower* and built log cabins was firmly ingrained in popular mythology. It was not until 1939, in the midst of yet another "rustic" vogue, that a man named Harold R. Shurtleff published his book explaining the true state of affairs. In *The Log Cabin in America*, Shurtleff pointed out that the first houses at Jamestown and Plymouth were not log cabins at all but crude replicas of the houses of England, made by amateurs.

The log cabin gained its foothold in the Delaware Valley, largely through the agency of Swedish settlers, who arrived in 1638 at the mouth of the Delaware, where they founded Christinaham. From there, the cabin's virtues of comfort and tightness led to its use by Scotch-Irish, German, and other settlers, and carried it into the

Pennsylvania hinterland west of Philadelphia, then down the Appalachian valleys into the southern frontier and East Texas and, simultaneously, across the mountains into the Ohio Valley and west until the wood gave out and sod became the material of necessity. In Alaska, the Russian version of the cabin left its mark.

We have all likely seen plenty of log cabins, *in situ* and in the National Park Service restorations at such places as the Great Smoky Mountains National Park. But the nearest thing to a proto–log cabin available is at Bridgeton, New Jersey, where the Swedes settled. From this well-made prototype, the variations over roughly three hundred years—for cabins were still being built in the middle of this century—seem less significant than the similarities.

The shape of a basic log cabin was almost geometrically solid: one cube and another cube sliced diagonally in half set atop it for a roof (about 16 or 17 feet on a side, or an English rod). Its size was limited by the size of logs that could be practically obtained, moved, and placed: weight kept them to a maximum of about 20 feet. Thus, expansion of the basic one- or two-room cabin required a second module, unit, or "pen." The double-pen cabin, with breezeway in the middle, is the "dogtrot" cabin.

The log cabin's variants: round logs versus hewn logs, variations of notching, were all subsumed in the abstract ideal of the cabin. The basic form held sway in the "upland South cultural hearth," as it is called by the cultural geographers who have attempted to map and categorize the cabin by such criteria as notching (half saddle, full saddle, dovetail, and so on).

While a cabin could be built by two men, it was generally constructed by community assistance—as so many things on the "individualistic" frontier were.* In the winter of 1792 my ancestor, James Patton, a few years off the boat from Ireland, built himself a house in Wilkes County, North Carolina. "I got the assistance of the neighbors who were very kind, and in two weeks had a comfortable house

---

* Quilts, too, were put together of available materials, found like the logs of the cabin. A quilt was often the product of the days women spent working together while the men raised a barn or house. The cabin was a structure just two or three men could raise together. A barn, with its heavy framing, or a post-and-beam house, required a good number of hands, building sections on the ground, then pushing and pulling them into position. The log cabin quilt pattern was also known as the barn-raising pattern. One of the most common American quilt patterns, it assembled long strips of cloth in a way that reflected the notched corners of its namesake.

a story and a half high to move into. It was built of pine logs and covered with clapboards," he recounted, in a memoir filled with invocations of the wisdom of Dr. Franklin and advice to his children and grandchildren to pursue a plain and simple life.

☆   ☆   ☆

BARELY TWO CENTURIES after its revival in the New World, the cabin had become a kit, too, of associations and social and political meanings. It was that form that all future American housing would follow.

Andrew Jackson was not only born in a log cabin but lived in one repeatedly while he was building and then renovating the Hermitage—a house of such size and so filled with visitors as to make a mockery of its name. Successive expansions also became renovations that brought the house from its original Federalist style (1821) to Palladian (1831) to Greek Revival (1836). During the rebuildings, Jackson retired to the cabin that was the original structure on the place. He did not mention it during his two presidential campaigns; his legend focused on his rise to the status of the Hermitage, not his low beginnings.

Double-pen log cabin, circa 1820, where Andrew Jackson lived before construction and later during renovation of the Hermitage.

The Hermitage, as it appeared after 1836.

But by 1840, the log cabin had become the dominant symbol of American politics. In the presidential campaign of that year, the Whigs used it to market their candidate as a man close to the people. The log cabin stood in contrast to the alleged corruptions of "the Presidential mansion." They defeated Jackson's heir, Martin Van Buren, by offering their own general with his own humble background and outdemocratized the Democrats.

General William Henry Harrison was the fairly colorless leader of the Battle of Tippecanoe Creek, an engagement in which Indiana militia burned an Indian village and then retreated to safety. But when a Baltimore newspaper controlled by the Democrats attempted to make fun of Harrison by speculating that "upon condition of his receiving a pension of $2000 and a barrel of cider, General Harrison would no doubt consent to withdraw his pretensions, and spend his days in a log cabin on the banks of the Ohio," the log cabin as political symbol was born. The Whigs jumped on the image, parading model log cabins mounted atop poles, donning coonskins and rolling cider barrels in torchlight parades. Abraham Lincoln, a young lawyer, made some forty speeches on behalf of the ticket. As the other half of their strategy, the Whigs succeeded in depicting Van Buren as an aristocrat of high tastes—just as Jackson had succeeded in depicting John Quincy Adams as a New England aristocrat.

Among the artifacts of the elaborate marketing of General Harrison is a whiskey bottle in the shape of a log cabin, produced by a certain E. G. Booz of Philadelphia. Mr. Booz not only contributed his name to the language, but created a memorial to the birth of packaging. Whiskey, like the hard cider of the campaign, was previously distributed from barrels into the owner's own tin cup or jug. The use of branded bottles was a recent innovation in 1840.

The same packaging that was beginning to be applied to everything from whiskey to soap was applied to William Henry Harrison, whose political beliefs remained a cipher not only through the campaign but through the droning, three-hour inaugural address during which he caught his fatal illness, and the single month he spent in office.

☆   ☆   ☆

THUS THE LOG cabin was an established motif when Abraham Lincoln's handlers picked it up. Birth in the cabin was an added twist—like birth in a stable. Any number of politicians were able to play this tired melody into the twentieth century. Lincoln himself was reticent about his log cabin days. He wasn't just modest about his rise; he was, like so many self-made men, genuinely ashamed of his origins. When he told country stories, it was as a lawyer who worked for railroads and real estate interests. In his campaign biography, written by a newspaperman from a long interview, he begged off telling about his background by half jocularly, half sentimentally describing his early days in a quotation from Gray's "Elegy": "the short and simple annals of the poor."

Although he did not deny it, Lincoln did not speak of being born in a cabin. Nor could he have done much more than tolerate the media experts of the age, who depicted him splitting rails with an ax—a goof that no doubt brought smirks anywhere outside the cities.

Thanks to the work of the campaign managers, surviving birth-places of the presidents or simple cabins have become bathed in nostalgic value. Consider the tiny log cabin set on a patch of grassy park amid the skyscrapers of Dallas, the one under a temple top in Salt Lake City, those in restored pioneer villages across the country—and the restored cabin of Lincoln's own alleged birth-place, inside a National Park Service building in Hodgenville, Kentucky.

In truth, of course, most cabins were miserable dwellings. The cabins Frederick Law Olmsted saw in South Carolina in the 1850s were made of barely hewn logs, unchinked. "Through the chinks," he wrote, "as you pass along the road, you may often see all that is going on in the house; and, at night, the light of the fire shines brightly out on all sides."

In this country, no one wanted to live in a log cabin who could live in anything better. The cabin was always understood as a provisional dwelling, and was built no better than it had to be. The owners might cover that rude cabin with whitewashed clapboards, as my ancestor did, or with asphalt paper simulacra of brick or stone, as later generations did. But it is also fair to say, if nobody ever lived in one a moment longer than he had to, nobody ever failed to become nostalgic for one less than a moment after he left it.

To stay in it longer was to risk being considered no account, a Mudsill. The cabin was somewhere you lived until the wilderness was tamed, the fields cleared, the barn built, and the big house finally finished. Writer Hector St. John Crèvecoeur built a cabin as a temporary measure, then built his barn, and finally built himself a substantial two-story residence. But—and this is critical—after you had moved to the big house, the cabin was not torn down. Occasionally, in the South, it remained as kitchen or storage shed—and as remembrance.

Or, on the Southern and Midwestern frontier, you might gradually build the cabin into a mansion. One cabin would be joined by another, separated by a passage, to form a saddlebag, or, if the passage was open, a dogtrot house. A verandah, perhaps even with classically trimmed columns, might go on the front. Clapboards could cover the logs as soon as the sawmill arrived, and whitewash could turn the whole thing into a mock Palladian villa. This was "classing up the joint"—class was an appliqué, an embellishment of basic virtue by the bravery and endurance that brought wealth and fame.

☆   ☆   ☆

THE "CABIN" IS frequently seen of course in contrast to the "mansion"—usually of elevated location. The term encompassed the big house and the quarters, but also the poor white shack— maybe a shotgun house—and the banker's residence. And it is in the cabin that traditional values survive, while they are lost in the avarice and detachment of the "mansion." That juxtaposition is the center of the use of the cabin in the 1840 campaign. The cabin not only proved the virtues of Harrison, but stood starkly opposed to the "corruptions" of Van Buren's presidential "mansion." Hawthorne's House of the Seven Gables, with its superstructure of Puritan guilt and oppressive aristocracy, was built on the site of the log cabin of the honest yeoman cheated out of its site. In Faulkner's *Absalom, Absalom!*, part of the unnaturalness of the self-made aristocrat, Thomas Sutpen, is that he leaps from obscurity right into the mansion—built by slaves straight from Africa and a "captive" architect straight from France—without pausing at the cabin stage. Hank Williams sings to his lost flame in a "loveless" mansion on the hill, and a kitschy Elvis poster depicts his progress from shotgun house to mansion on the hill—Graceland.

The cabin was the natural house for the natural man—that was the myth. Thoreau, writing in *Walden* five years after the Log Cabin campaign, still echoed the virtues it had objectified. He saw architectural beauty only in that which grew

> out of some unconscious truthfulness, and nobleness, without ever a thought for the appearance. . . . The most interesting dwellings in this country, as the painter knows, are the most unpretending, humble log huts and cottages of the poor commonly; it is the life of the inhabitants whose shells they are, and not any peculiarity in their

surfaces merely, which makes them *picturesque*; and equally interesting will be the citizen's suburban box when his life shall be as simple
and agreeable as the imagination, and there is little straining after
effect in the style of his dwelling.

This is the Thoreau who, with some seriousness, in effect advocated
housing the homeless in toolboxes he had seen by the railroad,
coffinlike structures with airholes drilled in the tops.

But even Thoreau, reducing things to basics, did not feel the
need to go so far back as to build a log cabin. His house at Walden
was a post-and-beam structure, assembled from second-hand lumber and beams, six inches square, he hewed himself with a borrowed
ax. ("The owner said it was the apple of his eye, but I returned it
sharper than I found it.")

Thoreau was either not aware of, or like Andrew Jackson Downing, landscape architect, was disdainful of the balloon frame. Even
though machine-made nails, by 1844, were selling for a nickel a
pound, he paid $3.90 for the handmade ones he used. Thoreau's
house at Walden, mortised and tenoned in the New England tradition, violated all principles of economy.

Later, in 1857, Thoreau in Maine rhapsodized over the naturalness of the cabins he saw there, "the projecting ends of the logs
lapping over each other irregularly several feet at the corners gave it
a very rich and picturesque look, far removed from the meanness of
weather-boards." He praised the cabin's

successive bulging cheeks gradually lessening upwards and tuned to
each other with the axe, like Pandean pipes.

It was a style of architecture not described by Vitruvius, I suspect,
though possibly hinted at in the biography of Orpheus; none of your
frilled or fluted columns, which have cut such a false swell, and
support nothing but a gable end and their builder's pretensions,—
that is, with the multitude; and as for "ornamentation" . . . there were
the lichens and mosses and fringes of bark. We certainly leave the
handsomest paint and clapboards behind in the woods, when we strip
off the bark and poison ourselves with white-lead in the towns.

These logs were posts, studs, boards, clapboards, laths, plaster, and
nail all in one.

Each structure and institution here was so primitive that you could
at once refer it to its source; but our buildings commonly suggest
neither their origin nor their purpose.

Thoreau's idealization was part of a general mythologizing of the cabin. The 1876 Philadelphia Centennial Exposition included a kitchen in a large log cabin with nice curtains and flowers in window planters. "Ye Olde" read the sign over the door. The "Ye Olde" inns and motels of the roadside had their birth here; so did the log cabin motif in restaurants.

This idealization of the naturalness of the cabin reached its height in the rustic movement of which Thoreau was so much an inspiration. By the beginning of the twentieth century, the rustic movement had apotheosized the cabin. The Old Faithful Inn at Yellowstone Park, built in 1902, was almost a parody of Thoreau's sentiments linking the natural and the ancients, with its huge columns made of single tree trunks with the bark left on, supporting two-story-high courts lined with balconies supported by forked limbs, also left in their natural state. Many a Maine and Adirondack camp adapted the same strategy, finding the "natural Gothic" in twisted limbs used for balusters, behind which sat chairs and tables in the same mode. One Maine camp, not long ago turned into a vacation community, now requires that all houses, old or new, be covered in spruce bark, the natural "ornament" in whose favor Thoreau spoke.

In the 1920s, the log cabin was especially mythified. A period of rapid mechanization, it produced increased emphasis on the past, in reaction. This was the time of Williamsburg and other restorations, of celebrations of anniversaries of Washington's birth (two hundredth) and Lincoln's (one hundredth).

A book by a semi-professional cabin builder named Chilson D. Aldrich, *The Real Log Cabin* (1928), was dedicated to Abraham Lincoln, "who has enshrined the log cabin in the hearts of his people." In his chatty, guide-for-hire tone, Aldrich tells how to build cabins, but also how to "camouflage these concessions to the softer life. . . .

"Did you ever stop to think that the reason we have so few great men nowadays is because there are so few log cabins for them to be born in? . . .

"Pioneering has become an art instead of a duty."

Aldrich advocates low pitch and wide eaves on roofs because they are picturesque. He talks of "square shooters," and celebrates the positive effects of the great outdoors—the cathedral of the forest—

on acumen in business. His pictured models and plans have such names as "Avon Bard," "Seven Glens," and "Trailsyde."

The contemporary revival of the log cabin in prefabricated kit form, as a vacation house, is closer to Lincoln Logs than to the original. Architect Frank Gehry, master manipulator of kit elements from the American building repertoire, included a mock log cabin on the roof of his Norton House at Venice, California. The counterculture log cabin evolved in the seventies into the serious business of selling kit cabins, all neat logs, barkless, and with expensive wood stoves and wall-to-wall carpeting. The cabin ended up a flattened image, reproduced from the quilt style in stencil and pastel-painted plywood "country-style" knickknacks, part of the "ducks on a stick" mode of boutique kitsch, along with ducks and cows and pineapples. The cabin has become a cliché.

The change in symbolism was clear in a cruel joke that had a ring of truth to it: that Jimmy Carter was the first president to go from the White House to a log cabin. Defeated for reelection, he hired an interior decorator for his vacation cabin, and to relax, built old-fashioned furniture there with power tools.

With remarkable speed, the cabin was replaced by what were called balloon-type houses for the poor. The ease with which cabins were simply covered with clapboards and sharecroppers moved into their new houses—dogtrots of frame instead of logs—showed just to what extent they had replaced the older models while retaining the same plan and function.

A few years after Andrew Jackson retired to his (by then) Greek Revival mansion, Abraham Lincoln purchased a Greek Revival cottage in Springfield from a local clergyman. He was to paint its white in clay colors, an amateur taking literally the advice of Andrew Jackson Downing to paint houses in the colors of the earth on which they stood. And as he became more prosperous, he added onto and renovated the house.

A year before speaking of the Union as a house divided, he ordered the builders to add a second story. The cottage had been constructed of local oak; for the addition, the wood was white pine, brought by water from Wisconsin on the Illinois & Michigan Canal. Fashionable brackets were added, and the humble cottage had become a prosperous mansion. Lincoln, returning from a three-month turn on the law circuit, is said to have claimed he did not recognize the renovated house as his own. The original cottage was made with a sturdy, thick-beamed structure called braced frame. But the second story was made of thinner, standard-sized lumber, yielding a frame as thin and rectilinear as the borders of the Western territories then becoming states. It was a balloon frame. Just as the iron rail replaced the split rail of Lincoln's youth, the balloon frame replaced both post-and-beam and log construction.

The log cabin was a kit whose parts could be found almost anywhere there was wood. The balloon frame relied on a kit of parts created by industry: nails and standard-sized wood pieces. The balloon frame used light pieces of lumber, joined only with nails, not pegs, and diagonally braced, to replace the heavier post-and-beam structure with its varying-sized members that went back to medieval times. Its skeleton relied on the external cladding to give it final strength, joining frame and skin in a new way. The new form was made possible by two developments: inexpensive nails—a balloon frame required lots of nails—and the standardization of lumber cut by circular saws.

On Mulberry Row at Monticello, where his slaves had their workshops, Thomas Jefferson had established a nail "factory," which he hoped would offset the erratic flow of agricultural income. Industry was not unfamiliar on the plantation. Indeed, the first foundries where iron or copper was smelted were often called "plantations," as if the metal were just another agricultural product to be coaxed and nurtured from the soil. The nailworks was different. There was some division of labor here, as one slave cut blanks from Pennsylvania iron for others to head and shape into nails. It was a system close to that Adam Smith had advocated in pinmaking. But it was already doomed.

Jacob Perkins, who in 1796 patented a machine to make 10,000 nails a day, was only one inventor of such a device. In the first quarter of the nineteenth century, the price of nails would drop dramatically as wire-cut nails replaced rod ones. The United States proved more amenable to the introduction of the nailmaking technology than did Europe. Samuel R. Wood, a Quaker, tried to introduce a nailmaking machine in England, only to see it destroyed by an angry crowd, as tailors had destroyed Thimonnier's primitive sewing machine in France.

The balloon frame was the invention (if one can specify so exactly) of George W. Snow, a carpenter, real estate man, and building-supply dealer—a lumber dealer, in other words—who knew that the price of nails had fallen from 25 cents a pound before machine-cut nails to 5 cents in 1833. He was also, significantly, a surveyor, at ease with the play of angles and diagonals. It was the diagonal portion of the balloon that made it work; the balloon frame applied the truss system to house building. The balloon frame also depended on the existence of sawmills, equipped with blades like those made by Disston of Philadelphia. And it depended on the absence of skilled carpenters to do mortise-and-tenon work demanded by the traditional post-and-beam structure; it was designed to be built by amateurs.*

But the first realization of the idea is generally attributed to one Augustus Deodat Taylor, a Hartford carpenter moved to Chicago, who designed St. Mary's Church in balloon-frame form in 1833. That church was subsequently taken down, moved, and assembled three times.

The balloon frame made possible the rapid building of Chicago and a thousand other towns in the West that thought of themselves as the next Chicago; many of them were dispatched, in prefabricated form, by rail from Chicago itself. It also allowed the use of younger and smaller logs and fir woods. The balloon frame could be erected easily, by one or two men.

Balloon framing did not bring radically new shapes of buildings, at least at first; the new method of construction adapted itself to the old shapes before new, the dogtrot before the ranchhouse. The house where Elvis Presley was born was a balloon-frame version of the traditional shotgun house—a form with African ancestry by way of the Caribbean by way of Louisiana.

Balloon framing has lent us the 16-inch spacing of studs and standard-sized plywood. It was the seed crystal around which formed the whole elaborate scale of the American building industry—off-the-shelf building materials, standard walls and stan-

---

* The difference between the balloon frame and the log cabin corresponded to the difference between the barbed-wire fence and the split rail, "snake" or "worm" fence. Machine-made products contributed to completely new sorts of designs, replacing a neo-primitive design based on the abundance of materials.

The split rail fence seems to have been a uniquely American feature, summing up the plenitude of wood and of land—the zigzagging fence took up much more space than a straight stone or post-and-beam fence.

But equally American was barbed wire, which depended on the machinery to make the wire—the first such machine was an adaptation of a coffee grinder to twist the barbs.

dard pipes and standard wiring, the stuff of Sweet's Catalogue, the 30-inch-wide appliance, the 33-inch-high counter—the whole kit of items one can buy at any building-supply store today. European émigrés enjoyed this sort of standardization. It was said of the Austrian-born architect Richard Neutra that "Sweet's was his Bible." The French designer Philippe Starck calls American architecture "an architecture of catalogues." Today, while new-age woodworking buffs swear by the post-and-beam house and disdain the balloon frame, even office buildings are frequently erected with a metal stud version of the balloon frame, later to be sheathed with glass instead of the plywood of their residential kin.

Frank Gehry, master of the low budget and low tech, appreciator of the unappreciated building-supply house material, plays off the balloon frame in a number of his buildings. Sometimes he will acknowledge it—in the best modernist mode of "acknowledging the flatness of the canvas" or "acknowledging the artifice of narrative"—by stripping it bare, exposing studs and joists as objects in their own right, or by organizing the face of wall around the four-by-eight plywood module.

To see a balloon-frame structure under construction, with its thin spars of bright new wood and sheets of plywood like sugar wafer, almost sweet against the sky, is to appreciate the genius of the form. A balloon frame is so lovely that the arrival of the wall panels and, often, a veneer of non-structural brick and the closing of the roof, can only produce disappointment. Such an unfinished house is the model of possibilities, with the specifics to be realized by the customer builder or the do-it-yourselfer. For that, one returns to the catalogues, the rest of the kit.

The great catalogues of building products are made literal in the American store type, once known as building-supply outlets, now more often "home centers." At such places, housed in structures composed of the basics of the industrial construction vocabulary, trussed girders and dangling fluorescent lights, the tools of Ames and Disston, of Stanley and Black & Decker, abut their Oriental competitors. Wood and wire and pipe, particle board and plywood, glass and corrugated tin, nuts and bolts, nails and tacks—all are here.

"Do-it-yourselfers welcome," the signs say. Here, Americans can find their homes-away-from-home.

CHAPTER SEVEN

Wonders of the Modern World, or Bridges and Plumbing

NEAR HIS CONNECTICUT factory, Eli Whitney in 1823 installed the first of Ithiel Town's truss bridges and gave its inventor a testimonial letter to help sales. As part of his advertising, Town quoted Whitney—who better?—in praise of the truss's simplicity, lightness, strength, cheapness, and durability; but most of all he noted the independence of its parts, and their interchangeability: "the construction is such as to afford great facility in taking out any piece of timber and replacing it with another."

The Town truss was a classic American approach to building, a token of how American design changed the way of looking at buildings, raising the utilitarian to grandeur by the simple addition of size—size, along with novelty, being the oldest American clichés. Americans looked at building differently: they presented it as tall-talk, Bunyanesque engineering, full of boast and bluster. The very word "skyscraper" has overtones, if not provable origins, in frontier slang for a tall man or a tall tree.

But that size was achieved with the interchangeable part, the democratic atom of construction, from the two-by-four to the T rail to the I beam, that fitted together in a great grid of structure. The classic American types of structures were all of this sort: grids of metal made into dramatic shapes with a skin. The balloon-frame house and the balloon railroad shed, the skyscraper, the giant dam

116

and the great bridge, the massive grain elevator and the huge concrete or steel-frame factory, the roller coaster and the stadium. The cast-iron building and the gas station were both built from kits.

Americans have built big, using little pieces. The framework for this sort of building—which is, at least in ideal, the engineering equivalent of democracy—is the grid system, the skeleton, the replicated truss. Scale in building was seen as a virtue in itself. American building has been a process of encasing kits of interchangeable parts in models that were also expressions of individual distinction and corporate image—luxury built on standardization, just as the Pullman car rode the bogies of a standardized rail gauge, to a high emphasis on individualism.

When we think of American bridges, we think of the great suspension spans—the Brooklyn Bridge or the Golden Gate. These are what Marcel Duchamp was likely to have had in mind when, in his defense of "R. Mutt and his Fountain," he said that "the only things America has given [to culture] are her plumbing and bridges."

Plumbing and bridges were negatives of each other: one contained water, the other spanned it. When he wrote of plumbing, Duchamp was poking fun at the American obsession with hygiene, at the elaborate bathrooms that appeared first in millionaires' mansions and then in hotels before coming home. But Europeans often saw wider cultural importance in American bathrooms. Austrian architect Adolph Loos called the plumber "the quartermaster of American culture." But it is hard not to think of Duchamp's plumbing and bridges in a wider sense: what Americans had been doing for a long time was bridging and plumbing the continent.

Over the huge beast of the continent, we cast nets—the grid of the land system, as regular as Thomas Jefferson's graph paper, and inspired by Cartesian plans he worked out on his desk. A network of internal improvements: canals and turnpikes first, then railroads and highways, telegraph lines, aqueducts and their pipe grids, then electrical and telephone grids. And at the center of these grids were the structures, with their grids of trusses and steel reinforcing their concrete. The skyscraper was a consequence of the city street grid, with its maximization of private land values. The grain elevator lay on the rail grid. The dam and its dynamos sat at the center of the irrigation and electrical grids.

In American public waterworks, plumbing achieved an unlikely heroism. Early nineteenth-century European visitors were proudly directed to the Fairmount Waterworks in Philadelphia, completed in 1822. The pumping machinery, housed in miniature temples designed by the architect Benjamin Latrobe, was juxtaposed with a "Bridge of Sighs" of Venetian inspiration. When New York's Croton Aqueduct system, debouching into an Egyptian-style reservoir on the future site of the New York Public Library, was completed in 1842, mass public celebration hailed the designer, John Jervis, also renowned as a pioneer engineer of railroads. In Chicago, the waterworks that later inspired Oscar Wilde to his remarks on the congruence of "the line of strength" and "the line of beauty" was also a monument to civic pride. The key to the Chicago works was a system of piping and a five-foot-wide, two-mile-long, brick-lined tunnel to the middle of Lake Michigan. The work of one of the country's leading sewer experts, Ellis S. Chesbrough, it was considered a triumph not unlike the Transcontinental Railroad, when both were completed in 1869. Plumbing as engineering became a modern wonder of the world.

☆   ☆   ☆

THE GREAT SUSPENSION bridges were grand models, European in origin and highly engineered, most of them by recent immigrants, often Germans or Swiss, from the Roeblings to O. H. Ammann, engineer of the George Washington, Verrazano-Narrows, and many other bridges, or Joseph Strauss, chief engineer of the Golden Gate.

But bridges made of networks of trusses (which Duchamp may never have seen) were more characteristically American, and far more common. The American trusses were necessary because men and equipment for masonry construction were virtually impossible to obtain. Wood, by contrast, was very cheap. They were the covered bridges of New England, built to the patents of Town and Pratt and Howe and Long and Burr and Bollman, or bridges such as the precarious scantling piles of hastily built Western railroad trestles. These bridges were packaged kits, based on patented truss systems. The same sort of engineering that went into bridges was used to cover the common spaces of public buildings: railroad stations, gathering halls such as the Mormon Tabernacle, and more recently the sports arenas and convention centers, such as I. M. Pei's Jacob K. Javits Convention Center in New York, with its Tinker Toy assemblage of roof elements and round junctures.

They were also spiritual models: idealistic, soaring, and dramatic. Until the skyscraper came along, the Brooklyn Bridge was the highest structure in New York and offered a new view of the city. The mundane truss bridges possessed their own beauty: Bach to the Beethoven of the great suspensions that came later.

They began from a vision that combined romantic engineering with prefabrication. In 1811, a builder and surveyor named Thomas Pope described his patented bridge design, a prefabricated twin-cantilever flat arch. In *A Treatise on Bridge Architecture, in which the Superior Advantages of the Flying Pendant Lever Bridge are Fully Proved*, Pope argued against mere academic theories of engineering, and asserted that in building there was no true standard except experience. The Pope bridge, notes Alan Trachtenberg, was not initially designed for any specific place but "was an invention in the broadest sense, a contrivance to be used wherever a bridge was needed. In this sense it can be spoken of as a 'pure' bridge. . . ."

But so grand was the vision in Pope's eyes that he felt compelled to resort to poetry to describe a plan for erecting a bridge of his type over the East River—a proto–Brooklyn Bridge—or over the Hudson. "The poem is a plea for the chance to build one model bridge. . . ." writes Alan Trachtenberg. This was a bridge that wanted to soar: a "flying bridge." Although shipwrights endorsed the engineering of the structure, which resembled some great inverted keel, or the rainbow to which Pope's poem compares it, it would not have been practicable to build at the 1,800-odd-foot length required to span the East River.

Far more practical was the reinvention of the truss by Timothy Palmer. Palmer, a New Englander, looked back to Palladio, whose works architects in the colonies were perusing, and to actual trusses proposed in his writings. His great achievement was the 1806 Permanent Bridge over the Schuylkill at Philadelphia, now gone. It was the first covered bridge—the backers of the project at first resisted the added expense of the covering—and therefore the original of all the covered bridges in New England.

The trusses—evolving steadily from Palmer to Town to Burr to Whipple to Wendell Bollman, and on—created the covered bridge,

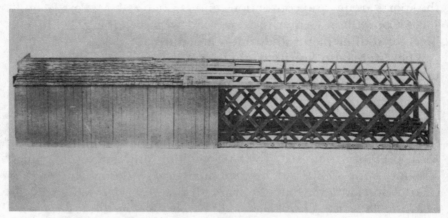

Covered bridge using Town truss system, cutaway model. (Smithsonian Institution)

and they were in construction what they resembled in outward appearance: long barns extended across a river. While the typical New England barn might have but one truss with its "king post," or two, adding a "queen," Theodore Burr's truss offered a whole series of equally treated trusses joined by a supporting arch. Patented in 1817, Burr's truss was used in hundreds of covered bridges. A couple of years later, Ithiel Town obtained a patent for his truss, the first truss system that made a bridge act structurally as a single beam. According to legend, Town, an architect, originally sketched out the plan on the wall of a Connecticut inn. The Town "lattice" truss, patented in 1820, did away with the arch and, topped by a gable roof, gave the impression of the stately marching of timbers across the waters.

Town's slogan was "Build it by the mile and cut it off by the yard," suggesting bridging as a kind of raw material. He charged a dollar a foot for the rights to his plan—a democratic schedule of fees, it was felt, that avoided penalizing the builder of the small bridge with a single fee. Town also sent scouts across the country looking for violators of his patent: if discovered, they were charged $2 a foot to avoid lawsuits.

Town, the designer of such buildings as the Congregational Church on New Haven Green (1814), was an excellent salesman, who traveled with pamphlets about his bridge. With the money he made from his patent, Town assembled the nation's finest collection of books on architecture.

The truss was the key to the New England covered bridge of calendar cliché. Look inside the long shed of such a bridge, and you will find an interlacing of beams and posts and arches; the cover protects the wood from damp and rot.

Town's truss, with its network of beams to create the dynamic effect of a single large beam, was essentially the balloon frame applied to bridge building. The covered bridge was a package: a shell to protect the kit of engineering that was the patent truss. Soon his truss bridges were stretched across rivers in the developing backcountry as well as in New England: over the Great Pee Dee near Cheraw, South Carolina, or over the Yadkin in North Carolina for the new Raleigh to Petersburg railroad, or on the national road in Maryland and Ohio.

This basic frame Town used everywhere. He was one of the country's first professional architects, designing state capitols, churches, and villas. For his architecture—including toll and gatehouses for the bridges—he moved freely among the images in his extensive architectural library, gathered on trips to Europe and purchased with the royalties of his bridge truss design, from Greek Revival to Egyptian and Gothic.

Town's truss was also used in architecture. Henry Grow, who had built a number of Town truss bridges in Utah, was called on to work out a roofing scheme for the Mormon Tabernacle, begun in 1863, a vast tortoise shell of a building, with a roof of arched lattice trusses.

Town's truss enjoyed only a short commercial life. By 1840, others who had seen his success patented improved trusses that were licensed and applied, first in wood and later often in iron. But as late as 1866 the Town truss was used in the 460-foot Cornish-Windsor Bridge, spanning the Connecticut River between New Hampshire and Vermont. Like most Town trusses, this one was constructed using only hammer, ax, saw, and auger. The trusses were assembled in a meadow before being fit together over the river.

Later trusses were also arrived at empirically, no more scientific than Town's charcoaled plans on the tavern wall. Stephen Long's truss was so perfectly engineered that admiring Europeans inquired what texts he had based the design on. It was not until 1847, when iron bridges using the Town and other trusses had been constructed, that a Squire Whipple published the first American text scientifically analyzing the stresses of bridges and introduced his "bowstring" truss.

The Town and other trusses present the eye with a pleasing, tinkling rhythm. They possess a sort of democratic engineering—a democracy of members; no king or queen posts, only a democracy of wooden beams, each beam bearing virtually the same role in the support of the structure. Such systems stood in contrast to the more dramatic engineering hierarchy of the suspension bridge.

☆   ☆   ☆

THE STRUCTURES THAT carried the Transcontinental Railroad across the country were, like the military scope and methods that constructed them, an extension of wartime practice. The builders of the Transcontinental Railroad brought many bridges in prefabricated form from Chicago, from a firm that built Howe truss bridges, of wood reinforced with iron. But many were also built on the spot, from huge amounts of lumber. The tunnels, snowsheds in the high Rockies, and above all the trestles, were in the American tradition of the impromptu and temporary. In their grand precariousness, the virtual delirium of their members and cross members, their obsessively repeated shapes and spaces, they were as dramatic as any engineering structure ever.

One of the greatest of these was the Dale Creek Bridge in Wyoming. Structurally crude, it succeeded by sheer mass. It was an overgrown successor to the basic wooden bridge exemplified in Concord and immortalized in Emerson's poem about the first battle of the Revolutionary War as "the rude bridge that arched the flood." That bridge featured trios of piles joined with X's of diagonal bracing. Each trio supported a beam, and the beams supported a gently, almost Orientally arching bed of planks. At Dale Creek, the work of the Union Pacific's British-born bridge engineer Arthur Brown, the piles have grown to a vast, almost buzzing network of beams and cross braces like a mad matchstick construction. Like many of the engineering works of the railroad, the Dale Creek trestle was built quickly and had to be replaced soon after completion. There was no subtlety to it, no innovative structure; just a relentless and rapid "bruteforcing" of the problem of the ravine to the solution of the structure with a mass of material.

Dale Creek Bridge, Union Pacific Railroad, 1868. (Smithsonian Institution)

It was a continuation of the Civil War bridging techniques of resourceful engineers such as Hermann Haupt, whose work inspired Abraham Lincoln to exclaim, "I have seen the most remarkable structure that human eyes ever rested upon. That man Haupt has built a bridge across Potomac Creek, about four hundred feet long and nearly a hundred feet high, over which loaded trains are running every hour, and . . . there is nothing in it but beanpoles and cornstalks." This was provisional engineering—just good enough and no better than it had to be—a structure whose mere adequacy was astonishing.

☆　☆　☆

THE GREAT SPANS, from Palmer's Permanent Bridge in Philadelphia to the Verrazano-Narrows, embodied that familiar ideal: the wonder of the modern world, the myth of the new pyramid, the new cathedral. They stood at the other extreme of the contrast between the structure as grand model, uplifting monument, single shape, and the structure as kit.

The question of suitable "style" for such structures was unanswerable. Did engineering demand architectural style? Working on the towers of the Brooklyn Bridge, John Roebling racked the style books, the kits of past styles, for the appropriate piece, before settling on softened Gothic. For future suspension bridges, the engineers would be joined by architects—Carrere and Hastings, Cass Gilbert—who were designing great public buildings at the same time. The bridge was no longer seen as a mere functional structure but a public monument, whose engineering needed to be suitably encased.

New York's George Washington Bridge was a famous example of the claims made for American functionalism by the proponents of the undecorated tradition, such as John Kouwenhoven. They never tired of telling how, when the bridge's intricate metal towers were finished in 1931, they immediately became so beloved by the public that a flood of letters prevented the bridge authority from erecting the masonry shells—designed by no less a figure than Cass Gilbert, architect of the Woolworth Building, the "Cathedral of Commerce"—that were originally intended to cover them. A second factor was that the bridge authority found that painting the towers regularly would be cheaper than building the masonry shells, which offered little structural (specifically stiffening) support.

The engineers, led by Othmar Ammann—an engineering genius who, like virtually all the bridge builders, was European by birth and training—had conceived its structure in the assumption that it would be covered with the concrete and granite shell. Dutifully, although with only partial conviction, one suspects, Ammann himself even argued in favor of covering his work with the planned Cass Gilbert shell. Le Corbusier praised the bridge as the best in America, and waxed lyrical about the play of light through the trusses of its towers and the ribbons of its cables. In this bridge, he said, steel architecture had at last found its own form. "Here," he wrote, "steel architecture laughs."

George Washington Bridge, steel towers minus planned masonry shell.
Completed 1931. (Port Authority of New York and New Jersey)

Artist's rendering of
George Washington
Bridge with masonry shell.
(Port Authority of New
York and New Jersey)

The skeleton was seen as "pure" engineering, untainted by aesthetic intention. Left uncovered it became a found object, a grand artifact of that tribe as strange to the average citizen as the Pygmy or Masai—the engineer. In this it recalled the Brooklyn Bridge, which despite the intended artistry of its Gothicized towers, figured as a plain object discovered and rediscovered by artists (Joseph Stella), photographers (Walker Evans), and critics (Lewis Mumford). The other great suspension bridges such as the Golden Gate (1937), the Bronx Whitestone (1939), and even the late Verrazano-Narrows (1964) treated the towers as "pure" engineering.

But whereas their towers were designed from the beginning to be seen, the George Washington was unique. The newer bridges offered towers that were less steel cages, with more appearance of solidity; the George Washington celebrated the contingencies of girder and truss, suddenly exposed, and their power to make great forms in their own right. The sheer repetition of these structures in public view cemented those forms on the consciousness. They became monuments and icons.

☆   ☆   ☆

THERE WAS A curious connection between bridge engineering, whose products at their best seemed to hover or soar, and aviation. One of the Wright brothers' key advisers and mentors was Octave Chanute, credited with engineering the great bridge across the Missouri at Kansas City. To Chanute, who had devoted himself to aviation after a career in civil engineering, Wilbur Wright wrote of plans for his airplane: "My machine will be trussed like a bridge."

Alexander Graham Bell, in the long anticlimax of his life after the development of the telephone, was also interested in aviation. For his huge "man carrying kites," he employed triangular truss frames, anticipating the "tensigrity" shapes of Bucky Fuller, and even the cellular matrices of wing structures developed by John Northrop and others. Bell's kites were strange and lovely things, painted red for visibility, futuristic in appearance, but ultimately unsuccessful in their ambitions of flight.

More important was the application of the triangular truss to towers and other structures. Bell was one of the earliest experimenters with what are now called "space frames," using tetrahedral shapes—the purest realization of the truss's division of structural forces into triangles, the strongest of geometric forms. These honeycombs of trusses were later to be used in buildings such as New York's Jacob K. Javits Convention Center. They are the truss driven to its extreme.

The shapes were a strange and striking echo of the telephone network Bell had foreseen as more important than the individual instrument he had developed. Bell understood from the beginning that the system was the solution—the system of telephone wires resembling, he said, pipes delivering gas or water, a utility. They were to be a giant net or frame in space—he foresaw long-distance calling early on—a frame across space. The man whose invention led to the nation's key communication network went on to develop the truss into a network of structure.

☆   ☆   ☆

THE COUNTRY'S FIRST basic transportation network was built on, not over, its waterways. Until the arrival of the railroad, transportation by sea from Boston to Charleston for example was far cheaper than by land. And the first great works of the civilizers—the first of many succeeding engineering "wonders of the world"—were canals linking the rivers. The rivers themselves had to be dredged, diverted, and directed, through such works as the jetty complex at the mouth of the Mississippi constructed by James Buchanan Eads, the creator of the great St. Louis bridge that bears his name. The canal boat, the flatboat, the application of steam to shipping in a variety of forms, were dominant objects and icons.

To make the deserts bloom, the rivers would also have to be dammed—replumbed to provide water for the new gardens of the Western desert—and to provide power for the new plumbing of electricity, the power grids. Hoover Dam was one of the great icons of the thirties. It had been begun as Boulder Dam by the engineer-president who was to lend it his name, but it became a symbol of government intervention in the economy of the sort associated with Franklin Roosevelt, who dedicated the dam. This was pump priming at its most concrete.

Dams brought electricity—and electric churns and washing machines and radios—to rural areas. They also brought a pride in "progress," that wonderfully, naively simple ideal. (Even today the town of Tupelo, Mississippi, boasts in the signs on its borders of being the first town electrified by the Tennessee Valley Authority— and not of being the birthplace of Elvis Presley.) Although debate about the Tennessee Valley Authority was sharp—its dams were the New Deal made physical—there was a strange measure of consensus about the great dams of the West. They were perceived as much as objects of nature as of political or social culture.

They were cathedrals for the dynamos that dispersed power over the massive new electric grids, stoppers for whole new lakes, great jars in the wilderness. They were models of the values we held in common, and figured in American culture of the thirties with all the iconographic power they did in Fascist or Communist states. Laborites saw them as monuments to the workers who had built them. Capitalists saw them as monuments to can-do private industry, succeeding despite onerous government inspection, recalcitrant unions, and bureaucratic bungling. New Dealers and the Hooverites, the Luce magazines and folk singer Woody Guthrie—who wrapped up the power of the structure and the power of the workers who built it in "Grand Coulee Dam"—could all wax enthusiastic. There was a shared human power implied by the dams that could "reclaim" the desert—the building agency was the Bureau of Reclamation, founded by Theodore Roosevelt in 1902—with power and water for Los Angeles. (Southern California would get half the kilowattage of the dam's generators and the All American Canal, feeding from the reservoir behind it, would water the fields and orchards of the Imperial Valley.)

But the controversy over the naming of the Boulder/Hoover Dam suggested an ideological struggle beneath the common admiration: were these the works of untrammeled free enterprise and Republican virtues, or of New Deal collectivism? The iconology of the dam, we would see, adapted with equal ease to communism (as in the cult of the Dnieper Dam in the Soviet Union) to fascism, and to nationalism of all stripes in Africa.

The dam was originally planned for Boulder Canyon and retained the name for a while even after the site was shifted to Black Canyon. Ray Wilbur, Herbert Hoover's Secretary of the Interior, whose department included the Bureau of Reclamation that built the dam, named it for his boss in 1930. FDR's appointee to the same job, Harold Ickes, undid the naming in May 1933. Boulder was, he said, a "fine, rugged, individual name." The name change, asserted Ickes, marked the claim that the dam was dedicated to the public and not to any private interest—a reference to the Republicans' use of federal force to break unions on the site. The final return of the title to Hoover Dam was made in 1947 by a Republican Congress.

The massive project of building the dam and its successors intrigued the press. So did the rough Old West–reborn life of the workers' towns where drunkenness and violence were common—twentieth-century versions of the rail towns that blossomed and wilted along the Transcontinental Railroad. Secretary of the Interior Wilbur had his pocket picked when he arrived for the formal beginning of construction, the driving of a ceremonial rail spike (echoing the one at Promontory Point in 1869) for the first step of the work, building access rails to the site. Later, these rails were to carry excursion trains loaded with tourists.

At 726 feet, the dam was to be the world's tallest. It was to be constructed of 230 huge blocks of concrete—enough concrete, the boosters of the dam never failed to note, to pave a highway from San Francisco to New York.

The pressure of the concrete under such weight produced heat; it would take a century to cool and harden, the engineers estimated, and would crack and break in the process. So a network of miles of one-inch pipe was laid in the concrete as it was poured, and the world's largest refrigeration plant, capable of freezing 1,000 tons of ice a day, there in the middle of the desert, generated cool water to be pumped through the pipes. The century was reduced to a couple of months.

The Hoover was followed by the Grand Coulee, Shasta, Hungry Horse—all of them grist to the mills of the newsreels and newsmagazines. Such engineering marvels fascinated the press; comparisons to the Great Pyramid—generally to its disfavor—were frequent. *Life* magazine's editors, meeting in the Cloud Club atop the Chrysler Building, decided to put Margaret Bourke-White's photographs of the Fort Peck Dam on the cover of their first issue. The story inside was not only about the grand concrete forms of the dam itself, the statistical marvels of cubic yards of concrete, the tale of the water-filled tubes buried in the gray mass to cool and harden it evenly, but about the rowdy nightlife in the shanty towns that grew up around the construction. This, another magazine proclaimed, was a sociological as well as an engineering project. Harnessing the power of the river was associated with harnessing the power of the people, en masse.

"Uncle Sam," went one account, was extending "a kindly but firm paternal helping hand to the workers." This included housing them in standardized cottages laid out in the gridblocks of Boulder City.

Zane Grey wrote a posthumously published novel called *Boulder Dam*. A film of the same title appeared in 1936. The theme is the recalcitrance of a worker who, "back east," has killed his boss in a fit of rage. Working on the dam, however, he learns to be part of the team and—in an episode freely adapted from the exploits of one worker, Louis "The Human Pendulum" Fagan—swings out on a crane hook over the canyon to save two fellow workers.

The dam's associations were Egyptian: a pyramid in the desert. Inside the powerhouse, wrote Joan Didion of her excursion to Hoover, she touched the turbine—and got carried away, finding "a dynamo finally free of man, splendid at last in its absolute isolation, transmitting power and releasing water to a world where no one is."

☆   ☆   ☆

BRIDGES AND DAMS were public works and, for all the jokes, not even J. P. Morgan or Jay Gould could buy the Brooklyn Bridge. But you could buy or build a skyscraper, if you were Frank Woolworth or Walter Chrysler, and attach the epic power of the big structure to your business and your reputation. With its network of steel girders encased in a dramatic form, the skyscraper became a means for advertising, for theater, for civic drama.

☆   ☆   ☆

ON THE TOP floor of the Chrysler Building, Walter Chrysler for years kept on display his tool kit—the set of tools he had assembled and in some cases made while he was still working as a mechanic for the Union Pacific Railroad. They were there as a token of how far he had climbed—from the roundhouse to the penthouse, so to speak—but they were also an apt symbol of function of the Chrysler Building itself, as a tool for self-assertion and monumentalization, as a public relations tool and advertisement for the company.

The Chrysler, of course, had elements of the company's cars incorporated into its design: the hubcaps and hood ornaments rendered as ornamental elements. The Woolworth Building, the famed Cathedral of Commerce, was mother church to all the little five-and-dime chapels on Main Streets across the land. It was inspiration—and advertising. And, as he did in his stores, Frank Woolworth made sure every detail reflected himself. He tested the elevators, approved the choice of toilets ("Sanitas" brand), and even considered the shape of the flush handles on the urinals in the men's rooms.

To usurp the spiritual role of the cathedral for lucre might seem American materialism at its most Philistine. Architectural historian Charles Whitaker, who loathed them, contended that "skyscrapers are buildings the first purpose of which is to enable the individual, or group, to cash in on land." They were speculative real estate, and imaginative architecture. But whatever motives of individual monumentalization or greed drove it, it was a characteristically American way to grand effects: what Thomas A. P. Van Leeuwen, the Dutch historian of the skyscraper, borrows from Johan Huizinga to call "transcendental materialism."

You didn't have to have your company's name written on the skyscraper, although General Electric did on its, amid a *Metropolis*-set swirl of steel lightning bolts and other devices; although McGraw-Hill did on the bridge of the ocean liner Raymond Hood designed to reflect the identity of the company and its boss.

The shape of future skyscrapers would be a company logo: the Chippendale top of the AT&T Building or the Citicorp Building's slash roof in New York, the Transamerica Pyramid in San Francisco, the juxtaposed rhomboids of the Pennzoil Building in Houston. Skyscrapers were models of the most dramatic American sort; icons of companies and commerce, but also of far vaguer and more grandiose emotions.

One of the greatest and least appreciated of skyscrapers is not the Empire State or the Chrysler or the Tribune Tower, but 70 Pine Street in Manhattan, a.k.a. 60 Wall Street, a.k.a. the Cities Services Building. What is wonderful about it is not its crystalline roof fixture, although it glows theatrically at night in the mist and haze of Wall Street, nor the mystery of its proud silhouette—mystery because locating the base of that silhouette, at street level, is very difficult. What is best about the building is that the architects, a firm

Model above entrance to 70 Pine Street, New York.

Architectural rendering (1935) of 70 Pine Street by John C. Wenrich. (The Landmark Society of Western New York)

called Clinton & Russell, left their model for the building, rendered
in stone, above its front door. That model is more romantic than
even the lit building at night. It seems softened by time and the
erosion of nearly fifty years of Wall Street drip and drizzle—partly
because the model omits the distracting detail of the railings atop its
setbacks, not to mention the window air conditioners of later addition.

Built in the Depression, like so many skyscrapers, the Cities Services Building falls halfway between the Woolworth and the Chrysler in style; it is a kind of Gothic Deco. It also suggests something
about the skyscraper's identity: located away from its nominal address, but connected to it by a passage/bridge that serves as an
excuse for its claim to the more expensive location. Cities Services
was a building known by its silhouette—a silhouette that was visible
only from a distance or from the model above its entrance.

Models had long figured in architecture, but they took on a new
importance in the age of the skyscraper. Cass Gilbert, the architect
of the Woolworth Building, appears in a sculpture on the facade of

Contemporary pedestrian's-eye view
of 70 Pine Street.

that building, holding a model of the shape he created—a reflection of similar sculpture in the real cathedrals. But it was only in the twenties that modeling became the dominant mode of designing skyscrapers such as the Cities Services Building.

These models, along with the romantic, hazy pastel or chalk renderings by such men as Hugh Ferriss and John Wenrich, summed up the essence of the American skyscraper in its golden era.

It is virtually impossible to find any discussion of the history of the skyscraper that does not within a few paragraphs begin to talk about "the expression of the steel skeleton." But people do not build buildings to express their skeletons. They build buildings for a purpose: to make money, to swathe the demands of commerce with aspirations more than merely commercial, or to attempt to assure their immortality. The skyscraper was a tool for dramatizing personal and/or corporate success. It was a model for aspiration— upward striving—encasing a kit of repeatable, rentable office spaces of standardized size and features.

The theme of the skyscraper as icon, as model for the American spirit, is its upward thrust. The steel frame was the standardized kit that made the modeling of the exterior into such dramatic shapes possible. The steel frame is the means to this end; it has drama, soaring aspiration, modern monumentality. The steel frame and the elevator were a means to get the building high for economic reasons and prestige, for commercial verging into spiritual reasons.

Louis Sullivan said as much when he famously declared that a skyscraper must be "tall, every inch of it tall." Just as the clipper builders had believed that their models, and their ships, should not only be fast but look fast, the skyscraper should look as well as be tall. For Sullivan, this had little to do with the steel frame, which was a means to construction efficacity and efficiency. He admired the pre-steel-skeleton Monadnock Building in Chicago, with its masonry walls like a tall armory. He called it "an amazing cliff of brickwork, rising sheer, and stark . . . with a direct singleness of purpose, that gave one the thrill of romance."

The skyscraper as kit of established building elements was a Chicago tradition, one aesthetically resolved by Sullivan with the classical column as model: the elements of his tall buildings correspond to the base, shaft, and capitol of a column. The outside shape tells the story in model form, with a beginning, a middle, and an end. Still, Sullivan's buildings were unified rectilinear structures to which the ornament was applied. They terminate with a cornice, not, as in New York, with a modern-day steeple—as the Empire State or Chrysler do.

Expressing the steel frame was incidental. Its predecessor, the iron skeleton, had been even more carefully hidden within traditional shapes—-in, say the dome of the U.S. Capitol, or for that matter of the Statue of Liberty, a model enlarged and supported by a skeleton designed by Gustave Eiffel himself. Appreciating the steel frame came after the fact: it was the old European appreciation of the functional and unornamented, of unintentional beauty, pursued in the face of all contrary evidence. The European modernists seemed to look right through the terra-cotta moldings and renditions of old campaniles and cathedrals that made the skyscraper's skin. They saw only the skeleton within. Even the word "skyscraper" made proper modernist architectural critics cringe; they preferred "high rise office building."

But for the developer, the skyscraper had to have an immediately seizable shape. It was a speculative enterprise, native to America because the real estate and zoning conditions made it possible. A skyscraper, said Cass Gilbert, is a machine for turning land into money. Far denser European cities never scraped sky because they limited building size and limited ownership of land.

By the twenties, the shapes of skyscrapers in New York were less concerned with expressing the skeleton than with expressing the "zoning envelope" created by the 1919 zoning regulations. That envelope was defined in shadow and shade in Hugh Ferriss's series of drawings. It constituted a block of possibilities from which architects modeled their skyscrapers.

"Modeling buildings" was a favorite phrase of Ferriss, who rendered a number of skyscrapers and created an imagery that transcended the vision of the individual architects. "This sketch"— reads one of his captions in the book *The Metropolis of Tomorrow*, which preceded Fritz Lang's film *Metropolis*—"was made with the fancy in mind that a number of such gigantic shapes were actually existing; that they were composed of clay; and that they were awaiting the hand of the sculptor. Indeed, the crude clay of the future city may be imagined as already standing. There must come architects who, using the technique of sculptors, will model the crude clay into the finished forms."

Just as the skyscraper sheathed the structural kit of the steel frame in dramatic shape, it sheathed the workaday kit of the modern office—with its phones, its typewriters, its office chairs and desks, and eventually its modular partitions—in the romantic model of soaring tower. It mystified the banality of bureaucracy in the theater of the skyline.

The steel frame, from one perspective, stood for Chicago, where it, like the balloon frame, had been developed. But the dramatic model stood for New York, where the tower was part of the grand theatricality of the city—the tower, gilded and chromed and washed like a star in footlights. When Irwin Chanin opened the Chanin Building in 1930, he inserted a brief notice in the theater sections of New York newspapers, announcing the "premiere illumination" of his new building, promising "212 floodlights of 30 million candle power painting a picture on the facade and towers of New York's newest wonder. A thrill for all."

The *New York World* responded in kind, sending its drama critic to cover the event in the form of a mock theater review, which praised the display's "real eloquence," "grandeur of tonal distinction," and "synchrony of shadow and light." Mayor Jimmy Walker, also on hand, averred with his usual glibness and sensitivity to his constituents that the building "was as beautiful as the River Shannon."

Chanin's previous real estate ventures had included several Broadway theaters; he recognized the skyscraper for the theater it also was. And every day for more than half a century, he went each morning to the office he had built for himself in the building, an elaborate Art Deco fantasy complete with bathroom, that was in its way the equivalent of Walter Chrysler's tool display. The owner was inextricably linked to the building that bore his name. The salesman was part of the address he was selling.

☆   ☆   ☆

WILLIAM VAN ALEN, designer of the Chrysler Building, was accused of being the Ziegfeld of his profession. It was a reputation ratified not only by the theatricality of the building he created for Walter Chrysler, but by the fact that he appeared at the famed Beaux-Arts Ball of 1929 dressed in a suit representing that structure. His virtuoso performance, however, was a building stunt. In 1930 he hid the "vertex" or "spike" of the Chrysler's top inside its summit until the builders of a downtown rival in height had complacently topped off their building. Then he lifted the shiny steeple through the top and took possession, for a little while, of the title of world's tallest building.

The Empire State Building, which seized the title away in 1931, combined the financing power of the Du Pont empire, the administrative skills of its key adviser John J. Raskob, and the blue-collar appeal of its hired front man, Al Smith, former Democratic presidential nominee. It was a speculative office building, whose sales began so unsuccessfully that it was known for years as the "Empty State Building." For the Empire State's designer, William F. Lamb, sheer scale was what transported the building to the level of a wonder; for its shape, his wife later claimed, he had taken as his model "the clean soaring lines" of the simple wooden drawing pencil. One imagines a gigantic Claes Oldenburg project, or a Saul Steinberg drawing where the tools of the drafting table become a skyline.

But the building's theatrical aspirations were summed up in the lobby, where stained-glass windows depict the wonders of the ancient world, sufficient in their implication that the structure they reside in is the wonder of the modern one. Hugh Ferriss at first saw the skyscraper just this way: as a successor in the line of evolution that ran back to the Assyrian ziggurat. But by the end of the great skyscraper era, in 1942, he viewed these soaring buildings instead as "monuments to the rugged individualism of the period; their topmost, unnecessary floors and gilded spires, to the conspicuous waste and rampant advertising; their long shadows across the slums at their feet, to the exploitation."

Even as the skyscraper was being created, the engineers of bridges and railroads were turning their talents to a similar kind of drama creating entertainment, spectacle. They made their own modern wonders to match the glories of the Old World through the particularly American strategy of turning mundane girders and trusses into near-operatic spectacle and lightness of construction— into the lightheartedness of entertainment.

☆ ☆ ☆

GEORGE WASHINGTON GALE FERRIS was a thirty-three-year-old
bridge engineer when he conceived the great wheel that today bears
his name. The scene was a banquet presided over by Daniel Burn-
ham, the chief architect of the World's Columbian Exposition, the
great world's fair that was to arrive in Chicago in 1893 amid a
depression and a year after its intended opening, to celebrate the
four hundredth anniversary of the discovery of America. Augustus
St. Gaudens thought the assemblage of architects, artists, and engineers represented the greatest gathering of talent since the Renaissance; but the buildings they designed for the "White City" that was
the center of the Exposition were constructed of plaster strength-
ened with fibrous aggregate and intended to endure only for the
four or five planned years of the fair's existence. Nonetheless, the
architects and artists—save Louis Sullivan, whose Transportation
Building sought variance from the Renaissance guidelines, and who
believed the whole complex a sham—were mightily pleased with
themselves. And at the banquet, they scolded the engineers for the
paucity of their contribution.

Ferris responded with the concept of the great wheel. It would
be proof that engineering could be as dramatic—perhaps as
aesthetic—as pure architecture. It was also one of the great monuments of engineering applied to entertainment in the best American mode.

The concept was not a new one; as early as the seventeenth
century fairgoers had ridden horizontal and vertical wheels. Children sat on large cartwheels that spun; the device replicated a
device for jousting games that was medieval and possibly Middle
Eastern in origin. Industrial machinery from water wheel to treadmill served as models—the water wheel, widely used in Europe by
the time of the Norman invasion of England, was the dominant one.
The fairs of the seventeenth century offered pleasure wheels, like
25-foot flax wheels, to which passengers clung and by which they
were turned. These, says David Braithwaite, historian of carnival
devices, derived from the treadmill and water wheel.

The Ferris wheel echoed the wheel of the river steamboat, with its
paddles corresponding to the Ferris wheel's cars and its slow, lit
rotation echoing some of the drama of the steamboat as showboat—
whether as festive racer or with fine food, dancing, and theater. It
was like a combination of the Corliss steam engine of the 1876

Centennial Exposition and the Eiffel Tower: a dynamo in which one could ride. Or it could be compared to a kind of circular railroad. The thirty-six cars, each 27 by 13 feet, were far larger than the sort of car bodies that make up standard carnival Ferris wheels today. Allowing for standing room, they held sixty people each.

The Ferris wheel, of course, was basically a carousel turned on its edge. With the advent of steam engines, the carousel could take on new and more elaborate forms, accompanied by powerful calliopes. Some had ships, which pitched and rolled. Carousels, roller coasters, Ferris wheels—all these devices exaggerated the physical effects of their industrial models: the wheel spun, the rails dipped and rose, in virtual parody of the originals. The railroad was rendered as roller coaster—one of the first, from 1884, was the LeMarcus A. Thompson Gravity Pleasure Switchback Railway at Coney Island.

The first Ferris wheel, World's Columbian Exposition, Chicago, 1893. Photograph by Charles Dudley Arnold. (Gernsheim Collection, Harry Ransom Humanities Research Center, University of Texas at Austin)

The Ferris wheel was designed to impress as a demonstration of the power of engineering. The promoters supplied key statistics to awe customers (as stadium and skyscraper boosters were later to do): weight, 2,100 tons; steel-forged axle, 33 inches in diameter and 45 feet 6 inches long; carrying capacity, 150 tons of people. Motive driver: two 1,000-horsepower reversible steam engines.

But it was an aesthetic construction as well. The lights and movement of the wheel lent their rhythm to the whole fairground. The wheel turned as if under water, and suggests, of course, the steamboat, the old mill writ enormous—and the bicycle. In 1893, the bicycle was a fad—Orville and Wilbur Wright's bicycle shop was so prosperous that they could take time off to attend the Exposition (although there is no record that they rode the Ferris wheel).

The Ferris wheel was erected not in the White City but on the Midway Plaisance, the commercial entertainment zone, to which the committee of architects and artists had agreed only grudgingly. There, proto-Disney international settings, from Spanish to Samoan, offered gauzed entertainment in educational wrappings. The culture of Cairo was represented by belly dancers doing "the hootchy kootchy."

Ferris did not find much support even among this sort of enterprise. He had to buy a license on the Midway for $25,000 and raise the quarter of a million dollars it took to build the great wheel through a stock issue. The stockholders did well: for 50 cents a ride (two circuits, twenty minutes) a million and a half passengers rode the wheel. The enterprise ended with a profit of over a million dollars. Many rode repeatedly; several couples were even married on one of the cottage-sized cars. Ferris himself died three years later, impecunious—a not uncharacteristic American last act, and one that suggests a comparison between the wheel he invented and the ups and downs of the American economic machine in the 1890s.

From a distance it was the part of the fair you saw first. It became the Eiffel Tower, the Trylon and Perisphere of the fair, though not by intention. Many have seen in the Chicago fair the conflict of the genteel and the popular rendered in physical form: the Ferris wheel symbolizing the popular triumph over the European simulacra of the White City which were to be the ruling forms of aristocrats and enlightened reformers in the Cities Beautiful of the next half century.

The fair celebrated four hundred years since discovery, and four hundred years implied a significant history—if you did not stop long enough to remember that nearly a century and a half had intervened between Columbus and the establishment of significant white settlements, and another century or so in varying degrees of primitivism. But to the visionaries of 1893, four centuries of history seemed enough to establish America as rightful heir to the classical tradition, usurping old Europe and standing on the verge of its own version of the Renaissance—even if rendered in plaster-and-straw models.

The Court of Honor centered on the administration building by Richard Morris Hunt, the first American to study at the Ecole des Beaux-Arts, the first of many, most of them represented in the buildings of the Exposition, who sailed back to the United States with the kits of classicism in their heads and portfolios, like highbrow versions of the builders' books of the early nineteenth century.

While the fair was under construction, the Panic of 1892 struck hard at the economy. Workers were striking the factories at Homestead, Pennsylvania, and President Cleveland was yielding to the local bosses to send in federal troops to break the strike.

The Ferris wheel itself offered diversion from the hard times, perhaps. But it also contrasted with the Crystal Palace or Eiffel Tower of past fairs, and with the White City itself: it moved. It was a machine offering successive and shifting viewpoints to the passive observer, by contrast to the White City's grand, linear perspectives—Baroque by way of Haussmann's Paris. The pedestrian had to move himself in the White City, with its deep views across "Lagoon" and boulevard.

There was never another Ferris wheel as large as the original, which was taller than the Statue of Liberty and at 286 feet offered by far the highest prospect to visitors to the World's Columbian Exposition. After the Exposition, there was talk of the Buffalo Bill Wild West Show bringing it to Coney Island. George C. Tilyou, the creator of Steeplechase Park at Coney Island, may have tried to buy it. Failing this, he built his own, at half size but nonetheless advertised as "the world's largest Ferris wheel"—a statement justified during those times when the original was disassembled. (That original, disassembled and rebuilt once more for the fair in St. Louis in 1904, seems finally to have disappeared.)

Coney Island's Wonder Wheel survives, and the slow rotating wheels of any number of county and state fairs and traveling carnivals keep the Ferris wheel alive. But in time, of course, much of the amusement park's entertainment was to be replaced—by the movies. Film, with its little sprocket-driven belts, the railroad in a box, the assembly line of action of the film camera, replaced what had been essentially individual experience of stage sets. What was left for the rides were the physical sensations of rushing and rising and falling and careening that no wild chase scene could do more than suggest.

But even while the engineering of entertainment was shifting to the electric and eventually the electronic, the application of civil engineering to leisure would survive in other entertainments: in the great stadiums that already, in 1893, were beginning to replace the more pastoral baseball and football parks.

☆   ☆   ☆

THE SEVEN WONDERS of the Ancient World can be found not only in the lobby of the Empire State Building, but also at the Astrodome in Houston, which from its opening in 1963 also advertised itself as the Eighth Wonder of the World, and at the Superdome in New Orleans, which boasts that it can contain the Astrodome. That stadium—a sign in the Superdome's offices runs—could fit inside the Superdome "like a car in a garage."

In 1904, when Henry James returned to Cambridge, Massachusetts, after an absence of many years, he walked along the Charles River and noted the new Harvard football stadium, the first of its kind in the country, gleaming in the moonlight—a futuristic silver if not golden bowl. For him, stadia were a thing of the ancient world, the ruins of the Colosseum where Daisy Miller could go at night and be infected simultaneously by the sins of the Old World— feeding Christians to lions and loose sexual behavior—and by symbolic fever, fever that broke down morality and class conventions simultaneously with the body.

While advocates of the still new boom in college athletics emphasized their classical virtues—a healthy mind in a healthy body— James saw a Harvard football game as proof of "the American capacity for momentary gregarious emphasis."

Harvard Stadium, built 1903.

Harvard Stadium recapitulated the mythology of athletics as of the classical virtues. Wrestling and the Olympics (revived in the 1890s) were somehow equated with Pudge Heffelfinger's lead blocks behind tackle.

The stadium was an early experiment in reinforced concrete, a technique advocated by one of Harvard's engineering professors. President Charles Eliot was skeptical of the material; his administration enlisted a number of students to conduct an informal stress test on its piers, using the weight of their own bodies. It was not until a few years later that the Doric colonnade was added, that most basic order of classical architecture providing a symbol of the purity of sports in its early days posing as classical revival. This is the feature that gives the Harvard horseshoe its special appeal and sums up the revival in America of a classical building form that had been neglected for around 1,500 years.

Yale followed its rival in 1914 with a different concept—"the bowl." Three hundred thousand square yards of earth were moved to form a depression and surrounding berm. Part hole, part wall, the floor of the stadium was set nine yards into the ground. This giant environmental sculpture was "part of the earth's crust," boasted its creators. It was a work of civil engineering as much as of architecture—like an artificial lake. Finished about the same time as the Panama Canal, the Yale Bowl seated 70,000, dwarfing Harvard Stadium, and provided parking lots, soon to be occupied by the Model T's of the raccoon coat set. The ancient Colosseum, boasted one Yale alumnus, would barely "come up to the armpits" of the New Haven structure.

The rivalry between institutions that inspired such grandiosity proved that the stadium had more in common with the Roman Colosseum than with the Athenian public spaces where runners and wrestlers competed. But it required the spread of college football by imitation from the banks of the Charles to the hills of South Bend and the prairies of Oklahoma for the principle to become clear. Across the country, ever larger stadiums were built, and the Ivy League schools that had given birth to the sport were relegated to minor status in national rankings.

The original sports creed was watered down and blown up with alumni contributions at the Midwestern and Southern football powers. In those huge stadiums, the gladiatorial values of big-time sports undercut the "larger spiritual" issues they pose as embodying: thus Notre Dame's religious figures just outside the football stadium there are cynically nicknamed "Touchdown Jesus" (both arms upraised) and "First down Moses" (one arm pointing to the Promised Land). And when the game became professional, the stadiums and domes became civic monuments and utilities, as well as private. No city was big league without a big league stadium, and one citizenry after another rushed to indebt itself to build a better facility than the one in the next television market over.

The stadium became a dominant form of American public space, far more meaningful than any single political, social, or leisure arena, more than the town hall or concert hall. Its only competition was the airport and the highway system. It was the part of a city seen most often on television, and so became an important shaper of national opinion of a city's spirit and prosperity.

☆   ☆   ☆

BASEBALL'S NATIVE GROUNDS were more pastoral: the first, most famous, formal field for that game was called the Elysian Fields, a high-falutin' title for a piece of real estate in Hoboken, New Jersey.

It had begun as a pleasure ground owned by the Stevens family, of engineering and yachting fame. There, on a closed track, the first railroad in the country had puffed out its circuits, more at first as amusement for the Stevenses than as prototype for practical transportation; engineering for entertainment.

The first baseball players were, like the first football players, upper-class members of an exclusive club. The Civil War, of course, spread baseball throughout the Union Army. Games were played in camp and even among Union prisoners in Confederate prison camps. The process by which baseball's greenswards acquired first fences, then bleachers, then crudely engineered stands, and ultimately joined football in great covered television studios is the story of American sports. The uneasy progress from "field" to "park" to "stadium" to "dome."

Baseball's appeal spread out from the upper classes. At New York's Polo Grounds, around the turn of the century, the last vestiges of high-class baseball literally abutted the insurgent popular version of the game. The Nationals played on the east side of the long field that was the Grounds, facing Fifth Avenue, and charged 50 cents admission. Behind them, facing Sixth Avenue, a second, scruffy field played host to the American Association team, who charged only a quarter and tolerated the sale of beer. Soon the two would be united in the legendary monuments of the new Polo Grounds stadium and, more dramatically, across the Harlem River in the grand baseball palace, Yankee Stadium, erected by beer magnate Jacob Ruppert.

Stadiums began on the wrong side of the tracks—or atop them. Long before it became institutionalized with the pieties of a "national game," baseball sipped at the same netherworld saloon as horse racing or boxing—a side of the sport brought in the open by the 1919 Black Sox scandal. Navin Field in Detroit was the creation of a bookkeeper who was also a bookie. Baseball still had its roots on the other side of the tracks: many early stadiums were built on supplanted railroad marshaling yards. The irregular shapes of these tracts lent such places as Fenway Park in Boston an asymmetry that was to become charming only with age, when it stood in contrast to the boosterish geometries of perfectly rounded city-financed structures with all the character of a shopping mall.

Baseball's emergence from low status was paced, beginning

around 1910, by an increasingly affluent audience. At the same time, repeated fires and structural collapses of wooden stands led to the adoption of steel-and-concrete stands. The new stadiums offered even bleacher bums a touch of class. Shibe (later Connie Mack) Park in Philadelphia, the game's first steel-and-concrete facility, built in 1909, was rimmed with a mansard cornice topping a dramatic classical colonnade. Its focal point was a miniature version of the dome of St. Peter's.

By the beginning of World War II, major league stadiums were already laden with enough tradition to inspire enough nostalgia to fill the Roman Forum. It was in the rough and rangy older stadiums that intellectuals began to appreciate the "velvet and inalterable symmetry" of the baseball field (Thomas Wolfe), seeing in the "emerald" grass primal evocations of Eden, regarding the simple playground as a formal garden with power alleys instead of poplar allees. Meanwhile, the new stadiums being cast to frame this magic space began to look more like airport terminals.

The move of teams west, made possible by air travel and demanded by television, resulted in major changes in sports.

One was the purely geometric stadium, built away from the center of town on a site unconstrained by the existing urban plan. The first was Dodger Stadium. The first stadium intended for an audience that arrived entirely by automobile, it was a circle surrounded by acres of parking.

After dickering with Buckminster Fuller about building a geodesic dome in Brooklyn, wired with cable television viewable in the local bars, Walter O'Malley gave up and moved the Dodgers to Los Angeles. There, the stadium was dug into Chavez Ravine and set on pillars like the supports of freeway overpasses (or like those supporting the houses in the Hollywood hills hovering over the gridded lights of the city). In announcer Vin Scully's words, the symmetrical stadium suggested "a wedding cake," with its regular rounded layers of seats. Its little zigzig roofs were the type one might have seen on a carwash or burger drive-in of the time. The herringbone-painted patterns of parking lots radiating around it were as much a part of the scheme as the stadium itself—in contrast to the USC Colosseum, with its WPA neoclassical overtones, across town. Meanwhile, back east, apartment towers took the place of Ebbets Field and the Polo Grounds.

☆   ☆   ☆

STADIUMS HAVE BEEN the scene of innovation in engineering, almost unremarked and unsung. While bridges and dams and other "merely functional" structures have been given their due, athletic facilities have been looked down on and their construction unappreciated, from the reinforced concrete of Harvard Stadium to the air-supported roofs of the latest major league pleasure domes.

In 1962, ground was "broken" for the Astrodome by a group of Texas plutocrats firing shots from Colt revolvers. Roy Hofheinz, the man behind the Astrodome, claimed he had been inspired to build the structure by seeing the Colosseum in Rome—and hearing from the tour guide of the awnings used to protect Nero and his retinue from the elements.

The Astrodome was fitted with 4,596 Lucite skylights, the stuff of gunblisters on World War II bombers. It was a great greenhouse—until the players looked up and lost the ball against the glare. Even orange sunglasses failed to solve the problem, and the skylights had to be painted. The grass died and Astroturf was born of desperation.

"More than a stadium . . . a way of treating people," was the slogan.

Air conditioning was a key part of the pitch: this, after all, was the ungodly humid climate of bayou Houston. Inside, the Astrodome favored a kind of riverboat eclecticism, boasting zebra-skinned chairs in its conference room and a barbershop entered through Moorish arches. There was a Gothic chapel and an 1890s bowling alley. The presidential suite featured a sculpted nude figure at the base of its dramatically turning staircase and a round rug bearing the presidential seal.

All this untrammeled kitsch, along with the first scoreboard, has vanished now, like the oil boom and heady early days of the space program, which inspired the change in the baseball team's name from the Colt .45s to Astros when LBJ moved NASA in.

Stadiums—even domed ones—age as fast as athletes' careers end, and acquire a patina of nostalgia. Twenty years after its completion, the Astrodome had already grown as ragged and worn inside as an ancient ruin. It looked like the kind of place where cattle shows and tractor pulls as well as baseball and football games were held.

Sky boxes or luxury boxes, not dissimilar to those enjoyed by the Romans Roy Hofheinz considered his inspiration, became a vital part of the economics of team sports. Rented by corporations or wealthy individuals, with catered food and liquor and closed-circuit television, these boxes were no less than motel rooms set atop the stadium. Here, guests could socialize, do business, look down on the mortals in the regular seats—even watch the game.

The sky boxes became a major source of revenue for teams. Franchises, such as the Oakland Raiders, would shift cities and become the Los Angeles Raiders, in order to obtain more luxury boxes. One developer went so far as to propose financing a new domed stadium near the existing Shea Stadium in New York by building not just sky boxes but apartments into the structure, for full-time living and game-day proximity.

Luxury boxes were like the family chapels set in the sides of the great cathedrals, and in real major league cities, no one of a certain station and prosperity could afford to be without one. Plugged into cable television, but shielded by glass from the hoi polloi in the cheap seats, the luxury box was a kind of private apartment. The stadium, once offering a democratic mix of class and incomes, was now divided, just as the railroad car had changed from a common car to different types of special car—Jim Crow car, immigrant mover, Pullman compartment, or private car.

Almost as soon as the Astrodome and its concrete-and-steel-roofed kin were completed, however, they were already dinosaurs. Their type of construction—using lamella trusses and compressed ribs—had been supplanted by another: air-supported, Teflon-coated fabric domes. It was perhaps the most appropriate architecture for an era in which the business of sport was built on high-pressure salesmanship and hot air.

The visionary behind the fabric-covered stadiums was the engineer David Geiger, who got his start with a commission to build the U.S. Pavilion at the 1970 World's Fair in Osaka. Since the Crystal Palace of 1851, fairs had been the place to show off new building technologies. For the 1967 fair in Montreal, Bucky Fuller had been called on to put up a huge version of his geodesic dome—the same sort of dome he tried to sell to the Brooklyn Dodgers. The success of the Osaka dome led Geiger to a commission for the Pontiac Silverdome, outside Detroit, whose translucent roof was held up by a pressure three and a half pounds per square inch—or about 20 percent—more than normal. The cost of the structure was about a third as much per seat as that of the hulking concrete Superdome. Soon even Tokyo had adopted the American technology and built its own version of the Geiger structure, even importing American college and professional football teams to play inside. This was the great American building tradition recaptured: lightness in the service of spectacle—the balloon frame made literal.

CHAPTER EIGHT

Ship Shapes
and Land Yachts

THE YACHT *AMERICA* swelled the American bosom with national pride when it defeated British boats in 1851 and lent its name to the America's Cup. The victory coincided happily with American triumphs at the Crystal Palace World Exposition in London, where McCormick's reaper, Colt's revolver, and American tools of all sorts won Continental admiration.

Designed by George Steers, *America* was built for a scion of the country's first industrial families. John Stevens—whose brother Robert invented the T rail for railroads—brought the locomotive *John Bull* to the United States, and laid out the most notable early baseball ground (Elysian Fields) on land his family owned between Hoboken and Weehawken, New Jersey. Steers's fast little sailboat *Martin Van Buren* had won a prize offered by the New York Yacht Club, of which Stevens was a founder, and in 1841, probably seeking his patronage, Steers designed a new type of racing shell drawing just four inches of water and named it after John Stevens. Evidently, the ploy worked.

Despite Steers's success with *America*, he could persuade none of the merchant firms to let him build a model of a clipper ship that would magnify the *America*'s lines into a great cargo vessel. It would be the fastest clipper ship of them all, he boasted, and its hull so finely shaped that no eye could determine its midpoint, "just like the well-formed leg of a woman," which inspired the shape.

150

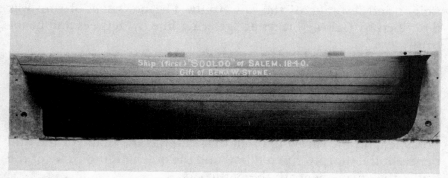

Builder's half or "lift model" for ship *Sooloo*, Salem, Mass., 1840.
(Peabody Museum)

A model survives—sole repository of his particular vision—a beautiful, sculptural thing, combining resonant line with the wood-grained detail of handcraftsmanship. It is a builder's half model, so called because it represents one longitudinal half of the hull. Such models were the way the great ship designers of the clipper era— Steers, Donald McKay, and John Griffiths—created the hulls that revolutionized ship design. Although no American clipper ship still exists, the models preserve their shapes as functional sculpture. But the clippers' names and legends are well known, from the stories of dashing voyages around Tierra del Fuego to Gold Rush San Francisco by the *Flying Cloud* and *Sovereign of the Seas*. Donald McKay and others not only modeled their ships but drove them and their crews to record passages.

For all the talk about its purity of shape and embodiment of the beauty inherent in function, the clipper ship was given its form by competition: it was capitalism made physical. Its success came not from the breaking down of production into individual processes, nor from the division of labor, but from a concentrated, plastic modeling of the form toward a single goal: speed. Speed, to reach San Francisco or the Orient in record time. Speed, to hold off the attractions of steam packets for a few more years.

For the sake of speed, the ruling factor of the design was not cargo capacity but the ability to slice through the water and to carry sail. The result was the characteristic sharp—or, as sailors called it, "smelt head"—bow of the clipper, and a hull that distributed the weight and leverage of the sail evenly. The new style of ship was described by George Steers as "growing fine by degrees and beautifully less."

"Our cathedrals," Samuel Eliot Morison called the great clippers. Blown along by the winds of legend, the clipper ship remains in our minds as a Currier & Ives print, billowing sail as complex as a crinoline hoop dress, towering above a bare slice of black hull between canvas and thundering main. But what made the speed possible, and the carrying of all that sail, was the shape of the hull beneath: the sharply tapered bow and the overhanging curve of the hull—the curves beneath the petticoats.

That we have nearly forgotten. The models remind us of how little of the great ships you could see and how strikingly modern the hulls appeared. The ship was the first mode of transportation to be streamlined, and to us it not only seems beautiful on its own terms but to foreshape the *Twentieth Century Limited*, the Lincoln Zephyr, or the Lockheed Electra.

Many naval historians agree with Howard Chapelle, one of their most distinguished number, that the clipper ship has been overpublicized, overcelebrated, and overrated. It was the Baltimore clipper, a small vessel dating from as early as 1730, that introduced many of the innovations of shape that made speed possible. The full-sized clipper ships, Chapelle argues, were not as fast for their size as smaller schooners and other yachts inspired by the Baltimore clipper, which was often adapted for privateering and smuggling. And the era of the clipper was limited because it sacrificed capacity for speed—a trade-off that became uneconomical once the demand for rare cargoes, such as tea or any number of items supplying the California gold fields, slacked off.

When the sculptor Horatio Greenough praised the simple unornamented beauties of the ship at sea, he understood that the need for speed had lent the vessel its shape. But it is highly unlikely he considered the type of ship that especially required speed, and did much in its development to streamline hulls. That was the slave ship.

Because of the high risk of capture, with loss of ship and cargo, slavers were built as cheaply as possible. "Therefore the slaver," wrote Howard Chapelle, "like the later 'rum-runner,' was very plain and cheaply built, without decoration or 'ginger-bread work.' " For a slaver, speed was vital to escape patrolling government cruisers and to cut the mortality rate among slaves. For such a ship, as for the clipper in the tea or California trade (or, for that matter, for the more contemporary "cigarette" speedboats of the Florida drug trade), "speed and profits went hand in hand," noted Chapelle.

Modeled on the Baltimore clipper, slave ship design, in turn, influenced the early schooner-yachts—the ancestors of the *America*. One American slaver was even converted directly into a British yacht. Such service wore clippers out even faster than the Horn; by the middle of this century, virtually all had vanished. But the builder's half models preserved their shapes—and the shapes of ships dreamed of but never built, like Steers's clipper.

For the designer, the half or lift model was drafting table and blueprint made solid. Self-trained by observation and study, these men found it easier to think in three dimensions than to draw plans in two. The classic half or lift model used in shipbuilding was made from wooden planks stacked up and held together by dowels, then carved as if they were one solid block. The usual scale was a quarter inch to a foot, making many of the models more than six feet long and as heavy as 200 pounds. Even more dramatic is the checkerboard model, in which the shape has been further highlighted by using long pieces of one-by-one, light mixed with dark, so the bow and stern curves come out as if graphed in solid wood.

When the carving was finished, the planks were unstacked and their lines traced onto graph-paper drawings, which were then expanded to full-scale layouts in the great mold lofts—as big as the ship itself—where the lines were rendered in colored chalk on a black floor, forming an area of vast curves like a Sol Lewitt drawing. From the lines, wooden molds or frames were fabricated against which white oak was bent to form the ship's skeleton.

Only one longitudinal half of the ship needed to be shaped, since the hull was symmetrical about the plane of its keel. When the shape had been transferred from the model, the planks would be reassembled and the model mounted on a board or against a mirror to hang in the designer's study or to present to the ship's owner or financial backer. Occasionally the models were given full rigging and set against a background or "shadow box," with results that range from downright kitschy to some combination of folk painting and Joseph Cornell's Surrealist boxes.

The half model was an American invention, developed in Massachusetts in the late eighteenth century and, with modifications, used to shape ships well into this century. The invention of the half model has been attributed to Orlando Merrill of Newburyport, Massachusetts, as early as 1795. One such model may have been used in the design of the war sloop *Wasp*, which earned renown in the War of 1812. It was one of the great innovations in nautical technology, "a proud emblem of American genius," said clipper designer John Griffiths. "The most important invention in the history of wooden shipbuilding," Carl Cutler, an historian of the clipper, calls it. It was a characteristically American combination of the intuitive with the scientific.

Models had been used before; English shipbuilders had even tested them in water tanks, a procedure also suggested by Benjamin Franklin. But previous models had been crude affairs and the economic impetus for applying their lessons was lacking. Errors frequently occurred in converting the model to full-scale plans. The new model type provided the shipbuilder with an expression for his intuitive observations: he could carve out the ship he saw in his head, much as Detroit auto designers later carved their imaginings out into clay models.

The great clipper ships could no more have been created without the half model than the F-15 without a wind tunnel; but the model was used as a means of rendering, not a testing device. A generation elapsed between the invention of the half model and its use as a test model in a tank. That was above all the work of John Griffiths. In 1836, when he was twenty-seven years old and a draftsman, Griffiths wrote a series of articles proposing radical changes in ship design.

He had dug up some thirty-year-old studies involving tests of geometrical shapes of wood performed in a brewer's vat by the Englishman Mark Beaufoy. From Beaufoy's *Nautical and Hydraulic Experiments* and his own insights, Griffiths developed new ideas: a slender shape, V-shaped in section, sharp at the bow and rounded at the stern, with maximum breadth located amidships. In 1841, he exhibited a model of the new type of ship he proposed; two years later he finished plans for the *Rainbow*, a ship for the tea trade commissioned by the New York merchant firm of Howland & Aspinwall.

The *Rainbow* proved Griffiths's points. Previous fast ships had been based on the intuitive designs of the Baltimore clipper. The *Annie McKim*, the first version of the Baltimore clipper to be fully "ship rigged"—to apply the shape of the Baltimore clipper to a full-scale cargo ship—had performed admirably for Howland & Aspinwall. But the *Rainbow* did better: it broke the record for a New York to China round trip, and paid for itself in a single voyage by racing back with the season's first tea and earning $45,000 in profits.

From then on, the process was one of refinement, as Griffiths's models progressed from intuition to a combination of intuition and science. He gradually modified his V-shaped hull section toward the U shape favored and proven by Donald McKay. The resulting hull, sliced amidships, bore the silhouette of a wineglass. Griffiths's praise for the model was tempered by his statement that "the man who builds one hundred ships by the same model, contracted or expanded, has had no more real experience than the man who has built but one. It is impossible to model vessels by eye, having no reference to known laws that govern the elements." Those laws were the laws of mathematical physics, in which the self-taught Griffiths alone of all the major clipper designers was expert. In the end, it became a mathematical model altogether, summed up in Griffiths's *Treatise on Marine and Naval Architecture* of 1850.

The shape of the clipper represented a revolutionary reversal of the traditional wisdom that the bow of a ship should be rounded and the stern tapered. The model had been the shape of the fish—"cod's head and mackerel's tail," was the builder's phrase for the ideal. But when such clipper designers as Griffiths began to read up on hydrostatic mathematics and test their models in tanks, they found that the immediate visual analogy was false. They went on to prove that a ship with a fine pointed bow, a center of displacement aft of center instead of forward, and a hull rising from keel to gunwales, could be driven faster than any sort of sailing ship in history. The clipper ideal was "clean, long, smooth as a smelt." The best models dramatize this ideal. The lifts or planks, especially in models of light pine and dark mahogany or cedar, create long horizontals that serve to accentuate the streamlined shape.

Clipper designers were notoriously jealous of each other and protective of their trade secrets; it was said no good designer ever gave away a true model of one of his ships. Early in his career, before he created the *Flying Cloud* and became world-famous, Donald

McKay had a partner. When the two partners split, they agreed to cut all their models in half—one taking the bows and the other the sterns—so that neither of them could individually exploit the work of their collaboration. In 1853, a dockside fire devastated the *Great Republic*, McKay's last clipper, and he never recovered financially. Soon McKay was reduced to keeping warm by chucking the half models of his clippers into the stove. Only three of his models survived him.

☆   ☆   ☆

UNLIKE GEORGE STEERS, with his talk of matching the beauty of the female leg, most of the builders of half models made no claim to aesthetic appeal—John Griffiths scoffed at those who called his ships beautiful. "We do not understand the import of the term beauty," he wrote. "We can give no other definition than the following: fitness for the purpose and proportion to effect the object desired." In this he echoed the Shaker elder who claimed that common ideas of beauty were "absurd and irrational."

The purpose and the object, Griffiths thought, were cash. The shape of the clipper, singlemindedly created for speed, was a summary of America's motive power:

> The American character seems to be but partially known abroad. It is only necessary for him to receive an affirmative answer to the question, will it pay? when he gathers up his scattered thoughts, and concentrates them into a single idea, or into the compass of a tele- graphic despatch; and then, as on wings of lightning, he is ready to circumnavigate the globe, or to embark in any enterprise within the grasp of thought, or the conception of the human mind.

Griffiths was a man in the grip of a contradiction. He scoffed at beauty, but he knew the effect of a correct shape: the viewer "is pleased with the appearance but cannot tell why. There is a certain something that makes an impression on his mind." He argued that a ship must look fast, even when at rest, a sentiment that suggests Louis Sullivan's decree that the tall building, to be beautiful, must be "every inch of it tall." On one hand, Griffiths raised up hydrosta- tics as the guide to the shipbuilder; on the other, he wrote: "we hold that no man can improve to any considerable extent either in ship-

building or any other branch of mechanism, whose volitions are not
the results of his own conceptions."

But in their search for speed and efficiency, the designers made
objects as smooth and modern as a Brancusi, with a presence as
mythical as any of Melville's or Poe's literary vessels. What appeals is
some primal pleasure inherent in their curves—the same pleasure
afforded by the clipper's streamlined descendants, the jet plane and
the racing car, or by the feminine forms that inspired George Steers.
In Currier & Ives prints, the *America* or the long, low shape of the
clippers was depicted in a way very much like the company's engravings of trotters and racehorses. And Horatio Greenough, the sculptor and proto-functionalist theorist, rhapsodized:

> Observe a ship at sea! Mark the majestic form of her hull as she
> rushes through the water, observe the graceful bend of her body, the
> gentle transition from round to flat, the grasp of her keel, the leap of
> her bows, the symmetry and rich tracery of her spars and rigging,
> and those grand wind muscles, her sails. Behold an organization
> second only to that of an animal, obedient as the horse, swift as the
> stag.

Greenough praised the "naturalness" of form suited to function,
and proposed an almost Darwinian theory of nature's abhorrence
of the decorative and redundant. Designers should evolve away such
ornament, he argued, as nature does. Griffiths, too, scoffed at
unnecessary ornamentation.

Although Greenough wrote before the clippers, he saw in the
Revolutionary-era frigate *Constitution* a contrast to the men-of-war
of the mercantilist age, which were encrusted with gilt and captains'
quarters with protruding bay windows, like countinghouses at sea.
The desire for improved function, he argued, had stripped off such
foolishness at the same time that it reshaped hulls. Greenough went
so far as to ridicule the carved figureheads of ships as a last vestige
of such tendencies.

Such highly personalized views were less scientific than spiritual.
While the hulls of the eighteenth-century merchantmen and ships
of the line were far from the sleekness of the American clipper or
frigate, it is doubtful that they paid much of a speed penalty for
their decorations. But Americans put their faith in what looked
right to them—and counted that it would be both efficient and

beautiful. In his *Treatise on Marine and Naval Architecture*, Griffiths wrote that a ship should look as if it would move even without the application of sailpower.

To be sure, the models look capable of movement, and of speed. The clippers' beauties live on chiefly in the form of these models, mantel-sized pieces of sculpture through which the designer's convictions—"the result," Greenough wrote, "of the study of man upon the great deep"—shine through.

Another equation killed the clipper: that of speed to capacity. In the heady days of the early 1850s, speed counted far more than capacity. Only the most exaggerated form of mercantile capitalism made the clipper successful; the ratio changed as things quieted down in California. Its reason for existence was to convert speed into currency. A barrel of flour that cost $6 in New York went for $100 in San Francisco, but supply and demand shifted so precipitously that a newly arrived cargo could end up worth too little to justify unloading. Tacks, for instance, were at one point in great demand but the price dropped so low so fast that an entire shipload of tacks was dumped overboard. By 1854, the extreme clipper was being succeeded by the more capacious medium clipper, and finally by the downeaster and windjammer, where size alone kept sail competitive with steam.

Slaves were one of the few cargoes whose weight-to-price ratio kept the clipper economical. Fast passages meant fewer deaths and a better chance to escape the authorities. Guano and coolies were the other cargoes with which the great record setters ended their careers. George Steers's only completed clipper design, *The Sunny South*, was turned into a slaver and renamed the *Emanuela*. Captured by the Royal Navy, she ended up ingloriously as a storeship.

The bust that followed the boom was brought on by the Civil War. Its blockade, the attacks of Southern raiding parties, the arrival of improved steam engines, and finally the crossing of the country by railroad, killed the clipper as a practical vessel. But by then the clipper and its lore were firmly established in American culture and especially language. The topmost sail of a clipper's rigging was sometimes known by the new term "skyscraper." And by the late 1860s Mark Twain, recently returned from the West Coast in the company of a clipper ship captain, noted of a New York theater that they "put beautiful clipper built girls on the stage . . . with only just barely clothes enough on to be tantalizing. . . ."

☆   ☆   ☆

IF THE CLIPPER ship's importance lay in the shape of the hull, another characteristic American vessel, the Western steamboat, was distinguished by the virtual absence of smoothness and shape to its hull. It was extremely shallow and loose-jointed, wide and flat-bottomed, almost a raft decorated with a huge edifice of super-structure. To build a hull this shallow for its length required clever technical innovations, and it was at least as formidable a task as shaping the clipper hull. To keep the long hull from sagging, it was often supported by "hogchains," which worked like a sort of inter-nal suspension bridge.

To many observers the Western riverboat suggested the trim

Mississippi River steamboat, circa 1850.

Eastern version turned inside out: inward-looking cabins and galleries now were arrayed toward the world around the ship. It was like the tall-tale frontiersman who boasted of being "half horse and half alligator"—it was half building and half vessel.

In his definitive *Steamboats on the Western Rivers,* Louis Hunter argues that "in the river steamboat the western designer-builders evolved a new architectural form, falling somewhere between the marine and civil branches of the art. The exigencies of river navigation in the west reduced the hull from a thing of potential beauty to a structure having too often the unprepossessing lines of a barge."

A *Harper's Weekly* article of the 1850s praised the steamboat as "one of the most striking, as well as most original forms of our altogether original American architecture." And echoing Oscar Wilde, it went on to claim that "whenever our people attempt to build public edifices, such as churches, state-houses, and private dwellings, after their own invention, they are pretty sure to make a frightful botch of it; but American steamboat architecture, which has grown out of the needs of our commerce, is not only original to us in its form of construction, but it is sometimes splendid in appearance."

Riverboats not only differed from ocean steamboats, their shapes were specific to individual rivers. We all know the Mississippi riverboat, emblazoned in memory via Currier & Ives; but boats for the less treacherous Hudson had fuller hulls, closer to those of seagoing craft, while those for the frontier trade and military operations of the Missouri were more compact than those of Mississippi boats and almost all stern wheelers. These were known as "mountain boats," designed for narrow channels and low waters, and drew as little as 20 inches unloaded. Custer's troops departed for the Little Big Horn in such a vessel.

If the clipper reflected a "singleness of purpose" in its parts and its evolution, the riverboat, observers noted, demonstrated "no common principles" in its construction. It was a makeshift kit, where such engineering tricks as the hogchains that supported the ends of the hull like the cables of a suspension bridge were mingled with grandiose and often eccentric cabin architecture.

Design ideas mingled in the riverboat like its crowds of people of all backgrounds and social strata. In Melville's *The Confidence Man*, the river steamer *Fidele* is described as an "arcade or bazaar," which "might at distance have been taken by strangers for some whitewashed fort on a floating isle. . . ." It is compounded of "fine promenades, domed saloons, long galleries, sunny balconies, confidential passages, bridal chambers, staterooms plenty as pigeon-holes, and out of-the-way retreats like secret drawers in an escritoire."

As in *The Confidence Man*, the riverboat figured as a sort of abstract social space for self-presentation: you could be anything at all on the boat, amid the theatrical scenery of its grand Gothic fittings. But while its passengers might mingle on deck, the riverboat was a sort of floating bazaar or hotel, a world in miniature, a social model in which the classes were neatly divided—including blacks, relegated after the Civil War to a separate cabin called "the Freedmen's bureau."

Steamboats were singled out by de Tocqueville as reflecting the impermanence of American design and construction. Faith in the progress of the technology discouraged solidity; the average steamboat lasted no longer than five years. But steamboats were also run by small operators, who had to get maximum profit out of them, were unencumbered by considerations of safety, and resisted long-term "business plans." The economy of the river, by contrast with the railroad to come, was radically laissez-faire. The steamboat offered a world in miniature, but a world constantly destroyed and recreated.

☆   ☆   ☆

MICHEL CHEVALIER, the trenchant French observer of America, was struck on his travels in the States in the 1830s by the sense of a people "encamped" rather than established on the land, and by the American's "passion for locomotion" which led him, even while sitting, to rock, whittle, chew, and otherwise remain active. Chevalier noted in particular the frequency and facility with which houses were physically moved.

One man made his reputation and modest fortune moving those houses: George Mortimer Pullman, whose business was abetted by the building of the Erie Canal near his hometown of Albion, New York, and the relocations it made necessary. Pullman's outstanding accomplishment, however, was the shifting of an entire hotel, complete with sidewalk, out of the way of a widened street. The hotel did not lose a day's business.

This profession was an apt foreshadowing of Pullman's ultimate success in turning the railroad car into something resembling a mobile hotel—and eventually a private mansion on wheels, a yacht for the land. In the 1850s, the American passenger railroad car looked to the riverboat for inspiration. While the European model and that of the first American railroad cars was the horse-drawn coach, even before the first rails were laid inventors were dreaming up independent "Prairie Steam Cars" and "land barges," self-propelled successors to the steam wagon that the inventor Oliver Evans had proposed in the 1790s. These boasted multiple "decks," and one even had a "Captain's office." They set the pattern of the railcar as ship, on which one could walk around—even transportation could not enforce immobility in America—and on which, when George Pullman was finished, one could eat, sleep, converse in a "saloon," even bathe and get a shave, in a self-contained world like that of the riverboat.

When Abraham Lincoln went to Washington to be sworn in as president, he arrived in a primitive sleeping car—in secrecy, on the advice of his security man Allan Pinkerton. Some political cartoonists inaccurately showed him disembarking from a boxcar. In fact, the car was not much more than that, differing mainly by the addition of three tiers of beds. There were no private compartments, no fancy fittings, no dining car. The Baltimore & Ohio had been experimenting with these since the late 1850s, at which time George Pullman was in Colorado. There, legend has it, he saw folding beds installed in tiny miners' shacks.

The development of the railroad car was part of the whole American adaptation of the railroad—one of the classic cases of a technology being redesigned to fit the landscape and society. English railroads were as straight as possible and their beds carefully constructed, set behind fences in built-up areas, run through tunnels or over viaducts. In America, where capital was scarce and land and materials cheap, the rails curved widely. They went around hills and valleys rather than through or over them. They ran right through town—practically every European visitor commented on the simple warning sign and bell that served to keep people and animals off the tracks. America presented the world with the "cowcatcher," first installed on the Camden and Amboy Railroad.

To keep the wheels of the railroad car on the uneven and curving rails, in 1832 Ross Winans developed the "bogie," a car beneath the car, a pivoting platform to which the wheels were attached. The bogie was a glorified version of the rotating front wheels of a soapbox derby racer. The front bogie might turn one way, the rear the other, while the car stayed steady atop them. Rails were modified, too. Robert Stevens developed the T-shaped rail, which replaced the English I, and to which wheels clung more tightly. In place of Continental stone bases, Stevens and others developed the use of wooden sleepers or ties to which the rails were attached with the familiar rail spike.

The bogie made possible longer railcars. Both in Europe and the United States, early passenger cars were based on stagecoach-style bodies, first lifted onto railroad wheels, later grouped in pairs on flatcars. In Europe, the stagecoach origins survived in the plan based on a row of compartments, each with its own door to the outside. But the United States favored a long wooden box of a car, containing a series of benchlike seats, with a stove at one end and a toilet at the other. Both these amenities were lacking on European cars—and so were the fires often caused by the stoves overturning in wooden cars. And while in Europe the baggage stayed on the roof—again, coach style—the American system exiled it to a separate baggage car. The black passenger often went there, too.

Dickens and others were struck by the absence of first- and second-class carriages, and the American railroad car has generally been seen as a democratic contrast to the closed compartments of European railroads. "It may be described as the democratic palace instead of a nest of aristocratic closets," was the characteristic comment of an observer of the American railcar, writing in Horace Greeley's *Review of American Industries*.

The story is not that simple. The evolution of the American railway car, from the open space that struck European travelers to the ornate chambers of Pullman and others, is an exemplary tale. The myth of the American railcar as a "palace for the people" suggests heads turned by the sumptuousness of elaborate interiors, with their carpet and carving and crystal. It ignores the question of which people were riding in such cars. It forgets the emigrant cars and the third-class cars and the colored cars. Dickens in 1842 had described the Negro car—"a great, blundering, clumsy chest, such as Gulliver put to sea in from the kingdom of Brobdingnag"—as inferior to white cars he already found miserable, "like shabby omnibuses."

When Lincoln left Washington for the last time, his body lay in state in his own private car, accompanied in the train by George Pullman's first grand car, "the *Pioneer*." The entire trip back to Springfield, through solemn crowds gathered at local depots, became a publicity stunt for Pullman. Platforms had to be widened and heightened to accommodate the new car. Soon after Lincoln's death, the Pullman car began to win the hearts of railroaders and the populace. It became a symbol of the Gilded Age, an era whose beginning was marked by Lincoln's journey to the tomb.

Nothing better symbolizes the seizure of public vehicles for private purposes than the private railroad car; the whole history of the robber barons is summed up in their décor—as bogus Continental as the name "Credit Mobilier" or the arabesques on its stock certificates. It was on his private railcar, *Idle Hour*, that William Vanderbilt issued what came to be the motto of the age of the robber baron, "The public be damned."

In fact, the Pullman car marked the gradual institution of class distinctions and separate compartments. Pullman was inspired by the canal boat and riverboat, whose histories had also seen a progression from an open space to closed compartments, and by the special train created for Napoleon III in 1857, with interiors by Viollet le Duc.

As with so many items, the story here is the creation of a classy or premium product, a reassertion of individual differences in wealth against the democratic grain. This was accomplished with decoration—Brussels carpet, ornate polished walnut, and acres of mirror—modeled on European lines, and bearing (as automobiles later would) all sorts of names in foreign tongues.

Efforts were made over the years to make the seats more comfortable; hundreds of patents for adjustable and reclining seating were issued. The railroad car—followed by the automobile and airplane—was one of the laboratories in which the modern easy chair was developed.

After the Civil War, the opening of the Transcontinental Railroad marked the development of more and more private cars, cut first into compartments and then into what were eventually called roomettes. Specialized cars followed. The dining car, for instance, was a response to the famous abuses of station restaurants, born of a conspiracy between conductor and restaurateur. The all-aboard signal came just as orders were served, depriving the customer of the chance to eat them. Grabbing what they could, the passengers still left much food, already paid for, to be recycled for the next group of passengers. Pullman himself patented the first two forms of the dining car. He named the first the *Delmonico*, after the restaurant. In swift succession came the smoking car and the parlor car. Soon the cars boasted bathtubs and barbershops. Because the American railroad often crossed unsettled or sparsely settled territory, and there were few restaurants or other facilities along the way, it made sense to put as many such services as possible on the train itself. Transcontinental trains became as self-sufficient as riverboats or ocean liners—and were fitted out in similar style.

☆　☆　☆

IN MAY OF 1870, the Union Pacific put Pullman's *Pioneer* back into public relations service, ferrying a group of Boston financiers cross country on the newly completed Transcontinental Railroad. The passengers shot the soon to be eradicated buffalo from its windows. The train consisted of the "hotel cars," two sleeping and drawing-room cars, two commissary cars, a baggage car with icebox, and a smoking car equipped with a printing press for publishing a newspaper en route. A crowd of fifty thousand viewed the train before it set off, and the whole city of Boston followed its progress via newspaper reports.

One of the drawing-room cars, the *Palmyra*, boasted an organ, which offered accompaniment for hymn singing and lighter entertainments. Passengers dined on antelope steak, golden plover, and prairie chicken (although later travelers told tales of being served what turned out to be prairie dogs, the cute little creatures they could buy live at station stops). When they reached the West Coast, the passengers ceremonially mixed a bottle of water from the Atlantic with one from the Pacific.

Such stunts were part of Pullman's method. His success was in large part due to his ability to sell luxury to the middle class. At first, it is true, the red carpet was designed to discourage patrons from spitting tobacco juice on the floor, and signs were posted requesting passengers to remove their boots before retiring. But gradually the fixtures of the railcars—Pullman's and others'—became more and more elaborate. And if even the middle class had these amenities, the rich needed more.

Pullman did not invent the sleeping car or the fold-down berth. Rows of beds, adapted from steamships and canal boats, had been in existence for twenty years or more. Many inventors contributed parts of the sleeping car concept, notably William Wagner, who outfitted the Baltimore & Ohio, among other railroads. Pullman bought him out. Nor did Pullman invent the terms "floating hotels" or "moving palaces," themes of his publicity. Both were applied to the steamboats and generally understood to be the goal for railcars as well.

What Pullman did was to combine and buy up patents and rival companies, and promote the idea of his cars so successfully that they became identified with his name. He wisely began by signing up Chicago railroads, of which he soon had the lion's share, to buy his cars and then let him operate them. He worked his way east, where Wagner and other cars were already in place. By offering him stock in the Pullman Company, he persuaded Andrew Carnegie to shift to Pullmans on the Pennsylvania Railroad. As the century progressed, Pullman swallowed his competitors, until he dominated the field almost entirely. In the process, he created the myth that he had invented the sleeping car—a myth as elaborately decorated as the cars he built.

Descriptions of the Pullmans read like catalogues of precious materials: red Turkey carpet, Circassian walnut, Honduran mahogany, tufted satin, ivory inlay, onyx columns, fringed valances, and beveled mirrors. Pullman boasted of his "marquetry room," where artisans imported from the Black Forest crafted his car interiors. There were Italian Renaissance dining cars and Louis XVI smoking rooms and Greek reading rooms. And one of Pullman's competitors, Colonel Mann, summed up the spirit of the railroad car in 1883, promoting his "all-room" sleepers under the slightly titillating title of "boudoir cars"—and naming them after operas and opera stars. The sleeping car was followed by the diner car, the parlor car by the compartment car. The ultimate realization of the Pullmanizing process was the private railroad car.

It was fitting that one of the first private cars should have been

Lincoln's. Although during his lifetime Lincoln would have considered such a car too undemocratic, he did more than anyone to create the great age of the railroad. While he was debating Douglas back in Illinois, Lincoln was known as one of the nation's leading railroad lawyers. He successfully defended the railroad in the landmark *Effie Afton* case. The *Effie Afton* was a steamboat that had struck a railroad bridge; its owners sued the railroad. Lincoln's arguments established the right of the railroad to bridge rivers, even at some danger to riverboats, and helped mark the beginning of the riverboat's decline and the railroad's rise. And Lincoln, of course, pushed through the Transcontinental Railroad as a symbol of union—a symbolism marred by the fact that only when the Civil War removed the voices of the Southern states could the railroad's course be chosen, a choice, of course, that favored the industrial North.

Soon, the leaders of the new economic order that Lincoln's administration helped establish had their own cars. No self-respecting robber baron would have been without his own "PV" (private varnish), as these cars were known in the business. Architect H. H. Richardson designed one. Rockefeller and Carnegie, Gould and Harriman owned them and named them, like their yachts, after wives or daughters. PVs were hitched to the train in the station— and generally carried free of charge if the owner had managed to avail himself of one of the many seats on railroad boards of directors. J. P. Morgan's car had racks for his favorite Moselle and a $4 a glass Schloss Johannisberger Auslese 1872. His greatest thrill in life, he once said, was crossing the Continental Divide in the shower bath of his own PV.

George Pullman went on to build a famous planned town outside of Chicago for his employees. He believed he could create housing and common facilities as well organized and upliftingly beautified as his railroad cars. Appropriately called "the town where everything fits," it was an expanded version of a Pullman railroad car, with one exception: no liquor could be sold.

Pullman, Illinois, was no piece of idle philanthropy. Pullman expected to make money on it; he even built churches and rented them to congregations. But in 1893 a depression led to cuts in wages while rents remained at their previous levels, and the famed Pullman strike of 1894 was suppressed by the National Guard. After

Pullman parlor car *Barbara*, circa 1892. (Smithsonian Institution)

that, any sense of benevolent cooperation between owner and workers was destroyed. The town was never the same again, and Pullman soon sold it.

What Pullman had failed to take into account was that his workers were like him: they wanted to *own* their houses, and as much as they were awed by the Pullman cars, they all dreamed of owning their own PVs. He could not foresee that his black porters would create one of the most powerful organizations dedicated to removing blacks from such roles—the Brotherhood of Sleeping Car Porters— nor could he imagine how the automobile, the trailer, the camper, the recreation vehicles to come would be everyman's private cars and "land yachts."

☆   ☆   ☆

WALLY BYAM, the creator of the Airstream trailer, printed business cards in the mid-1930s showing the huge ham-can-shaped trailer being pulled by an Airflow car. Streamlining helped distinguish the Airstream from all the other backyard trailers of that era—the time of the "tin can tourist"—and after World War II helped make it into a cult item, at first straightforward and then camp, like a pink flamingo of the road. Parked in the midst of the Mongolian desert, it was a silver jar radiating American resourcefulness through the wilderness—or so its proponents saw it.

Byam's Airstream was one of the few survivors of hundreds of trailers, with such names as "Covered Wagon," generically known as "road Pullmans." Henry Joy, head of the Packard Motor Company, built one for himself around 1915; William Kellogg, the cereal magnate, lined his with mahogany. This was "private varnish" re- born and conflated with the romance of the Old West.

In Sinclair Lewis's *Dodsworth*, published in 1929, Lewis's hero is a symbol of two decades of trends in the auto industry. Sam Dods- worth has achieved success by developing a basic, Model T car, a good solid everyman's car called the Revelation. He insists that the largest profits lie in selling automobiles as cheaply as possible to as many customers as possible. Finally considered a radical for "raving about long 'stream-lines,' " Dodsworth sells out to the "Unit Auto- motive Company," a GM-like combine, and is at loose ends in his life.

His boss tries to discourage him from a planned pilgrimage to Europe to see the great works of art. Forget it, the boss sniffs. "More art in a good shiny spark-plug than in all the fat Venus de Mylos they ever turned out." (Picabia had already created his *Portrait of a Young American Girl*, a mechanical drawing-style portrait of a spark plug that could have come from a Ford spec book, and Marinetti had already cried that a racing car was more beautiful than the *Venus de Milo*.)

Now Dodsworth dreams of "land yachts," of

> a very masterwork of caravans: a tiny kitchen with electric stove, electric refrigerator; a tiny toilet with showerbath; a living-room which should become a bedroom by night—a living-room with radio, a real writing desk; and on one side of the caravan, or at the back, a folding veranda. He could see his caravanners dining on the verandah.

A year after *Dodsworth*, Wally Byam left the advertising world to create the Airstream trailer. Byam not only realized Dodsworth's vision but provided the sort of superficial American travel that the novel mocks: sightseeing through a mobile picture window.

Byam's first products were kits, plans and specifications sold through the mail for building your own $100 "torpedo trailer." By

Airstream trailer with Lincoln Zephyr automobile, circa 1940. (Airstream, Inc.)

1931 he had begun manufacturing the "Silver Cloud," a more advanced version. After 1935, when it became clear the Airflow was a loser, the Lincoln Zephyr stepped into place on Byam's business card as the tow vehicle of choice. And by 1936, as Pan Am was introducing its "Clipper" flying-boat airliners, Airstream began offering its own "Clipper" model, built of riveted aluminum on a tubular steel frame, and including an experimental system of air conditioning. The Clipper looked a lot like Buckminster Fuller's Dymaxion Car of 1934 and made its owner feel like he possessed his own DC-3. Airstream was getting ready to take on the deserts not only of America but of the world.

That had to wait for the end of World War II and the beginning of the pax americana. In 1942, with aluminum dedicated to the war effort, Byam shut up his factory and went to work for an aviation firm. When peace returned, Airstream sales increased again. In 1951, a caravan of Airstreams headed south through Central America. Four years later, the Wally Byam Caravan Club was founded, a sort of combination of the Boy Scouts and the Kiwanis Club. Loaded into and out of the holds of cruise ships, Airstreams by the caravan visited Europe, Asia, and Africa, assembling, for instance, at the pyramids of Egypt, like Napoleon's marshals.

Like the chuck wagon, the Airstream implied a complete way of life, a cartoon domesticity mingled with a caricature of adventure, and wrapped in the sort of all-American camaraderie of the joiner that de Tocqueville would have recognized. Byam died in 1962, but his organization survived. Nearly 4,000 Airstreams gather at the annual "homecoming rally" each August in Jackson Center, Ohio.

The streamlined shape of the trailer—typically drawn by a huge Cadillac—was the symbol of travel made easy and penetration of the foreign accomplished without the friction of discomfiting encounter. Wearing the blue beret that Byam decreed (artists wear berets; the French, a foreign people, wear berets), Airstreamers learned to "feel at home wherever they go."

☆   ☆   ☆

THE AIRSTREAM CONTINUED a long tradition of bringing home along with you: consider the chuck wagon, that bit of home on the range. The first chuck wagon I ever saw, at round-up time on the

Four Sixes Ranch (the name came from a memorable poker hand that enabled its owner to buy the place) east of Lubbock, Texas, no longer moved at all. It was the genuine article all right, only grounded a hundred feet or so from a cinderblock bunkhouse. From the bunkhouse electrical wires and hoses provided light and water, instead of the open fire and lanterns and casks mounted on the side of the wagon bed. It was the Old West in its last throes, rigged up on life-support systems.

An electric light shone on the chuck box, with its drop shelf and drawers, but the cooking was still done over a fire, using that miraculous device, the Dutch oven, which a skillful cook can turn into a steadily heated bakery in the middle of a chaos of coals.

Charles Goodnight created the first chuck wagon in 1866, from a tough old Army wagon, for his cowboys heading up the Goodnight/Loving Trail along the Pecos. Immediately establishing itself as an American tradition, the chuck wagon has since lent its name to restaurants and its style to football stadium parking lot tailgate parties.

"Chuck," a wonderfully various and solidly Anglo-Saxon word, meant food or a meal by the end of the Civil War, the *Dictionary of American Regional English* asserts. The chuck box was a folding pantry on the back of the wagon, with a cooking board that dropped down on chains, Pullman style, or on a single folding leg. Above, the storage was almost nautical. Drawers as neat as in a Shaker desk held everything in its place: flour to upper left northwest, pinto beans below.

The staff of life was biscuits, beans, and bacon—the last called "chuck wagon chicken." Only when a cow was injured or killed did the cowboys eat beef, and every morsel was used, producing such dishes as "son of a gun," a combination of entrails. Bread on the chuck wagon was made with the sourdough method: a standing joke around chuck wagons was that the cook kept a frog in the sourdough starter, to keep it constantly stirred and "working" with his swimming. A key cultural geographic line could no doubt be mapped showing the juncture of Southern yeast bread culture and Western sourdough culture. I would be surprised if that line did not also cross the most vital locales for country-and-western music.

The chuck wagon was not just for cooking; it was home base for the round-up or trail drive. Bedrolls were stored inside. It was a symbolic, mobile home, a hearth on wheels. The cook himself was a figure of domesticity: the mess sergeant, the figure of wisdom and restraint who figures in western films—Walter Brennan trying to defuse conflicts between John Wayne and Montgomery Clift in *Red River*, Paul Brinegar as the cook "Wishbone" cooling down Clint Eastwood in the television show *Rawhide* (1959–66).

If the chuck wagon could be called ancestor of the lunch wagon and its heir, the diner, it was easiest to see the modern chuck wagon as the recreational vehicles observed at stock car races, national parks, and space launching, with the same power cords trailing out their backs. Roadside lunch stands would eventually call themselves chuck wagons with as much pride as irony. But the chuck wagon was the ancestor of the diner as well, despite the stainless-steel and aluminum streamlinings of that now much iconized institution. In between chuck and diner, the pedigree ran through the lunch wagon—the horse-drawn ancestor of the canteen trucks, with sides of quilted stainless steel, you find at construction sites. The diners we wax nostalgic over today were prefab units, designed to take advantage of the romance of the railroad diner—to look as if they had been unloaded from the rails and dewheeled, bringing with them a lingering sense of the elegance of the Gilded Age.

CHAPTER NINE

Model T to Tail Fin

ONE DAY IN the mid-1960s, at the base of a hill facing an intersection in Raleigh, North Carolina, the blue Chevrolet Corvair of which my mother was the driver and I the passenger was struck gently from the rear by a Cadillac some five years its senior, driven by a local North Carolina country music and television personality who bore the professional name of Homer Briarhopper. The Caddy's protruding bumper cones (known as Dagmars after a bosomy starlet of the fifties) struck the compact high, and dented the vented rear deck that covered its air-cooled engine. Mr. Briarhopper was profusely apologetic and somewhat distressed—a manner that suggested a previous driving record no more distinguished than his musical ones. Cash payment was graciously offered in lieu of the time-consuming process of insurance adjustment.

That collision neatly represents the conflicting traditions of the American automobile: utility versus luxury. Chevrolet succeeded the Model T as everyman's car, and Cadillac succeeded Packard as pinnacle of the staircase of automotive status.

The Corvair had replaced our aquamarine 1957 Chevrolet—a car with its own, smaller tail fins which would today be worth several times not only its original sticker price but that of Mr. Briarhopper's Cadillac as well. The Corvair represented a European influence on American auto styling. *Car & Driver* called the 1965 Corvair, the second generation of the model, the most beautiful car to appear in the United States since before World War II. Designer Ron Hill had looked for its shape to such Italian sports cars as the Cisitalia, the only car in the MOMA design collection. The resulting shape was smooth, and perhaps more elegant than that of the Lincoln Zephyr in which the boxer Jack Johnson had been killed, in the forties, near Raleigh.

Ralph Nader, of course, would make the Corvair famous for its danger. But, with its aquiline, grilleless front, its air-cooled aluminum engine, and four-wheel independent suspension, the compact was a far better and safer car than Mr. Briarhopper's overpowered, underbraked Cadillac. It was, its defenders say, Detroit's last daring car, but it was also fun. Elvis Presley gave his mother, who could not drive, a pink Cadillac. But he gave his fiancée, Priscilla, a red Corvair.

The Corvair represented one of the last, most elegant gasps of the American "everyman's car" tradition. There would be others after it—the ill-fated Ford Pinto, for instance, and the pathetic Chevette. The "compact" Corvair, like the Ford Falcon and Dodge Dart, was a response to the Volkswagen Beetle and a turning away from the big car with tail fins—the Cadillac type.

In the automobile—probably the most powerful American icon of the twentieth century—these two traditions warred from the beginning and war still.

☆   ☆   ☆

No PRODUCT MORE clearly illustrates the strange relationship between model and kit, between mass production, the Ford system, and flexible mass production, the General Motors approach. Henry Ford offered one model for everyone. At GM, Alfred Sloan provided a car "for every purpose and purse," a kit most of whose parts would remain the same, though a critical few would change—from inexpensive car to expensive car and from year to year. His line ranged across incomes and across time. In this process of marginal differentiation, the customer could virtually design his own car. (An often cited calculation of the possible permutations of General Motors models and options of the fifties claimed that their number exceeded the estimated number of molecules in the universe.) Years after GM began the system, Oldsmobile and Buick customers of the seventies would wax indignant upon finding Chevrolet-made engines under their hoods. But in the twenties, you could have found an Oakland engine in your Buick. It had taken the public half a century to catch on to GM's kit system.

The T, in retrospect, appears almost an aberration in the history of the American automobile, a great exception that proved the rule of aesthetics and fashion in the most basic twentieth-century machine. From Ford's T to GM's automotive array, the car evolved from a universal, basic appliance to a personal fantasy machine. Along the way, the auto designers showed how a particular form of functionalism could be transformed into fantasy—through streamlining, borrowing shapes from the only mode of transport that inspired dreams more fantastic than did the car, the airplane.

☆   ☆   ☆

THE MODEL T is forever fixed in our memory as if in a silent film, bouncing through mudholes, one wheel high on its crude suspension, the other dropping and rising, or leaping open drawbridges in Keystone Kop movies, being carried up on fire ladders or crushed by a steamroller, only to be resuscitated with an air pump. This was the Little Tramp of the American road, the Mickey Mouse, cheerful under all adversity in its standard black suit.

The Model T has vanished into legend. (Not a single contemporary American in ten, one suspects, could identify a T from a lineup of its contemporaries.) But nothing better summed up its immense popularity than the spate of Ford jokes it engendered. Collected in dozens of cheap paperback books, a staple of toastmasters and banqueteers, Ford jokes mingled pride and contempt for the Model T. A Ford without oil or gasoline "ran thirty miles on its reputation" went one of the more positive ones. Ford did not, as the rumor had it, begin the Ford jokes, but he enjoyed them and the company encouraged them. The one he liked to tell most often—on one occasion to President Woodrow Wilson—was of course highly favorable: A man arranged to have his T buried with him, "because it had gotten him out of every hole he'd ever been in." The jokes about the Ford's toughness and durability were cited by Wall Street analysts as one of the company's assets.

But as the twenties went on, the jokes grew more sardonic. They focused more on the T's rough ride—Henry Ford was said to have "shaken the hell out of more people than Billy Sunday," the brimstone evangelist. Virtually every new owner of a T was advised to get a tray to put under the car to catch falling parts. One T, it was said, ran over a chicken, which emerged unscathed, saying "cheep, cheep, cheep." Tin was a recurrent theme, with its associations to the disposability of the can. (In fact, the T's fenders were made of a fairly sophisticated alloy, vanadium steel, and its hood of aluminum.) A man sent a box full of tin cans to the Ford plant and received a T by return mail: "Although your Ford was badly damaged, we have managed to repair it." The T had no doors, but can openers were available from the company. "Lizzie Labels," slogans painted on the T, were popular in the mid-twenties: "I do not choose to run," "Barnum was right," and the durable "Don't laugh, it's paid for." The nicknames summed it up: a "flivver" was a good-for-nothing in the 1890s; "Lizzie" was the generic name for a black maid.

The tone is clear: the T was regarded the way soldiers regard any standard item of issue—a helmet, a uniform, a mess kit. At one point in the late teens, half of all cars sold were Model T's: the T was G.I.; the T was a car-toon, produced by the same processes of reduction to essentials, abbreviation, and ease of reproducibility as the great Mouse himself, or the gestural vocabulary of the Little Tramp. (When the Italians under Mussolini later attempted to create their own people's car, it was called the *Topolino*—the little mouse.) Mickey, Charlie, and Lizzie: all three were little black multiples, produced by moving belts and gears, abbreviation and repetition.

This is not the carefully groomed, belovedly washed and waxed car of American advertising legend, full body gleaming with soap in the soft suburban sunbeams, the swirls of wax polished down to the sound of the baseball game on the transistor radio. This was a car known for the abuse it could take, not the care it deserved at the hands of the adoring owner, whose identity it ornamented like so many pointed strips of gleaming chrome.

The T's virtues were its relative power—twenty-two horses capable of climbing steep hills and escaping the deep mud of country roads—and its ease of use—anyone, it was said, could learn to drive it in five minutes, and there was no complex gear shifting as it slid from forward to reverse without a pause. It was reliable and easily fixed, due to the huge network of mechanics and parts dealers.

You bought the T as a basic stripped-down car, and an industry grew up to decorate and improve it—tops, seats, reflectors,

ornaments. Some owners used the T for plowing; others put truck
and delivery beds on its chassis. Sears sold a top for the T at under
$12. There was no speedometer or windshield wiper.

The conservative look of the car appealed to the buyers of "motorized buggies," a class of simple, frail, retrograde vehicles built in
the early years of the century. These buyers were rural people who
might never have imagined themselves owning a serious car before
the T. The T took no interest in the European stylistic affectations
that were already leading to talk of "torpedo" and "streamlined"
bodies. The driver of the T, wrote E. B. White in his famous elegy to
the car, "Farewell, My Lovely," sat up high like a man enthroned.
The windshield was "high, uncompromisingly erect. Nobody talked
about air resistance."

But above all the success of the Ford had to do with its price—40
percent cheaper than any other real automobile. And that price, of
course, came from its production in huge numbers, due to interchangeable parts assembled with moving apparatus to deliver them
to the worker.

The T was introduced in 1908, but it would be a while before
Ford could institute what came to be called mass production. The
first parts of the car to be assembled on an assembly line were
magnetos, the alternators that converted mechanical power to electricity for the ignition spark. More and more parts were incorporated into the assembly-line system, and at each stage the price of
the car went down.

Because of its association with the assembly line, the T added
more to American life than cheap basic transportation: it was the
image of its own production, for better and worse. It was the image
of parts moving past workers on conveyor belts, the image of *Modern
Times*. That those workers, after Henry Ford's decree of the five-
dollar day, seemed to be well paid did not make up for the abuses
Chaplin lampooned in the film. The methods for manufacturing
the Tin Lizzie were adapted in part from those for making and
filling tin cans, in part from mechanized cattle- and hog-processing
plants—the assembly line was inspired by the disassembly line.

When, early in his career, Henry Ford contemplated the mass
production of cheap watches, he knew the problem was finding
enough people to buy them: he calculated he would have to sell
600,000 to break even. With the automobile, a virgin market, de-

mand to keep up with supply was no problem. One model—that, Ford said over and over again, was the key; find one thing that everybody wanted, and make that and nothing else. And so Ford applied some of the lessons of the watchmakers, and more of those of the gunmakers, sewing machine makers, typewriter makers, and bicycle makers, to the automobile.

The T was instantly associated with Henry Ford himself, and Ford instantly associated with the term "mass production." His methods not only produced a standard product in large quantities through use of machinery, interchangeable parts, and—the real contribution—mobility of parts on the assembly line, but the masses themselves. There are few stories more familiar: how Charles Sorenson and others created the modern assembly line and the price of the car fell as sales rose to almost half the total market. In a year's time, during 1913 and 1914, the man-hours required to assemble a Model T fell from 12.5 to 1.5. (In the late eighties, the

Joining body to chassis of Model T Ford, Highland Park Plant, Ford Motor Company, 1913. (Ford Motor Company)

figure worldwide was 25.8 hours, and the most efficient Japanese firms spent 18.2 hours per vehicle.)

But just as no worker likes to be thought of as simply another machine on the assembly line, no one really likes being tossed into the category of "mass." Ford promised that he was going to "democratize the automobile." "When I'm through, everybody will be able to afford one and about everyone will have one." He delivered on his promise; but the unforeseen result was not only that everyone was able to afford a car, but that no one could afford *not* to have a car.

The ambivalence with which the T was regarded—softened later by the mists of nostalgia and at the time by its true benefits—was the ambivalence felt toward mass production as well as mass products. Ford promoted his new production methods with the innovative public relations efforts of James Couzens, who distributed Ford promotional films free to local theaters. Unsympathetic observers noted the contrast between the frenzied drama of the assembly line and the glum faces of workers. Even bluesmen, who had once dreamed out loud of leaving the cotton field and the sharecropper's oppressions for the freedom of Detroit, began to worry about "keeping up with Henry Ford."

The five-dollar day, in fact the brainstorm of Couzens (the parsimonious Ford countersuggested $3.50), was needed to help cut employee turnover, which at times exceeded 200 percent. Ford had to hire 963 workers for every 100 it eventually retained, training each of them at a cost of about $100. More than a great publicity stunt, as a means to enable workers to buy the products they made (evidencing an understanding that the masses were now workers and producers at once), the five-dollar day was a recruitment tool. It brought so many job applicants to the factory gates that they had to be turned away with the use of firehoses, and it cut down on absenteeism and desertions. But as a measure of paternalism, the five-dollar day had its limits. Ford's "Sociological Department" rode close herd on the characters of its beneficiaries. When in 1927 he ended production of the T to switch to the Model A, Ford coolly laid off 100,000 workers. With the infamous battle of Dearborn, in which Ford thugs beat Walter Reuther and other labor leaders, all pretense of Henry Ford as a friend of the workingman vanished.

Ford virtually invented the term "mass production" (and characteristically overabbreviated it; shouldn't it be *massive* production?).

There was an unconscious irony here: it turned out that what was being produced, ultimately, was not just the car, in great numbers, but the mass. "Any color you want as long as it is black" was the twentieth-century successor to "The public be damned."

In 1925, the *Encyclopaedia Britannica* asked Ford to write the article on a new entry, "mass production." The author of the piece that appeared under Ford's name was William Cameron, who doubted that Ford himself ever read it. It was an influential statement of the new methods, reprinted in full in the *New York Times* and widely cited. But the piece focused on Ford's systems, without discussing their predecessors: Oliver Evans's automatic grain mill of the late eighteenth century, with its system of belts and Archimedean screws, for instance, or the meat-processing and canning factories that moved carcasses on overhead rails as early as the 1840s, or Cyrus McCormick's reaper factory.

Although the parts of McCormick's machines, which were made mostly of wood, were far from interchangeable, he did make much use of mortising machines, screw- and nut-cutting dies, boring machines, and metal presses, and apparently also moved parts among these machines. A vivid account published in the Chicago *Daily Journal* in 1851, the year when the reaper won awards and admiration in London, suggests that McCormick (like Ford after him) adopted some of the techniques for moving parts that meat-packers in Cincinnati—nicknamed "Porkopolis"—and Chicago had developed to move carcasses through successive stages of dismemberment on overhead rails. The account sums up the excitement observers would continue to feel watching modern factories at work:

An angry whirr, a dronish hum, a prolonged whistle, a shrill buzz and a panting breath—such is the music of the place. You enter—little wheels of steel attached to horizontal, upright and oblique shafts, are on every hand. They seem motionless. Rude pieces of wood without form or comeliness are hourly approaching them upon little railways, as if drawn thither by some mysterious attraction. They touch them, and presto, grooved, scallopped, rounded, on they go, with a little help from an attendant, who seems to have an easy time of it, and transferred to another railway, when down comes a guillotine-like contrivance,—they are morticed, bored and whirled away, where the tireless planes without hands, like a boatswain, whistle the rough plank into polish, and it is turned out smoothed, shaped, and fitted for its place in the Reaper or the Harvester.

☆   ☆   ☆

IGNORING HIS PREDECESSORS in industry was a symptom of Ford's curious attitude toward the past. He assembled the famous collection of historical Americana that became the Henry Ford Museum and Deerfield Village, and talked of reading objects like books— but he rarely seemed to read books. The Model T reflected Henry Ford: he had turned his mechanical bent into a means of escape from the drudgery and isolation of work on the farm, but it still somehow embodied the somber and scrimping virtues of that life in its resilience, economy, and technical stinginess. It looked backward to the rural landscape he despised and yet memorialized its values.

But just as the rural life was fast fading—more people lived in cities than in the country by 1920—the T was doomed to fade. It was a nineteenth-century product in the twentieth century— especially if you believed, as some argued, that the new century had only really begun with the Armistice. By 1915, it was already being described as looking like a maiden aunt. The shape of its roof suggested a bonnet. The front retained the dashboard of the carriage—devised to keep mud and manure off driver and passengers—as did many vehicles of the time. But the T's end did not come until 1927, when Ford sold only a third as many units as in 1926. Before Ford finally ended production, 15,458,781 of the cars had poured from the factories.

Was the T a beautiful object? Two opinions will serve as a sample of the change in thought about the T. The first is from Jane Heap, editor of *The Little Review*, who in 1926 organized a Machine Art survey and surely appreciated what was then taken to be mechanical beauty:

> Utility does not exclude the presence of beauty . . . on the contrary a machine is not entirely efficient without the element of beauty. Utility and efficiency must take into account the whole man. Let us take one of the simplest and most obvious examples . . . the Ford car. It is a by-word, for utility and efficiency, to the unthinking. Yet the thousands of jokes at its expense, the endless jeering on the part of the millions of owners, who also brag about its efficiency—is the

evidence against it. The lack of rhythmic balance in its organization, its stupid, sterile vertical lines, frustrate all consciousness of horizontal motion and velocity. It is justly considered a freak.

The second is from Jay Doblin, the industrial designer, writing in the 1960s:

A strong case can be made that the Model T is the most beautiful car ever built, a classic illustration of what an automobile ought to be. Even though little conscious aesthetic effort was expended on its design, the car is a dramatic, graceful, and pure example of what a wheeled machine should look like.

But if the T looked at least acceptable to its first buyers, it looked crude and antique to its last ones, who had the choice of fully closed, stylish competitors. More than anything else, the death of the T was caused by the popularity of the closed body. In 1919, only 10 percent of cars had closed bodies; by 1925, 56 percent did. In *Babbitt*, Sinclair Lewis has the ladies, in particular, clamoring for a closed car.

The closed body was a symbol: of ease of operation, of the self-starter, of less care—the Model T's spark plugs had to be cleaned every 200 miles—of adjustable seats and other conveniences. People wanted to be able to forget about the mechanism, even more than the planetary gear of the T allowed, to forget about the crank of the starter—especially women, for whom the required strength was often a severe limitation to freedom. (Self-starters had appeared on American cars as early as the 1912 Cadillac.) They also wanted color: tired of Ford black, they aspired to bodies painted in the non-fading color provided by General Motors' new line of fast-drying Duco enamels.

The closed body linked the production car to another tradition—that of the coachbuilder, who in many cases had begun actually building coaches, then switched to producing customized bodies for standard chassis. This was the way, for instance, that future General Motors head Alfred Sloan, then a prosperous executive for a ball-bearing concern, bought his first car, a Cadillac, in 1910. The coachbuilding business was restricted to the wealthy, of course, but 1927, the year that the T died, was also the year General Motors tried to bring something close to it to the masses. It introduced the first production car with body designed by a coachmaker, in the coachmaking tradition—Harley Earl's LaSalle.

Earl's father was a coachmaker in Los Angeles, who moved from the horse and carriage to the horseless carriage trade. Harley Earl entered the family autoworks in 1908—the year of the Model T. The Earl coachworks catered to the movie star crowd. Tom Mix and Fatty Arbuckle—for whom young Earl did his first car—were clients before the company was bought out by Don Lee, the West Coast Cadillac distributor, in 1919. Now the ordinary man—at least the man who could afford a LaSalle—could enjoy the benefit of the same genius that had ensconced Hollywood stars.

Earl was a director—he sat in a large canvas director's chair and boomed out commands with a volume designed to disguise his natural stammer—and picked up ideas from his crew of designers almost on whim. He was the first to use clay models; he had a full-size model built of one car just to study the effects of the grille. Earl was also a huge, charismatic bear of a man, sometimes likened to a football coach. (His physical size, he admitted himself, was translated into the dimensions of GM cars.)

The closed bodies Earl designed were the key to GM's variation on Ford—flexible mass production. With the auto market now larger, it was possible to produce a variety of cars and still produce them in quantities large enough for mass production. With the arrival of the full metal body, only at the end of the thirties, the designer's material was defined. Stamped metal lent itself to a variety of rounded shapes, limited only by the need to slip the stamped form from its die. And the number of cars of any shape was fixed by the life of the die itself. Changes in materials and manufacturing fit neatly with changes in design and marketing.

GM marketed by strata—not just one model for everyone but a kit of models, each suited to a particular purpose and purse. In fact, the strata fitted neatly into a ladder of price and prestige on which the buyer could climb as high as his means would allow, using time payments and trade-ins, the familiar pattern. This sort of marketing dictated a special kind of design—in this case, shaping the shell of the car. That was Earl's job, and the job of the Art and Color Section GM created in 1928.

The direction styling took was toward a lower, longer car. The inspirations it drew were from the world of the coach, at first, and later from the world of aviation, where the shape of the shell affected the function of the object.

Before beginning work, however, Earl went to Europe, whose Bugattis and Hispano-Suizas and Daimlers and Peugeots were still the source of many fashion trends in American cars. The man who would create car shapes that were to be considered the dreams of heartland America made physical came back, he said later, with five full notebooks filled with long, low, fast cars—rich man's cars he would provide to the not-so-rich.

Earl began the process with the 1927 LaSalle, the "sub Cadillac" that was the first car he completely designed for GM after being hired. There was a gap in the General Motors line between the high-priced Cadillac and the Buick; the LaSalle was intended to fill it. "On the line we now call the beltline," Earl wrote later, "running around the body just below the windows, there is a decorative strip something like half a figure 8 fastened to the body. This strip was placed there to eat up the overpowering vertical expanse of that tall car. It was an effort to make the car look longer and lower." The dominance of the horizontal line was Earl's legacy. The beltline drawn in chrome and creased sheet metal was emphasized in every way possible, controlling the swellings of streamlining in fender and hood and the flights of fins. Earl dictated that a single highlight should run the length of the car, like a theme or plot. The beltline was also reinforced by the painted stripe—the last touch of craftwork, the last vestige of the coachmaker's trade, which persists to this day.

In Charles Willeford's aptly named crime novel *Sideswipe*, there is a character who used to apply these stripes. Stanley Sinkiewicz has retired from Ford, one shoulder dropping three inches below the other, after decades of painting these lines:

> As a striper, Stanley had painted the single line, with a drooping striping brush, around the automobiles moving through the plant as they got to him. These encircling lines were painted by hand instead of by mechanical means because a rule line is a "dead" line, and a perfect, rule line lacked the insouciant raciness a handdrawn line gives to a finished automobile. Stanley's freehand lines were so straight they looked to the unpracticed eye as if they had been drawn with the help of a straightedge, but the difference was there. . . . The line was as vibrant as a tightly stretched guitar string.

Willeford has Stanley's line, in fiction, replaced by a method of painting the stripes with masking tape. "Of course it was a dead stripe, but it saved a few seconds on the line." In fact, that never happened; those lines are still done by hand.

This is a truth apparently missed by MIT researchers, including the appropriately named Stuart L. Inkpen, a graduate student, who recently patented a "method of painting customized pinstripes on automobiles by bombarding metallic paint with ion beams." One of the inventors said that the method "made it possible to paint designs without ever touching the car's surface and could be used by robotic equipment. . . . Moving the ion beam across the surface creates a line that can be expanded into any figure desired."

Those stripes continue the horizontal orientation of the American automobile, a fifty-year-old axiom of design which even foreign cars must face up to. It is hard to buy a car without such stripes; they make the car look longer and lower, and you will see them even when they quarrel with the curves and masses of the thing. During the "downsizing" of cars in the seventies and eighties, these horizontal elements became more and more pronounced, as a measure to combat the stubbiness of the new, smaller cars. Like the painted line, it was a last vestige of the handmade coachbuilding tradition.

☆   ☆   ☆

THE MOVE TOWARD the horizontal had been the evolutionary tendency of the car from the beginning, as the carriage, with its seat set high so the driver could see over the horses, was adapted to the shapes made possible by the engine. But under Earl's influence it became more pronounced. Earl agreed with Jane Heap: the car should show vertical lines of motion. Besides, vertical lines made a car look longer and therefore bigger. Over the years, GM in particular would play every possible trick of chrome and metal rippling to make even its large cars seem larger than they were. By 1932, a writer for *The New Yorker* was protesting that the typical car of the time "is built so low to the ground that we wonder why it is not infested with moles," and grumbled that to get in and out, you practically had to get down on your hands and knees and crawl. Earl justified the change on sheerly aesthetic grounds. "My primary purpose . . ." he wrote in a 1954 article for the *Saturday Evening Post*, "has been to lengthen and lower the American automobile, at times in reality and always at least in appearance. Why? Because my sense of proportion tells me that oblongs are more attractive than squares, just as a ranch house is more attractive than a square, three-story, flat-roofed house or a greyhound is more graceful than an English bulldog." That lower and longer in his cars went along with smoother and more rounded, incorporating a stylistic vocabulary drawn from the airplane, Earl did not say; but that change, too, had been part of his mission.

After flirtations with "torpedo" shapes and "streamlining" as early as 1908, after the streamlining efforts of Reo and Graham in the 1920s had failed to make those companies into stars, by the thirties streamlining the automobile seemed not only inevitable but overdue. The big question was, would it, should it, happen gradually, or in a single great leap forward?

The Chrysler Airflow of 1934 proposed the latter choice. The Airflow helped inspire the design of many later cars, including the Volkswagen, first called the *Kleincar*, which was on Dr. Ferdinand Porsche's drawing board as the Chrysler appeared. Take the silhouette of the Chrysler Airflow, fold it over itself in a couple of places, and you get a profile that looks almost exactly like that of a Volkswagen Beetle. The Beetle was one of the most innovative and successful cars ever; the Airflow was one of the most innovative—and unsuccessful. The first streamlined production automobile was a commercial failure, selling some 11,000 examples in its first year, 1934, but tapering off before being discontinued in 1937. The fortunes of the Airflow serve to illuminate the power and the ideal of streamlining.

But both cars had predecessors. The Czechs were turning out a streamlined car called the Matra until Hitler took over their country and put them out of business. The Airflow and the Volkswagen both owed a debt—later redeemed in the form of royalties paid—to Paul Jaray, a German Zeppelin engineer who worked out his ideas in the same wind tunnel where the *Hindenburg* was tested. Jaray patented a basic streamline shape in the United States in 1927.

The Airflow was introduced in 1934, the year that marked the high tide of streamlining. Showcase railroad trains such as the *Twentieth Century Limited*, the *Santa Fe*, and the *Hiawatha* bore the classic streamlined shape that helped them to compete in speed and appearance with the emerging threats of competition from airline travel. The first Douglas airliners were flying, all metal underneath, sleek riveted Duralumin on the surface, the epitome of streamlining. And at the Chicago World's Fair, Buckminster Fuller was showing his bulbous three-wheeled aerodynamic Dymaxion Car.

Tests by Glenn Curtiss, the motorcycle and airplane pioneer, had suggested that a streamlined automobile could increase speed by 25 percent. To get the most out of his new high-compression, eight-cylinder engine, Walter Chrysler turned to the possibilities of streamlining. He gave a free hand to Carl Breer, his chief designer, and a team who went to Orville Wright for advice about using a wind tunnel to test automobile designs. Supervised by William Earnshaw, those tests, using blocks of wood carved in the shapes of contemporary vehicles, revealed that most of them provided less wind resistance when driven backwards. Aerodynamic modeling suggested other possibilities: "negative lift," Breer thought, might be used to hold the car against the road. The body, in effect, might function as a "spoiler," a wide fin like those that keep the rear ends of racing cars on the pavement. Like Ferdinand Porsche, Breer originally considered a rear-engine scheme, which had always made mechanical sense because it eliminated the drive shaft and evened out weight distribution, but also lent itself naturally to the streamlined teardrop shape.

The Airflow's technical innovations lay not only in its shape, however. Its basic structure was a "unibody" or "monocoque" cage of steel members, a first in an era when the skeletons of cars were still wooden, and an aviation-derived concept much talked of at the time. Its body, except for fenders and hood, was a single piece of stressed steel.

Part of the idea of the new design was to smooth out the ride as well as improve aerodynamics. The wheel base was very long, to even out the weight on the two axles, and the passengers were placed more nearly in the middle of the car than previously. The Airflow also gained about ten inches of interior room due to its extension over the wheels. But all those factors contributed to lending the car a "funny" look—some early commentators compared it to a rhinoceros, and from one angle it looked like a bald man with a walrus mustache, as drawn by James Thurber.

The seats were externally framed in chrome-surfaced steel tubing, like seats from an early airliner or a department store knockoff of the Bauhaus. The tubing of the rear seats even swelled at their elbows to accommodate ashtrays. This was all a bit much for Mr. and Mrs. Consumer of 1934.

But it was the grille that seemed to give people the most problems—the coursing waterfall of metal streaming over the smoothed front like a chromed whale baleen, between the eyes of the headlights. The standard Chrysler had a noble square brow, and even the raciest of cars took on the wind with a squared engine compartment, faired to a sharp prow rather than curved. The signal that the Airflow had failed came when Walter Chrysler called in "styling" experts to fix the grille. They tried four distinct schemes to hide the unconventionality of the automobile behind a more conventional grille, but none worked.

The importance of the Airflow lay not only in its technological advances, which designers borrowed for years afterward, but in its basic design ambitions—to integrate the shape of the car into a whole, to meet the eye as smoothly as it met the air. In practice, its legacy was the principle it demonstrated: people would not buy a car for a merely functional reason, however sound. The Airflow was too strange a car for them. And it showed the fascination with streamlining for what it was: a highly charged symbolic issue, not a practical one. Streamlining was a set of motifs, an air, a mood—one auto designer aptly called it "the badge" of the up-to-date car. Woe to those who took its practical benefits literally.

The Airflow became a lost classic. Walter Chrysler failed to mention it in his autobiography, declining to claim even that he was foresighted in backing it. The Airflow may have gone too far too fast, as some designers suggested, but its effects were lasting. Even the Japanese imitated it in the Toyota A-1 of 1935, the first car produced by that company. And by 1940 its lessons would have been mostly incorporated into production cars. Its problem was aesthetic, not technical.

Norman Bel Geddes lent his praise to advertisements for the car, but he was well aware of the limits to public taste. Bel Geddes and most of the other industrial designers were politicians of form—they moved very carefully and by gradual steps into the grand future they envisioned. Raymond Loewy, who produced a streamlined Hupmobile in 1934, spoke of the "MAYA" principle, an acronym for "most advanced yet acceptable." Bel Geddes projected an ideal car and then worked backward toward existing models "by twenty per cent at a time," until he had a projected ideal evolution of eight cars. His car number five, for instance, projected in 1928, was to be produced in 1933. The eventual goal, reached by annual model changes, looked something like Fuller's Dymaxion Car or William Stout's Scarab—a modified teardrop.

Other designers adapted the streamlining aesthetic to auto fashion without nitpicking about function. Harley Earl applied streamlining very differently in his 1934 LaSalle; he took elements from aviation and used them decoratively. By contrast with the Airflow's cascade grille, in the 1934 LaSalle the hood and grille were given a vertical emphasis. If the Airflow looked like a locomotive of the streamlined era, the LaSalle looked more like a ship with its prow cutting through the waves. In addition, the radiator was placed well behind the sharp grille so that it could be wider. And although, as the auto design historian Paul Wilson has noted, people were extremely reluctant in the mid-thirties to accept any car whose radiator seemed to sit in front of its wheels, the LaSalle's sharp nose lent it acceptability.

So did that of the Lincoln Zephyr, designed by John Tjaarda and Eugene Gregorie for Edsel Ford, which appeared in 1935 with a European air to its version of streamlining. When he drove the prototype around Palm Beach, the wealthy wintering there were enthusiastic, and despite his father's skepticism, Edsel Ford managed to have it produced for the wider market. The "next best thing to flying," trumpeted the first ads for the Zephyr, which showed it in front of one of the new DC-3s. "You're skimming straight for the horizon!" it cried. "A warm wind softly fans your face as you soar to the crest of the hill. And stretching below you are ribbons of roads and trees that dwindle into pencils . . . watch out—or you'll bump into one of those big, billowy clouds! You're driving a new Lincoln-Zephyr, mister, and that means you're riding high!"

But when it became clear that the Airflow had given streamlining a bad name, the aerodynamic aspect of the Zephyr's styling was deemphasized. Its sales were steady but not dramatic. By 1940, the public was increasingly skeptical toward auto design—and toward all industrial design, especially the streamlined. There was a growing public sense that auto streamlining was simply superficial, slick packaging; Disney satirized it in a Bugs Bunny cartoon based on the Tortoise and Hare fable in which the tortoise attributes his success to "a streamlined body." Chaplin made fun of it in *Modern Times*.

But streamlining had not just gone out of fashion; it had grown so familiar that it had entered the collective unconscious, becoming tied to a more generalized set of ideas about speed and power that would never thereafter leave the American sensibility. So ads for the 1942 and last prewar version of the Lincoln Zephyr, whose body shape had become less overtly aerodynamic, spoke of streamlining that "starts way down deep," and depicted the car beside a leaping sailfish. "This modernness starts at the very core of the car!" the ads went on. "The Lincoln-Zephyr is *naturally* streamlined." The flight metaphors had grown vaguer, more ethereal, more evocative of simple power and natural force.

After the war, the Zephyr was discontinued. The fastbacks that survived the war were soon replaced with longer, lower, finned cars. Harley Earl continued slowly and surreptitiously adapting the science and the styling of the Airflow and the Zephyr to General Motors cars. His basic manifesto was the 1937 "Y Job"—his first "dream car" or "concept car." Its low body, with long, sweeping sides, and body panels tucked under instead of grounded in running boards, was accented by nine parallel chrome streamlines on the fenders. The beltline had vanished; the ruling line was lower, at wheel level. More chrome bars, almost striations, were vertically arrayed to form a hungry mouth of a grille that would show up in postwar Cadillacs and Buicks. It also came equipped with "power everything"—power windows, door locks, and steering—a harbinger of the standard equipment of the big cars to come. By the 1940 models, his "torpedo" look had made the fastback and the rounded front standard, as well as widening the car and moving the front wheels forward in the manner of the Airflow. The silhouette of the 1940 Oldsmobile, argues Paul Wilson, was virtually indistinguishable from that of the Airflow.

But in adapting elements of aerodynamics to automobiles, Earl and GM's Alfred Sloan always understood that the effect would be strictly cosmetic. Earl fell in love with various airplanes—Clarence "Kelly" Johnson's twin-tailed P-38 becoming his chief inspiration during the forties. Later, he was obsessed with the swept-wing Douglas Sky Ray and other jets, the source of inspiration for the

Firebird series of prototypes. It was Earl's vision of what the air-plane had to lend the automobile—symbolism rather than practical streamlining—that prevailed, not that of Carl Breer.

By the 1960s, the Airflow had taken on the dimensions of legend. Claes Oldenburg, a friend of Carl Breer's son, created a number of art works on the theme of the Airflow, including a giant soft sculpture of the car and a plastic relief of its profile. In death the strange car enjoyed a success it never did in life.

☆   ☆   ☆

AMONG SUCCESSFUL PRODUCTION cars, the streamlined body was replaced by the tail fin. And the tail fin, even after it appeared on Chevrolets and Chryslers, was forever to be associated with the Cadillac.

The Cadillac, ultimately to become the most effulgent flower of American auto culture and the superficialities of body shape, was deeply rooted in the American system of manufacture. Long before the Model T, the Cadillac had been the first automobile built of interchangeable parts. It could trace its ancestry back to the developments of interchangeability at the U.S. armories in Springfield and Harpers Ferry, and less successfully at the armories of Whitney and Colt.

The Cadillac's father was Henry Leland, who had begun his career building textile machinery. During the Civil War, he built machine tools at the government's Springfield Armory. Laid off when peace returned, he moved to Colt, where he built machine tools, and then to the classic machine shop of Brown & Sharpe, where he began applying armory methods to the production of the Willcox & Gibbs sewing machine, a major brand of the time. Leland standardized screws and small parts, and began the modern practice of keeping replacement parts in stock instead of manufacturing them as needed—an innovation so basic it is hard to imagine the time before it existed.

In 1890, Leland established his own shop for machine tools in Detroit, where the leading business was already the production of cars—railroad freight cars—thanks to the proximity of supplies of iron and lumber. His shop of Leland & Falconer produced equipment on contract: guns, typewriters, motorcycles, an automatic

Lincoln Zephyr advertisement, 1941.

chicken feeder, and a pencil sharpener, employing advanced grinding machines and tolerances measured in thousandths of an inch. Leland became known for his insistence on precision—he called it an art. "The art," he said, "consists in always setting the limit [of tolerance] as large as the best practice will permit." Under such a system, noted one observer, the key factor was that "the workmen in charge of the machines need not be men of the highest skill."

Once set, the tolerance was relentlessly enforced; that was the key to making truly interchangeable parts. A young sales executive for a ball-bearing concern once confronted the wrath of Leland when some of his bearings failed to meet standards. He was Alfred Sloan, who was later to bring Leland to General Motors.

Leland helped develop the necessary precision to make the Westinghouse Air Brake successful. He excelled in making gears for the booming bicycle industry, supplying its leader, the Pope Manufacturing Company. His first engines were built for motorboats; before long he was turning out transmissions for Ransom Olds and finally engines with interchangeable parts. Leland owned an Oldsmobile that Olds delivered personally to the Leland home.

In 1902, Leland and other investors bought the liquidated Murphy Automobile Company and reorganized it under the name of Cadillac Automobile Company, named after Detroit's founder, Antoine de la Mothe Cadillac. It was the first automobile company in Detroit itself. Leland originally planned only to take responsibility for the engine, but by 1904 he had assumed direction for the design and production of the whole car. It was a critical year for the new industry: Henry Joy moved the Packard Company from Warren, Ohio, to Detroit the same year. Buick was chartered and Henry Ford began a new company.

The Cadillac was at first promoted by the sort of stunt that had become a staple of the auto business: it climbed the steps of the Capitol in Washington. In 1907, Sir Thomas Dewar offered a cup worth 100 guineas for the first car with standardized parts. In a famous demonstration at Brooklands, England, in 1908, three of the one-cylinder Cadillacs were run seven miles in the snow, disassembled, the parts shuffled together, some replaced at random from stock spares, and the cars reassembled using only wrenches and screwdrivers.

After two weeks, the three new cars were rolled out, each a jumble of body parts, and with the colors of the original trio rearranged now into a motley—a purple lake hood meeting a Brewster green side panel. They quickly became known as the Harlequins, but they ran the 500 miles through the English countryside required by the Auto Club's Standardization Test and won the trophy, into which Henry Leland promptly dropped his tiny granddaughter for a triumphal photograph.

The standardized parts of Leland's Cadillac were still assembled by hand. At the same time the Cadillacs were bouncing over the course at Brooklands, however, Henry Ford was readying with the Model T to combine interchangeable parts with assembly-line production. Meanwhile William Durant was buying up auto companies hard hit by the economic slowdown of 1907 and assembling them in a large combine he called first International, then General, Motors. By 1909, Cadillac had become part of General Motors, where it enjoyed two decades of prosperity at the top of Alfred Sloan's price and prestige ladder.

In the thirties, however, the Cadillac was nearly killed. The Depression virtually ruined the old coachbuilding business, and dealt such premium brands as Packard—then "the standard of the world . . . serving America's aristocracy"—blows from which they never recovered. It was the car of old money, the kind driven by Southern aristocrats in Peter Taylor's Memphis stories and the Los Angeles aristocrats in Raymond Chandler novels. Packard, in short, was what Cadillac was to become. Packard responded to the Depression by moving downmarket, with its model 102 of 1935. While gaining sales, it fatally lost cachet.

Cadillac, supported by the structure of General Motors, took a different approach. In June 1932, the depths of the Depression, the car was selling badly, and Alfred Sloan assembled his board of directors to consider whether Cadillac should not be scrapped, with the LaSalle becoming the top of the line. The conclusion seemed foregone when a young Swabian immigrant, a former mechanic with the Mercedes racing team who was unknown to most of the board members, asked for ten minutes to make a special presentation. Thirty-three-year-old Nicholas Dreystadt was a warm, absent-minded man, who burned holes in his tweed jackets with pipe ash and occasionally arrived at his office wearing one black and one brown shoe. But Peter Drucker, consulting at the time for General Motors, has recorded how Dreystadt noticed something that, he claimed, could help save Cadillac.

It was not company policy to sell Cadillacs to blacks, yet more and more, Dreystadt had noticed that well-off blacks—entertainers, athletes, entrepreneurs—were paying whites a premium to front purchases of the big luxury car. A fine car, Dreystadt argued, was one of the few trophies with which a black man could celebrate his success. Expensive houses were as forbidden to him as membership in exclusive clubs; luxury travel and society, restaurants and night-clubs were restricted. Why shouldn't General Motors quietly go after this potential market? It was small, to be sure, but Cadillac was a relatively low-volume, high-profit item, and blacks could make the difference. The board gave Dreystadt eighteen months to make it work. By 1934, Cadillac was profitable again; by 1940, its sales had increased ten times.

Dreystadt had carried Alfred Sloan's market stratification to a new extreme—to niche marketing. It was not only the black market, of course, that did it. By 1933, Cadillac advertised its V-16 model with an Anton Bruehl photograph of the engine, a highly shad-owed, Hollywood deification in a mode normally applied to the visage of Gary Cooper. The advertisement warned that production of the car—with an engine twice the size of the standard V-8 pioneered by Cadillac in 1914—would be limited to four hundred, and early orders were encouraged to avoid disappointment.

The Cadillac became a status symbol of a special sort, and in the process confirmed a separation of parallel "ladders of consump-tion." Clothing, housing, automobile, and income no longer moved in step. The big car parked in front of the little ranchhouse or bungalow became an American image. All those seeking the affir-mation of their success and Americanization seized on the Cadillac as the most accessible symbol. In 1955, *Fortune* magazine visited the unveiling of the new Cadillac to dealers and asserted that "probably never before has one material object become so much the focus of so many of the aspirations that propel the American ego. . . . In a society where their own and others' positions had been shifting so rapidly, [people] needed a fix—a visible symbol that would affirm, to themselves as much as others, where they had got to. . . ." Blue-collar buyers would strain to attain the dream of a Caddy, and eventually the outright poor might still manage to own a used "welfare Cadillac," commonly in imminent danger of repossession. Even the workers who built them aspired to ownership of a Cadillac. Johnny Cash sang about an assembly-line worker who acquires his own model one piece at a time, smuggled out of the factory, year by year, until he has a whole car. The only problem is that, with annual model changes, the resulting car is an amalgam of tail fins from one year, wheels from two others, and so on—a "harlequin" Cadillac. The status symbol was a moving target, thanks to the planned obsolescence fostered by Harley Earl's Art and Color Section.

☆   ☆   ☆

EARL CREATED ANOTHER dramatic shift in the look of automobiles in 1946, when he lowered the beltline, ran the headlight shapes back along the body—and, with some trepidation, added the tail fin.

The tail fin was a quoted, or "found" form. Most everyone by now must have heard of Earl's fateful 1941 visit to the Lockheed hangar where he saw an early test model of the twin-tailed P-38 Lightning fighter. (The P-38 was powered by Allison engines, and Allison was a subsidiary of GM, hence Earl's advance look.) He spent the rest of the war, safe from the pressure of annual model changes, contemplating the cars of the future, and the twin vertical stabilizers—the tail fins—of the P-38 never left his mind. While aces Joe Foss and Dick Bong used their P-38s to knock down record strings of Zeros, Zekes, and Georges in the South Pacific, while P-38s sent Admiral Yamamoto to his death, while P-38s over Germany earned the nickname "twin-tailed devil," Earl dreamed of the cars of peacetime to come.

Aerodynamically, the P-38 was to help make its own shape obsolete. In dives it could reach speeds approaching that of sound, and was therefore the first airplane to encounter what was then known as the "compression factor"—the outskirts, the prelude to the sound barrier, beyond which airplanes would have to assume a new shape. But to Harley Earl, on the eve of World War II, the P-38 stood for the future, and it played muse to his designers throughout the war. The cars of the fifties were being in large part shaped as the battles at Midway and Guadalcanal and Anzio and the Bulge raged.

By the time he got the chance to try it out, Earl was nervous about the new look. He had made his mistakes, such as the 1929 "pregnant" Buick, derided for its bulgy body. At first Earl was afraid that the tail fin, which began to arrive on the Cadillac starting with the 1948 models, might turn out to be another pregnant Buick. He nearly canceled the design. In retrospect, the first fins seem tentative, simple swellings bearing the taillights.

The word "fin" was not used; the look was called simply "rudder styling." But gradually the fins grew from shapes of the prop-driven age to the space age, to longer, almost marine fins with taillights set in them like retrorockets. By the end of the fifties, they bore no relation at all to the fighter plane that suggested them.

The rest of the P-38 motif showed up in the front bumper, first as two gentle bulges echoing the twin propeller spinners of the Lightning, then expanding by 1951 into full-blown "Dagmars," which in a collision with another car, like our Corvair, could function as rams as effective as a Roman trireme's.

The first version of the prop spinner had belonged to Studebaker and was freely borrowed by Ford. In 1946, with the old man a few months in his grave, Ford was nearly moribund itself. Both Ford and GM purchased Raymond Loewy's 1947 Studebaker, the first postwar car, to measure and study, but Ford borrowed designers from Studebaker for an entirely new look by 1948. They coolly took the main elements of Loewy's Studebaker to give the 1949 Ford its famous front, with a prop spinner shape set on small vanes inside a ring.

The tail fin, wrote Earl, "started slowly because it was a fairly sharp departure. But it caught on widely after that because ultimately Cadillac owners realized that it gave them an extra receipt for their money in the form of a visible prestige marking for an expensive car."

The aeronautical shapes that inspired the 1948 fins were already obsolete by the time they appeared. Earl and his staff, having spent the war years in variations on the P-38 theme, were confronted at its end with a new look—Kelly Johnson's new plane, the P-80 jet. That plane's swept wings and round nose would inspire his next dream car, the Le Sabre, a name that echoed another fighter plane, the F-86 Sabre jet, famed for its Korean War duels in Mig Alley.

Conceived in 1948, first shown in 1953, the Le Sabre retained the twin prop spinner imagery in its bumpers, but above them added an oval chrome snout echoing the air intake of jet aircraft. It also included the "wraparound" or panoramic windshield, whose source was the airplane canopy. It took the power of a GM to make such features succeed. Earl's whims dictated whole industry trends—responses to his themes or exaggerations of them.

By contrast with his achievements of the thirties, the fifties were excess, with the exception of certain superbly well-proportioned cars: the '53 Buick (for all its gesturing), and even the '59 Cadillac, the apogee of long and finned. In 1959, Earl retired. In 1960, the tail fin shrunk. It had already "trickled down" the GM ladder, bottoming out in the classic Chevrolet of 1957, where the aeronautical image became almost nautical: the Chevy's fins lent the car an appearance suggesting a fast sloop beside the man-o'-war of the Cadillac.

But by the late fifties, too, it was clear that Earl's original misgivings about the fin were well founded. The tail fin became the object of rivalry among auto makers; in the mid-fifties even Earl realized that it had gotten out of hand. Virgil Exner, formerly of Ray Loewy's studio, as well as Ford and GM, was now at Chrysler, and he was out-tail-finning Earl's boys. By 1956, General Motors designers began panicking. They added great chunks of chrome to counter Chrysler's fins and abandoned the slow process of working gradually with models.

The tail fins, which aimed to do for the back of the car what the spinner did for the front—i.e., signal aeronautical speed and grace—were what Detroit designers, well aware that they were impresarios of a kind of theater, called "cues." The tail fins were the most famous, but the use of such things went back to the days before World War I, when Packard's lozenge-shaped radiator became a brand sign that the company would repeat in modified form for several decades. General Motors spread the idea of brand marks down its whole line. For years the side of the Buick's hood was characteristically punctuated with round "ventiports"—miniature, chrome-lined portholes. Harley Earl claimed that they had originally been placed there to cool the brakes, but that, unbeknownst to him, the brakes had been "improved so much" that the holes became unnecessary and were filled in. How about that, he fairly chuckled in the *Saturday Evening Post*. Jokesters claimed the ventiports were there for the comfort of little creatures who lived in the engine. The ports became known as "mouseholes."

☆   ☆   ☆

WHILE THE SIXTIES had their highpoints in the strong, squared shapes softened with "hips" made standard by General Motors under Earl's successor Bill Mitchell, the auto industry seemed to have lost nerve for theatrics that climaxed in the fifties. Younger designers, who mocked the classic Detroit treatment as "the Big Package," came under the influence of the Italians and their idea of auto design as rolling sculpture—particularly Sergio Pininfarina, creator of the Cisitalia. And one handy source for the sculptural theme was at least a nominally functional one—streamlining.

The return to streamlining came suddenly, under the force of the crisis that followed the oil shocks of the seventies and the invasions of Japanese products at the bottom of the market and European ones at the top. Chrysler would have to be rescued by the federal government and Ford was in serious danger of losing its independent existence in the early eighties. Its annual loss for 1980 reached nearly $1 billion. Its image with buyers was at a historic low.

In search of a dramatically new look, chief designer Jack Telnack created a futuristic car that at the same time recalled its classic model of 1949. That '49 model had saved the company, when the mismanagement of Henry Ford had left it nearly bankrupt. The new '86 car, counted on to do the same thing, was to make a clean break with the downsized and slab-sided models of the seventies. The "aero look" claimed to reduce fuel consumption, but it also asserted a new company "culture," a new level of quality control, and a new brand of customer. It proved that functionalism did not disqualify a car style from showiness. "Form follows function, but also flair," Telnack liked to say.

Ray Everts, whom Telnack assigned as chief designer for the Taurus/Sable, wanted to achieve the same simplicity as the 1949 Ford, with its prop spinner front borrowed from Ray Loewy's Studebaker, which had borrowed it from airplanes. "A photo of the post-war model," he recorded, "was on my desk: the floating circle within a circle grille design was a prominent feature." At first two different fronts were planned, in the best Detroit tradition: one for the high-end models, and a more traditional one for the lower end. Everts persuaded the bosses to adopt one grille:

"My notion for the Taurus was to float the blue Ford oval in an oval hole, and I had this modeled with Bondo and tongue depressors for a big meeting. Not everyone shared my enthusiasm, but by the time we went into production, management insisted that we build the Taurus with only one grille—the one with the '49 Ford heritage."

The ads for the new models showed wind-tunnel-style smoke-streams flowing over the oval logo.

The Sable—the Mercury version of the car—similarly sported a band of vertical striations that had been the leitmotif of the marque since the forties and was a key motif of the '49 models so favored by customizers. Only, on the 1986 car, the striations were carried out in a back-lit plastic "laser light bar."

The Taurus/Sable had several bloodlines. It was quickly termed "Eurostyle," as if by some association with Bauhaus, Braun, and the School of Ulm. But the European tradition from which the new American streamlining learned was in part that of the luxury sports car, in part that of the scientific study of aerodynamics in the wind tunnel—the work of Jaray and others. It was also part of the long-standing mutual interaction between American and European automobile traditions. Ford and Lincoln had long been the companies most involved in borrowing from the aerodynamic work of Europeans, although Harley Earl began at GM admiring pre-

1949 Mercury, Eugene Gregorie, chief designer. (Ford Motor Company)

Ford designer Jack
Telnack and 1986
Mercury Sable. (Ford
Motor Company)

streamlined Hispano-Suizas. The Lincoln Zephyr and Continental
grew out of Edsel Ford's admiration for the European look. The
European emphasis on aerodynamics, in turn, grew out of the
Treaty of Versailles: the ban on military airplanes diverted aeronautical engineers into such interests as auto streamlining and left
them with wind tunnels in which to experiment on cars. The result
was a European heritage of streamlined cars, from the Volkswagen
Beetle to the Saab to the Citroën DS.

The new streamlining also reached back to the "torpedo" look
and fastbacks of the forties. The reappearance of automobile
streamlining in the late eighties was in keeping with the vogue for
collecting machine art of the thirties. An Airstream was featured in
the influential 1986 show "The Machine Age in America," which
toured several major American art museums about the time the

Taurus, the Chevrolet Beretta, and other streamliners were appearing. It helped dramatically to raise the price of such streamlined thirties collectibles as irons, toasters, and radios.

But there is another critical bloodline: that of the American— and particularly the California—tradition of car customizing. Customizing did the "let us build you one" motto of General Motors one better: it was the means by which the car owner literally shaped his own body style. In this way, it was a return to the California (i.e., Hollywood) private coachbuilding world out of which Harley Earl had sprung.

Jack Telnack was a former customizer. His first car was a 1941 Mercury to which he personally applied an acetylene torch (which he was later to display in his Dearborn office), then lowered and lengthened the body so it resembled the Lincoln Continental of the same year. A favorite model of customizers was the 1949 Mercury, of which a "chopped and channeled" specimen is driven by James Dean in *Rebel Without a Cause*. The purpose of customizing was to give the car a full, rounded look, but also to lengthen and solidify it.

Both of these cars showed the hand and the style of Eugene Gregorie, one of the designers of the Lincoln Zephyr and the key shaper of the '49 Mercury, who, like Telnack, had designed boats, eschewed the golf-playing camaraderie of most Detroit designers, and lived on an island in the middle of the Detroit River. Raymond Loewy was also a customizer, constantly (although via other hands) doing up stock Cadillacs and Lincolns with his own custom bodies—some of them quite grotesque and a far cry from his sleek designs for Studebaker. When Telnack was studying at the Art Center College in Los Angeles, the Ecole des Beaux-Arts of the automobile world, he bought a 1950 Studebaker Starliner, a low, sleek car with a prop spinner or "bullet nose" prow—son of the same look Ford was borrowing in the late forties.

The key element of aero was not the European look, not the futuristic tone of the shape, but the fact that it made the cars look bigger and fuller, and carried customers back to the good old pre-oil-shock days of big solid American cars. The buyers were not taken into the future as much as into the past, straight back to the cars of their childhood, their father's cars, in which the excitement of the streamlined future was still alive.

☆   ☆   ☆

THERE WAS ANOTHER sort of vehicle whose popularity rose and fell with that of streamlining—but as contrast and counterpoint to it. That was the Jeep and its kin; a car virtually without a body, and with no pretensions to smoothness.

The World War II Jeep was the Model T updated and become self-conscious. It was so much a cartoon vehicle as to be named after a cartoon character. The Jeep got its name from a conflation of "GP" for general purpose, the official designation for the quarter-ton truck in question, and "Eugene the Jeep," a character in E. C. Segar's comic strip "Popeye," officially entitled "Thimble Theatre." The Jeep was a goggle-eyed, somewhat feline, vaguely hedonistic-looking character, able to break the laws of physics.

To compare something to the Jeep was to declare it the most durable, functional, the simplest and toughest of its type. In this respect, the Jeep hailed back not only to the Model T—a renowned "mudder" of a car—but to the Studebaker wagon and the Conestoga wagon before it. In its toughness and ability to take on mud and inclines, the Jeep became a symbol of individual choice: it could

U.S. Army Jeep. (U.S. Army)

go anywhere, at least in theory. It could take you off-road—off the beaten track. The Jeep began as a modern military equivalent of the Model T, a general-issue transportation appliance. The Army started looking for a vehicle like the Jeep in 1923, when it put oversized aircraft tires on a stripped-down Model T chassis; one batch was even turned out, like the T, at Ford's Highland Park plant.

The Jeep's ancestors were practically toy cars. In 1929, the American Austin Company began producing a U.S. version of Sir Herbert Austin's small British car, with new body style by Alexis Sakhnoffsky, designer of custom autos and commercial buses.

The Austin was a joke, a mascot, beloved of Hollywood stars. Buster Keaton was one of many actors who owned one, and Will Rogers used Austins as mechanical horses in his film version of *A Connecticut Yankee in King Arthur's Court*. An Austin led the parade that opened the 1939 New York World's Fair. But the little car was merely a novelty, even after reorganization under the more aggressive name American Bantam. In 1939, however, the American Bantam Company, which was based along the still abuilding Pennsylvania Turnpike in Butler, Pennsylvania, lent the state's National Guard several of its cars for use as scout vehicles on maneuvers aimed at exploring the new dimensions of mechanized warfare. Their utility impressed the brass in Washington.

In 1940, the Army invited 135 manufacturers to provide proposals loosely based on the Bantam, for a four-wheel-drive vehicle that would weigh no more than 1,300 pounds and carry 500 pounds. Only two small motor companies submitted bids: American Bantam and Willys-Overland Motors, headed by John North Willys, an auto salesman from Elmira, New York, and one-time ambassador to Poland. American Bantam got the contract, with a slightly lower bid, but since its meager production capacity was soon outstripped, Willys was brought back into the competition.

Bantam's entry was underpowered and unreliable; it was Willys that perfected the vehicle with which the world would become familiar. The man who had most to do with the process was one Delmar G. "Barney" Roos. A classic American tinkerer, Roos had been chief engineer for Pierce-Arrow and Studebaker before he went to work for John North Willys in the thirties and created the light, powerful four-cylinder engine that became the steady beating heart of the Jeep. The original Willys prototype was too heavy and slow to meet military specifications, so Roos pared every fraction of an ounce possible from the engine, frame, and body, even thinning the paint. He excavated a minimal vehicle from the more complex original. The Army granted Willys the first major production contract in 1941: it paid just $740 for each of the original Jeeps.

The rest of the story is familiar: during World War II the Jeep went everywhere Kilroy went. (The Kilroy face and the face of the comic strip Jeep bore a family resemblance.) What made the Jeep special was that Barney Roos's engine drove all four of its wheels, enabling the vehicle to brave mud and sand, slope and gulley. This ability turned the Jeep into the GI's motorized mule and mechanical mascot, ennobled by Bill Mauldin and Ernie Pyle. More than half a million Jeeps were manufactured during World War II. Just eleven feet long and slightly over four feet high, with open body and folding windshield, the Jeep was rockbottom basic. It had no door or windows, only an optional canvas top. Its four-cylinder engine produced a mere 60 horses, but they were quite adequate to move the light body up steep slopes and through shallow streams, thanks to four-wheel drive.

Peace brought the Jeep home almost immediately; Willys introduced its CJ "civilian Jeep" version in 1946, and by 1949 the company was pushing its Jeepster as "distinctly personal" and a "dashing sports car." The Jeep in mufti was sportswear. Even though it posed as the old Army tunic you brought home from the front and wore around the house, it soon turned into the designer leather jacket you wore downtown. It was blue jeans gone from rodeo to Rodeo Drive.

Just as basic jeans were succeeded by designer jeans, Jeep's simple four-wheel drive was succeeded by slicked-up versions—Jeep Cherokee and Wrangler, Ford's Bronco and Chevy Blazer. And by the 1980s, beginning with Audi's Quattro, there were full luxury cars with four-wheel drive. In 1988, the use of the name "Jeep" was licensed to Murjani, Inc., the jeans and sportswear manufacturer.

The military bought its last Jeep in 1982. In 1985, American Motors, which had absorbed Kaiser, which had absorbed Willys, announced that it would stop production of the basic Jeep model, the CJ, whose sales had fallen from a high of 80,000 in 1978 to half that in 1984. The public outcry was tremendous; the company was forced to back down.

It took the man who introduced the Mustang, the universal car turned sporty, to understand how the Jeep had changed. In 1986, Lee Iacocca of Chrysler bought the Jeep trademark and line along with the rest of American Motors, just as the Jeep and its new Japanese competitors, which by the 1980s controlled a third of the market, became chic.

With the Mustang, Iacocca had turned a basic compact car, similar to the Corvair or Valiant, into a sports car. The market for the Mustang had been almost accidentally revealed by the Corvair, whose most successful model was the sporty Monza coupe. When Ford executives saw how the Monza had snatched victory from the jaws of early defeat for Chevrolet and the at first slow-selling Corvair, they realized better than Chevrolet did what Chevrolet had created: a car as personal and exciting as a sports car, but more practical.

Introduced in 1964, the Ford Mustang under its skin was a basic car, with many parts taken from the Ford Falcon compact. But what made it a success was that it was touted as "the car designed so you can design it yourself." The Mustang's buyers purchased an average of $1,000 worth of options—the list of choices would previously have been found only in a luxury car—above its $2,300 base price, to "design it themselves." They turned the universal car into a personal car, the plain car into a sports car. Impossible to envision except in red, with literal hips for the dawn of the hip era and a crisp little rear deck like a miniskirt, the Mustang was a sports car the way clothes are sports clothes. And for Iacocca and other marketers, the Jeep would become work clothes worn for fun. (No wonder there was an Eddie Bauer "edition" of the vehicle.)

The Jeep and its like were work vehicles as little used for work as work clothes were worn for work. Like the Mustang, the Jeep succeeded in making the ideal of basic transportation emotionally attractive in a way the compacts such as the Corvair—beautiful and innovative as it was—had not. With air conditioning and high-end stereo systems, today's "off-road sport utility vehicle" is a long way from the vehicle of World War II. "Off-road" is a promise of escape from the constricted arteries of city and suburb—a promise of adventure, a romantic aura of distant theaters of action and exotic climes. One Suzuki ad claims that "something happens to people . . . they get in a Samurai, they disappear, and the next thing you get is a postcard from Burkina Faso."

For the military itself, the vehicle changed, too. In the early

1980s, after forty years, the Army Jeep was gradually replaced by two vehicles. The "Hummer," for HMMWV (High Mobility Multipurpose Wheeled Vehicle), and the CUCV (the Commercial Utility Cargo Vehicle or "cuc vee"). The latter was a Chevrolet Blazer purchased "off the shelf" or, as the Pentagon liked to put it, "as a non-developmental item." For the Army, this notion of the non-necessity of reinventing a wheeled vehicle was novel. That such vehicles were commercially available at all was also a sign of the extent to which the Jeep ethos had penetrated the consumer market. By 1991 the Hummer, a star of the Gulf War, was for sale to civilians. Actor Arnold Schwarzenegger bought the first one.

☆   ☆   ☆

THE JEEP, TOO, could be set against the tail-fin Cadillac to sum up the two poles of American design—the one rough and basic, unemotional, male; the other sensual and theatrical and female; the one asserting basic universal transportation needs; the other asserting the means of transportation as drama and dream.

The difference between them was indirectly pointed out when astronauts flew to the moon. For years, science fiction had taught us to expect a trip to the moon on a single great rocket, landing tail first and settling onto extended fins. When we actually made the trip, it was with a series of discardable "modules," of which the one that actually touched the lunar surface was akin to the Jeep.

But it was a method that was quintessentially American: employing "modules" all of which were discarded in a demonstration of waste for the sake of efficiency, until all that was left of a 360-foot-high package of rockets and fuels and equipment was a small cone that would fit in a bedroom, into which three men were crammed.

The LEM or Lunar Excursion Module ("Excursion" as in "excursion fare" was a wonderfully casual/serious choice of word) was the most personable of these modules. It was originally designed as a compound of geometric shapes: flat cylinder landing stage, topped by the sphere of the crew compartment, the body systematically sliced and faceted for weight reduction so that its twin windows took on the appearance of hooded eyes, with the exit hatch to the lunar surface set beneath as nose. It was a head on a flat abdomen, without thorax, with legs and antennas.

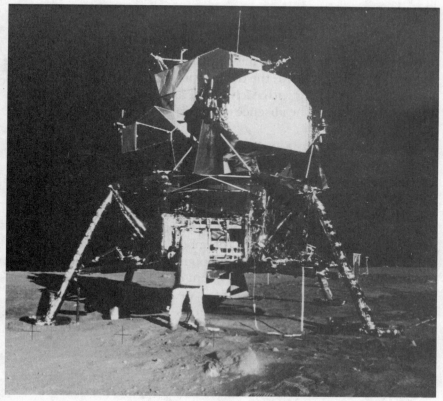

Project Apollo Lunar Excursion Module (LEM).

It possessed the anti-aerodynamic shape of earth satellites, carried to orbit beneath conical shrouds. Its geometric miscellany corresponded to the satellite's juxtaposition of solar panels, instrument boxes, and antennas. It was a shape worthy of the Russian Constructivists, only descended into zoomorphism: the bug, or a Cubist mask, with the triangularly notched eyes. And over it all, like an afterthought, wrappings of crinkly gold foil—clothing.

LEM, too, was a potentially wonderful name, rich with associations those at NASA would not likely have made and if made, would have rejected in horror: a name out of Swift (LEMuel Gulliver, traveler) by way of Beckett, and by wonderful coincidence, Stanislaw Lem, the great science fiction writer for whom space exploration was problematic and who saw our appliances mastering us.

LEM had the resonance of HAL, the acronymic computer of Stanley Kubrick's *2001: A Space Odyssey*, a device grown finally malevolent.

No wonder NASA would have nothing to do with its own name. Instead of calling it LEM, the device was swathed in patriotic names such as Eagle, varying with each flight, as if to assert the high-flown ambitions which, in the absence of soaring smooth lines, it could not itself express.

The buggy suggestions of the device were picked up by the advertising agency for Volkswagen, which, despite such American compacts as the Ford Falcon and Chevrolet Corvair (last manufactured in the year of moon landing), stood for the tradition of the Model T. In a famous series of ads the company lost no time in drawing a comparison between the people's car and the device by which astronauts commuted to the moon. Indirectly, the ad was an explanation of why going to the moon quickly lost its romance. It was one thing to land in the equivalent of a Model T, a Jeep, a Bug. It would have been very different if we had landed—as the early visions of things had it—in a rocket as glorious as a Harley Earl Oldsmobile Rocket 88, riding down atop the fiery blast of thrusters to settle softly on tail fins.

CHAPTER ☆ TEN

Airlines

AMERICANS IN THE twentieth century have always considered aviation something like our own national technology. The railroad and the automobile, however radically transformed in the United States, were European inventions, but aviation was ours.

But it has been in translating—thoroughly but never quite completely—what began as the most romantic of technologies into the banal that Americans have exercised their obsession with aviation. That obsession has centered on two ideals: the ideal of enabling everyone to fly, and the ideal of creating the single, perfect wing.

The strange mythical shape of "the flying wing"—the pure wing, with no extrusive fuselage or engine or tail—is an image that has obsessed the best American aviation minds the way the ideal clipper ship or streamlined automobile did their makers.

Almost as soon as the airplane was invented, Americans took it for granted that it was only a matter of time until everyone would be able to fly. Hadn't it been that way with the railroad and the automobile? Clearly, flying, like driving, was destined to become the privilege of every American. The idea of a plane for everyone, a universal aerial appliance, has been with us since the twenties. By the thirties, the dream changed focus from the individual craft to common carriers. The flying flivver never arrived. Everyman's airplane turned out to be a seat.

☆　☆　☆

211

NO DEVICE SO romantic has ever been created by more mundane procedures than the airplane.

Before their first powered flight, the Wright brothers carefully set up a camera. They arranged to forever freeze the tilt and bend of a certain fraction of a second as their moment of success. As a child, I assembled a plastic model kit of the Wright Flyer which came complete with a figure of Orville, in the exact posture rendered by the brothers' tripod-mounted Korona, and with the oil can and battery and tiny bench that figure in that photograph, snapped by a local named John T. Daniels. The photograph is a token of the Wrights' calm anticipation of the success of their do-it-yourself research and development program.

Another photograph taken with the Korona shows the inside of the balloon-frame shelter the Wrights built at Kitty Hawk. Its beds folded up, Pullman style, its shelves are rowed with canned goods— a line of tomatoes as neat as a Warhol print. Of all the storied American inventors, the Wrights stand the furthest from the stereotype of the clever impecunious toiler with an inspiration—the Edison with iconic light bulb of insight perched above his cranium. A whole nexus of services made their work possible: the Smithsonian Institution, to which the brothers wrote in 1894 for information on previous researches; the weather bureau, which suggested the Hatteras area of North Carolina as possessing the winds requisite to their testing; the train system that carried them there, the standardized system of lumber that allowed them to buy parts for their planes and planks from which to erect balloon sheds for their quarters ($4 worth of two-by-fours made the launching track for their airplane, which they jokingly named "Junction Railroad"); canned foods to eat on the test site and photographic materials to produce the first documented moment of a major invention—and a telegraph system to wire home to their sister and their father, the good bishop, the news of their success.

Then, of course, there is the bicycle-parts system that enabled them to support themselves in the shop in Dayton—at least until the bicycle craze of the nineties abated—by assembling their own models of bicycles, named after local historical figures, from off-the-shelf frames, handlebars, chains, and wheels. From the bicycle, as much as from their reading about the German glider pioneer Otto Lilienthal, they seem to have formed the idea that an airplane

First powered airplane flight, 1903. Photograph set up by Orville
Wright, snapped by John Daniels.

would have to depend on the pilot's control as a bicycle depended on
its rider, and would not have to be as inherently stable as, say, a farm
wagon.

Bridge structures, as explained by their correspondent and adviser, the bridge engineer Octave Chanute, confirmed their idea
that the airplane's structure should incorporate the classic American truss. "My machine will be trussed like a bridge," wrote Wilbur
Wright early in the project. For this purpose, the Wrights employed
Roebling wire, the steel wire developed by the creator of the
Brooklyn Bridge. Future wings would surround complex structures
with simple shells: every wing, no matter how sleek and curvilinear
its exterior, would reflect the same basic approach to building light
and strong that could be found in the Town bridge, the balloon-
frame house, the steel skeleton of the skyscraper.

For a while, Americans looked to Henry Ford to provide the aerial
equivalent of the Model T. The public eagerly anticipated some sort

of "flying flivver" that would do for aviation what the T had done for automation. Ford tried to oblige. He developed a single-engine, single-seat plane, but it was never produced. Yet for the most part he directed his attention to commercial aviation—to the airplane as railcar, not road car. Commercial aviation in the United States got its real start right after World War I, when the Guggenheim Foundation sponsored a "model airline" flying Dutch-built Fokker trimotors between Los Angeles and San Francisco. The planes, scarlet with an arrow and full headdressed Indian silhouetted on their sides, boasted mahogany interiors. Cold chicken lunches were served en route, but even the trimotors could not escape rough weather and the resultant airsickness. After some flights, the planes had to be hosed out.

One thing Henry Ford understood very well was that fear was the big obstacle to commercial aviation. In 1925 he set up a Reliability Tour, based on early auto endurance rallies rewarding range and safety rather than speed. Tony Fokker, the Dutchman who built planes in New Jersey, added two engines to one of his existing wooden aircraft and invented the modern trimotor. Fokker and his three-engine plane won the reliability contest.

Ford noted the success of the Fokker trimotors, as did William Stout, a former newspaper editor who had spent the war years in the aviation division of Packard Motor Company. Stout created the first all-metal passenger plane, the Duralumin "Air Pullman," as he called it, and went to Ford for support. The idea intrigued Ford, who liked new metals: Duralumin appealed to him as the possible equivalent of the vanadium steel he used in the Model T. He put the Air Pullmans to work ferrying parts among his Detroit, Chicago, and Cleveland factories on a daily schedule. In February 1926, Ford Air Transport also began carrying the U.S. mail under contract from the government. On its nose, the Air Pullman bore the boldly lettered legend "Made in Detroit" to foster the local vision of Detroit as the center of aviation as well as of auto manufacturing.

Ford gave shelter in his hangar one night to Lt. Commander Richard Byrd's Fokker F-7 Trimotor, and before morning his engineers had swarmed all over the plane, measuring it and noting its structure. Ford ultimately bought that Fokker Trimotor and lent it back to Byrd for his North Pole flight of 1926. Ford bought out Stout's business and had him design an unsuccessful trimotor of his own.* Ford then brought in his chief engineer, William Mayo, and James McDonnell, later chief of McDonnell-Douglas. Their plane, the 3 A-T, was succeeded by the Ford 4 A-T, or the Tin Goose, a plane with a range of 250 miles, carrying twelve passengers in a cabin large enough for a man to stand up in. The Ford Trimotor began flying in 1927, the year of Lindbergh's flight. But the next year, Ford's friend and test pilot, Harry Brooks, was killed in a crash of the Ford single-seat plane, and the automobile man lost interest in aviation. By 1933, Ford was out of the airplane business altogether.

The Ford vision of the airplane as flying railcar competed with another model: that of the airplane as oceanliner, which attached its hopes to the dirigible or to vast seaplanes. Streamline designer Norman Bel Geddes envisioned huge, flying wings with multiple decks, inspired by the German Dornier DO-X of 1929. That airplane's fuselage was a virtual ship's hull. Six pusher engines were mounted on top of its single wing. With three decks and portholes, it was as close as possible to a ship that flew.

Bel Geddes's airliner number 4 of 1929 was even larger: a multideck flying wing set on two huge, bulbous floats. It was designed for aerial refueling on a forty-two-hour trip from Chicago to Plymouth, England, and listed among the crew were a masseur and masseuse and a manicurist. Designed with the help of the German aeronautical engineer O. A. Koller, Bel Geddes's plane required twenty 1,900-horsepower engines, ten pushing and ten pulling, to get off the ground—or rather, water. The engines could be maintained and even swapped out in flight from inside the wing. Its wing spread was to be 528 feet. Bel Geddes's scheme might have worked—the economics of the trip were carefully calculated, and for a while a group of Chicago businessmen was seriously interested.

The nearest thing to Bel Geddes's scheme that ever flew was the apogee of the development of Pan Am's flying boats—the Boeing

---

* After Ford turned his Pullman into the Trimotor, William Stout turned to automobile design. The result—which he called Stout Scarab—was far more streamlined than his airplane. (Presumably he did not associate the scarab with the dung beetle, as some observers of the ovoid car did.) The Scarab was never to become a production car; Stout eventually lost another idea to Ford. His name was given to a junior high school in Dearborn, Michigan, a couple of miles from the design studios where, in the 1980s, Ford was to turn out aerodynamic cars—still not quite as streamlined as Stout's. Ford's airport is now the track where those cars are tested.

Sikorsky S-42 Pan Am Clipper "flying boat," on Miami-Havana run, 1939. "Very comfortable but a bit noisy," commented the sender of this postcard.

314 Clipper, developed in 1939. The Pan Am Clippers of the 1930s were few in number, tailored to their specific routes, and grew in size from early Sikorskys through the Martin 130, and finally to the size of the Boeing 314, with a wingspan comparable to a 747's. It could carry as many as seventy seated passengers, or thirty in sleepers on a San Francisco to Hawaii run. The interiors were "yacht-club Deco," in the phrase of design historian Donald Bush, with club chairs and small cabins and meals served by nautically garbed stewards. Norman Bel Geddes was hired as a consultant for Clipper interiors.

The Pan Am Clipper, which was in fact part boat, was the highest development of the ship model for air travel. Customers were the wealthy, service was valued, and the necessary refueling stops were treated as ports of call. Pan Am boss Juan Trippe used Charles Lindbergh as a figurehead, as many airlines were to do, even trying to enlist him to fly the first trans-Pacific flight.

But even without Lindy, the Pan Am Clippers provided a dashing and romantic image of air travel that prepared the public for more mundane airlines. Humphrey Bogart appeared as the captain in a film called *China Clipper*. Pan Am Clippers took off under Japanese fire and bombs on Pearl Harbor Day. They rescued deposed dictators from Caribbean and South American capitals. Henry and Clare Boothe Luce rode them across the Pacific to visit the likes of Chiang Kai-shek, envisioning in their magazines the American century when flight to the distant corners of the U.S.-dominated globe would be not only possible but essential.

In one way the name was ironic: the Clippers were the farthest thing from the sleek, streamlined shapes of the ships for which they were named. In another way, however, it was apt: the commercial life of the clipper-type flying boat was even shorter than that of the sail-clipper ship. The sleeker, more efficient model of the "airliner"—a direct adoption of a term for a fast railcar—emerged triumphant.

Marking the shift from rail to air was a company called Transcontinental Air Transport (TAT), formed by a consortium including the Pennsylvania Railroad. TAT's route was half rail, half air. Service on the air legs was modeled on passenger car service (pilots joked that the initials stood for "take a train"). In July 1929, TAT established its first mixed rail and air link from coast to coast, called the Airway Limited. Passengers rode the train from New York to Columbus, Ohio, where they boarded Ford Trimotors and flew on to Waynoka, Oklahoma. There they were shifted back to the rails (the Atchison, Topeka & Santa Fe) and went airborne again in Clovis, New Mexico, on the last leg, to Glendale, California. The torturous schedule called for a two-day trip, still a day less than the best purely rail time. On board, the amenities were the best Pullman level. But after sixteen months and the Wall Street crash, the Airway Limited was canceled, never having earned a penny. Reorganized, TAT went on to become TWA—the airline that led in providing the new type of aircraft that would make flying as common as riding the rails, the DC-3.

Knute Rockne and Franklin Roosevelt were the two men most responsible for the creation of the DC-3. On March 31, 1931, a TWA Fokker F-10A crashed in a field near Bazar, Kansas, killing Rockne, the famed Notre Dame football coach. A farmer saw one wing separate from the craft before it hit the ground; failure of the airplane's plywood wings was cited as the cause of the crash.

Then, in the spring of 1934, Franklin Roosevelt and his postmaster general, political boss and patronage czar James Farley, canceled mail contracts the Post Office held with private aviation, handing the job over to the Army. A disaster of crashes and errant airplanes followed, and finally the government had to place the business back in private hands. But the move had shown the airlines just how vulnerable they were to the whims and vagaries of politics. Passenger service was not yet profitable; only the mail kept the scheduled airlines flying.

FDR employed aviation as a symbol of the new approach he wanted to represent. Roosevelt had made a storm-tossed flight in a Ford Trimotor to the 1932 Democratic Convention in Chicago, where he broke precedent by delivering an acceptance speech in person. He seized on the romance of flight that Lindbergh had created, and demonstrated the do-it-now fearlessness that was to be part of his appeal.

By 1933, Boeing had developed its twin-engine 247, which can rightfully claim to be the first modern airliner. But conservative design decisions meant that the wing of the 247 still ran through the cabin, producing an ungainly foot-high bump that passengers and stewardesses had to step over. United Airlines had ordered enough Boeing 247s to tie up the production line for two years, and its competitors had to look elsewhere for competitive aircraft.

On August 2, 1932, a month after Roosevelt's flight to Chicago, Jack Frye of TWA invited aircraft manufacturers to design a new airliner for his firm. The specs called for three 500-horsepower engines to achieve a cruise speed of 150 miles per hour, a range of more than 1,000 miles, a passenger capacity of twelve, and the ability to take off or climb from TWA's highest-altitude airport (Denver) with one engine out.

One of the solicited manufacturers, Douglas Aircraft, responded with a bold decision: to give up the established trimotor configuration in favor of just two engines. They would use 750-horsepower Wright Cyclones, an evolution of the Wright Whirlwind that bore the *Spirit of St. Louis* to Paris. The resulting DC-1 first flew in July 1933. The DC-2 was an expansion, to fourteen passengers. But the DC-3 was the one that did it. By the time it was designed, engines of 1,000 horsepower were available, allowing the fuselage to be widened. The DC-3's capacity—twenty-one passengers, or fourteen in a Pullman setup—allowed the airlines to break even on most of their routes for the first time with mail revenue. (It even had a small private cabin that quickly became nicknamed the "honeymoon hut.") Its speed made it easier for airlines to adhere to something

Douglas DC-3. (Piedmont Airlines)

approximating a regular schedule, and maximized use of each aircraft. After the arrival of the DC-3, airlines reduced their total number of airplanes during a period when the number of passengers increased four times.

The success of the DC-3 and its successors depended on a new kind of wing construction, a modified version of a design developed by John Northrop: stamped aluminum sheet in a channel pattern for strength. The double wings of the Wright planes, "trussed like a bridge" with their spars and guy wires, had functioned together as a sort of sandwich. In the wing John Northrop pioneered, the two halves of the sandwich were joined. The skin, instead of simply being applied, like the old fabric cover, now played a structural as well as aerodynamic role—stiffening the wing as the skin stiffens a balloon-frame house. Another key design decision was the shaping of "fillets" to merge wings and fuselage with minimum drag, removing the pockets of rough air that formed when the juncture was a sharp angle—an innovation for which Douglas could thank Dr. Theodore von Kármán, who developed the idea in the wind tunnel at Cal Tech.

The DC-3 was put together from existing technologies, and the end was the happy result of both accident and the overdesign of ignorance. The elements were: radial engines with shaped cowlings and variable-pitched props (varying the pitch of a propeller was important for maintaining maximum efficiency at higher speeds), set into rather than over or under the wing; a retractable landing gear; and a wing based on the "cellular" structure of John Northrop.

Of all American airplanes, none has ever elicited quite as much affection as the DC-3, which by shape, style, and sheer ubiquitousness created for itself a following as intense as the "cargo cults" its appearance overhead were said to inspire among primitive societies, who took it for the conveyance of returning gods. The DC-3's low wing and nose-up stance lent an eager anticipation to its curves even on the ground. These curves and its smooth silver skin, in contrast to the boxy bodies and corrugated skin of the trimotors, announced something totally new and modern.

The DC-3 was modeled into lamps and clocks, and its lines were abstracted into objects of every shape and use. It was in many ways the single ideal shape of the thirties. Walter Dorwin Teague, the industrial designer, wrote:

> In a Douglas plane, the form of the fuselage is repeated in the form of the motor housing, and the horizontal fins of the tail recall in diminished size the contours of the wings. Everywhere in the plane we find a recurring contour-line, consisting of a forward thrust and a long backward sweep: it begins with a relatively short parabolic arc that quickly straightens out into a long backward-streaming curve which gradually and gently descends to the level of its starting point. This line defines the form that bodies take when they have adapted themselves most perfectly to the flow of air or water currents, and it has become, as much by its own character as by association, expressive of forward-straining flight. We find this form and its expressive line recurring throughout the whole structure of the plane, as we find them in fish and birds and falling drops of water. I do not know where in modern design to look for an example of rhythm of line composed more perfectly than in these transport planes. Exigencies of function have forced every form to be modeled after the same pattern, a peculiarly significant pattern, until from any point of view we find in the whole a harmonious consonance approaching perfection.

In addition to its novelty, the DC-3 inspired by its durability, its ability to fly on after damage and engine failure and, eventually, sheer age. It has continued to fly for a half century, slipping easily from front-line service on major airlines to universal wartime service to the milkruns of small commuter airlines. It was the first airplane that many people flew.

A DC-3 lasted so long because stress was shared among many structural members, so that no single one of them became vulnerable to cumulative wear. It lasted so long in imagination because it endured the stress of wars and crises of all types, from the Berlin airlift to Vietnam—where it was converted into a gunship—and surfaced in many guises, like Indiana Jones.

When the DC-3 is compared with its predecessors and successors, even jets, the differences seem minimal.

It set the pattern for airline interiors as well as external shapes. Chief engineer William Littlewood looked into the DC-3's "human factors." He took particular care choosing colors for the interior: dark for floors, suggesting solidity, and innocuous grays and calming blues above. An acoustics expert, Dr. Stephen Zand, was called in to reduce interior sound levels.

The reality of the "aerial railroad," however, was not completed until after World War II, with the invention of "air coach" class, to fill seats. As late as the forties, assurance that the passenger would not get airsick, would be properly fed, or would take off and land safely and on time at the specified airports was nothing that could be taken for granted. To become a success, air travel had to reassure and pamper the passenger, ensconce him in a cocoon of safety and leisure. For this, time would show, he would put up with limited space, and even with mixing of classes, as long as his seat was comfortable.

The seat—its width, comfort, and accompanying amenities such as food—was what mattered to the passenger. The industry tracked its success in "seat miles." Boeing spent nearly a quarter of a million dollars designing the seats for its Stratocruiser; such seats served as models for fifties recliners. Circulars from my local department store still regularly advertise the Stratolounger easy chair, as bulbous as the Stratocruiser.

The industrial designers were called in. Walter Dorwin Teague established an office in Seattle to handle his work for Boeing. Niels Diffrient of Henry Dreyfuss's office created flatware, dishes, packaging, seating, and interiors for American Airlines in the fifties, using "human factors" and ergonomic studies to lend specific

Lockheed Constellation airliner, Clarence "Kelly" Johnson, designer.

numerical and technical support for the designers' recommendations. In the new world of statistical marketing and seat/mile efficiency calculations, such numbers finally obtained respect for a designer.

While the airlines, by the fifties, had succeeded in taking much of the fear out of flying, they also succeeded in making it almost banal. By 1955, the airlines carried more passengers than the railroads. The airliner had at last achieved the goal envisioned by William Stout: it was as unextraordinary to the passenger as a Pullman car.

☆   ☆   ☆

As the DC-3 gave way to the larger but basically similar DC-4, DC-6, and finally the jet-powered DC-8, and air travel became more mundane, the fuselage assumed prominence over the wing. The airline passenger rarely knew what kind of plane he or she was traveling on, save whether it was or was not "a wide body"—a Boeing 747, 767, or Lockheed L-1011 or Douglas DC-10. The wings of the jets, to the few who noticed such things, could look stubby; the engines seemed to be strapped onto the back or wings. These were becoming trucks—the 747 was like a tractor trailer, with its buffalolike hump. While the new symbol of the romance of aviation was the Anglo-French Concorde, hundreds of more mundane Boeings—overstuffed 747s and stubby 727s and 737s—were flying around the world.

The last romantic American airliner was the Lockheed Constellation, designed by Kelly Johnson just before World War II broke out. The Constellation took the shape of its wing almost directly from that of the P-38 fighter; its tripart tail was a relative of the fighter's twin tail. With that punctuation mark of a tail, its long-limbed wings and fuselage, the Connie was a lovely shape that sat upright and proud on full landing gear. The P-38 had inspired Harley Earl's tail fin; by the time the Constellation was in its heyday, Cadillacs sported rear ends that suggested its stabilizers. The Constellation was the last plane that Orville Wright, then seventy-two years old, snaggle-toothed and balding, flew. It marked the end and beginning of a number of eras. The Connie competed with the Boeing Stratocruiser, with airframe and engines based, like the Constellation's, on the B-29. The Constellation was graceful and efficient, the Boeing Stratocruiser bulbous-nosed, fat, and ugly, as well as a mechanically troublesome aircraft. But the Stratocruiser offered comfortable seats, with a spiral staircase leading to a "conversation pit." Its temperamental engines and props were difficult to maintain, but its large interior, designed by Raymond Loewy, made it attractive to passengers.

The Constellation, its fuselage lovingly tapered, revealed something else: the industry pattern was expanding, extending airplane lifespans. For this, and for replacement parts, a straight fuselage was best; the efficient kit won out over the perfect model.

☆   ☆   ☆

WHEN THE AIRLINER turned flying into an everyday experience, the romance of flight was left to exotic experimental and military airplanes. There is always one plane the public knows well and aviation buffs adore—the hottest, most sensual craft—the iconic plane of the day, the model of the "leading edge in aviation." The Lockheed Sirius Monoplane held that position for a while. So did the *Spirit of*

St. Louis, which succeeded in replacing the biplane with the monoplane as the public's mental illustration of the generic aircraft.

During World War II, various planes struggled for this position. The flying-wing bomber assumed the mantle in the late forties. In the fifties, the swept-wing test planes, like Chuck Yeager's X-1, and the Korean War–era jet fighters, such as the F-86, held it. But for the last twenty years or more, until the arrival perhaps of the Stealth bomber, this special status has been held by a plane whose design began at Lockheed's "Skunkworks" advanced development facility, named for a moonshine still in Al Capp's "L'il Abner" comic: the A-12, or, as it is better known, SR-71 reconnaissance plane.

It looked, from different angles, like a manta ray, a sword, or a racing boat. One engineer compared it to "a cobra that swallowed two mice." It was officially called the Blackbird, but when heated in flight, its black paint actually glowed blue against the deeper blue of near space. It was also called the Habu, from the name of the cobra that lived around one of its bases in the Philippines and in testimony to the hoodlike "chine" shapes flaring away from its cockpit.

The SR-71 was another of the great Kelly Johnson's designs. From the P-38 and the Constellation, Johnson had gone on to develop the P-80 Shooting Star, the first U.S. jet; the F-104 Starfighter, with wings the thickness of two razor blades; and the U-2 (in just eight

Lockheed SR-71 reconnaissance plane, Clarence "Kelly" Johnson, designer. (Lockheed Corporation)

months) with the same speed, efficiency, small staff, and absence of bureaucracy. In an industry that literally measures engineers on a big project by the acre, the Skunkworks employed only 135 to create the SR-71. There, one motto was the famous acronym KISS—"Keep it simple, stupid." "Simplificate and add lightness" was another.

Despite the obvious differences in appearance and speed, the Blackbird was a descendant of the twin-tailed and twin-engined P-38 Johnson also developed at Lockheed. The twin tail was a sort of Johnson trademark. His career had begun in 1933 when, as a young engineer fresh out of school, he was called in to help solve the major aerodynamic flaws of Lockheed's Electra, the company's first all-metal airplane. After wind tunnel tests, Johnson replaced the original single horizontal stabilizer with a twin tail and the airplane went on to become a commercial success. Amelia Earhart flew one, and the airliner to which Bogie bravely entrusts Ingrid Bergman in the misty final scene of *Casablanca* is an Air France Electra.

The SR-71 possessed a complex appeal—much of it incidental to function. The plane was secret, exotic in technology and shape. It was the first airplane to employ elements of what later became known as "Stealth" technology, radar-absorbent materials and a shape—long and narrow, with wings and fuselage integrated by chines—that reduced its radar image.

On the one hand, the Air Force affected coyness about the plane's performance—listing it as "Mach 3 Plus"; on the other hand, it proudly used the SR-71 to set world speed and altitude records. Its black color, its composition of exotic material (titanium), and its appearance, made it unique. The titanium skin had to be corrugated to allow for expansion at high temperatures—a technique that reminded old-timers of the corduroyed aluminum of the old Ford Trimotor. On the ground, cool and unexpanded, its parts sagged apart and the whole airplane leaked—fairly gushed—fuel.

The airplane retains its mystique despite the fact that the last SR-71 was assembled a quarter of a century ago. Jealous of its performance and seeing a possible rival to his B-70 and TFX fighter projects, Secretary of Defense Robert McNamara ordered production stopped and the tooling destroyed. When the last SR-71 was retired, in 1989 (to be replaced, it was rumored, by a new secret superplane called the Aurora), it still looked not only brand new but ahead of its time.

☆　　☆　　☆

SUCH AIRPLANES APPEALED because they were thought to be examples of pure functional beauty. But it had never been so simple.

For the real aeronautical engineers, more and more familiarity with the science, more and more time in the wind tunnel at higher speeds, brought home the same truth that had struck John Griffiths: what looks right is not always right. Streamlining is often counterintuitive—or transintuitive.

It was the same question the clipper ship creators had raised: the ship should be fast, it should also look fast, but does looking fast also make it fast? Griffiths had answered in the negative: he pushed mathematics as the most important tool. The first clippers and other fast ships did not look right to the eye of the time.

The key point was the approach to the sound barrier, and speeds of more than 500 miles per hour. The result was known as "the compression problem"—the building up of successive pressure waves on the surfaces of the plane. The first plane to confront the problem, Kelly Johnson's Lockheed P-38, took its shape not from pure aerodynamics, but from the fact that at the time of its design American engines lagged behind those of the Europeans. To reach the speed needed for combat, two engines—supercharged Allisons—would have to be used. After Johnson had been given specifications, he said disingenuously that, to use two engines meant that "the shape took care of itself."

In fact, his drawings show that he looked at a number of configurations: one engine pushing and one pulling, two engines set in a wing without a twin tail separate, and so on, before deciding to continue to tie the engines back into the tail and leave the pilot alone in the middle of the wing in his teardrop-shaped pod. The P-38 was less sleek than other fighters—the P-51, say. It could at times look awkward, but with elements of grace—especially in the pilot's pod and the engines with their booms. The fighter was like a football lineman whose speed and agility surprise despite the chunkiness of his build; but it was extremely maneuverable and could stall, spin, and recover beautifully.

It could also dive, and in a high-speed dive, the P-38 encountered waves of air that built up as the aircraft approached 600 mph or so, threatening to tear it apart. Johnson's team solved the problem by avoiding it. They put dive brakes—hydraulically operated panels that unfolded like the palms of hands—on the plane to pull it out of

Lockheed P-38 fighter, Clarence "Kelly" Johnson, designer. (Lockheed Corporation)

the dive whose speeds created the compression problem. But it was clear that in the future, aircraft would have to be reshaped. Soon, the compression problem would be known as "the sound barrier" and it would shift the ideals of "streamlining"—what "looked right" even to the layman's eye—into a more complex and far less straight-forward realm. The breaking of the sound barrier and the arrival of "swept-back" wings once more changed the ideal of the plane that looked "naturally right."

☆   ☆   ☆

JOHN NORTHROP WAS obsessed with the purity of wing design, and his greatest dream was to create an airplane that was pure wing. "The flying wing," he called it—an exclamation point of a name—and it was a more than technical aspiration.

Northrop said that he had "always been a nut about streamlin-ing," and from the twenties to the late forties he worked to achieve his life's ambition, if not obsession, creating an airplane that would be all wing, with no fuselage. It was an idea that seemed inherently right to him. He had begun research on a flying wing in the twenties. In 1939, he created a balsa-and-tissue-paper model. The

vision of a full-size version goaded Northrop through his distinguished career to reinvent conventional wings, turning out the
Sirius, the Lockheed Vega, and other great airplanes.

The flying wing is, in theory, a more efficient wing because it
"span loads"—spreads the load more evenly across the wing—and
because it lacks the drag of a tail. It is also much more difficult to
control, awaiting the day of computers to control the surfaces of the
unsteady wing, a wing in its purest form.

Northrop worked for Douglas from 1923 to 1926, then went to
Lockheed. In 1928 he formed the Avion Corporation, which was
bought by United Aircraft. In 1931, he left to form his own Northrop Corporation, financed and controlled by Douglas. Northrop's
Lockheed Vega appeared in 1927, the year of Lindbergh's flight. It
looked something like a larger *Spirit of St. Louis*, with the characteristic cylinder heads of the radial engine projecting from the front
cowl; but it was sleeker and more slippery. The fuselage was molded
plywood, and John K. Northrop built the whole thing in six months
at a cost of $25,000. Amelia Earhart would fly one; she posed for
pictures in front of a Cord 310 and a Vega, and each vehicle made
the other look lovelier.

Small improvements make large differences in airplane performance. The NACA cowl for engines was developed by the National
Advisory Committee on Aerodynamics in the late twenties at the
new government wind tunnel at Langley Field, Virginia. Its installa-
tion on the Vega increased cruising speed from 135 mph to 155
mph. The creation of engineer Fred Weick, the NACA cowl was one
of the unsung but critical innovations of modern aviation that made
the modern airliner possible.

In 1930 came the Northrop Alpha, which TWA flew loaded with
mail and a handful of passengers, and the Boeing Monomail, which
had a retractable landing gear. In 1931, the Northrop-designed
Lockheed Orion appeared, with "pants" for its landing gear. Its
224-mph top speed was faster than that of any military production
airplane.

Northrop's dream reached its climax in a series of flying-wing
bombers, including the YB-49. After flying cross country in 1949,
the YB-49 soared past the Capitol, looking like something out of
the film *The Day the Earth Stood Still* (in which it actually made a
cameo appearance). President Truman went aboard—although not
for a flight.

The original flying wing was highly unstable, and suffered several crashes. (The death of one of its test pilots, Glen Edwards, lent the famed Air Force Base, Muroc, in the Mojave where it was tested, its present name.) Northrop's flying-wing bombers turned out to be a failure—perhaps ahead of their time, perhaps a case of romantic fascination with shape getting ahead of practical application. To find such romance in the heart of a supposedly clear-eyed engineer was revealing. It was also not rare. Something that looked right should work right. Wrote the equally clear-eyed aerodynamicist Theodore von Kármán: "I have always thought it a shame Northrop's wing failed. He believed that if something is beautiful it is right. Visionaries with his daring and imagination should succeed."

Eventually, Northrop—the company, not the man—would succeed in creating a flying wing. In January 1981, deep within the Pentagon, the aging and exiled John Northrop was invited to the office of the Secretary of the Air Force. Northrop had been a shadowy figure since the Pentagon canceled his dream project—the flying-wing bomber—and he lost his company in 1952. Now, after signing a security oath, he was briefed on the new Stealth bomber, and watched as a model was taken from a wooden box: it was the son of the YB-49. The old man burst into tears. "Now," he said, "I know why God has kept me alive for the last twenty-five years." Within a month, he was dead.

Meanwhile, the company that still bore his name went on to build the strange plane in the model. The "pure wing" of the Stealth bomber was the ultimate descendant of the Northrop fascination with flying wings. The flying wing baffled radar with its multiple curves. To accomplish this, the plane's surfaces all had to be soft enough—obtusely angled—to disperse the radar waves. The resulting batlike shape is almost Gothic in the complexity of its curves.

Designed on a huge computer, it lived as an ideal in silicon, behind the flickering screens in tiny bits on the tapes and discs of the giant system where every part was modeled, drawn, sized, and stored before a single piece of material was cut. The system was said to remove the need for building most of the traditional mockups and models—they were in the computer, where the "numeric control part verification system" tested designs and handed information off to programs to drive the machine tools.

As long as it lived inside the computer, the Stealth was a wonder.

Its appeal lay ultimately in its mystery, as proven by the Air Force and contractors' efforts to hide it long after everyone except the public could have seen all there was to see of the plane. It was the last refuge of the mystery and romance of the air. Because of its secrecy, its great expense, its presumed technical sophistication, the Stealth fascinated people. There was a whole group of buffs— Stealth watchers—who gathered around Edwards Air Force Base and other Air Force test sites eager to catch glimpses of the plane. Northrop's obsession was shared by Stealth watchers, for whom the tiniest glimpse of or hint about the airplane was titillating. They traded bits of information like gossip—groupies of the flying wing.

But Northrop was not the only company with a Stealth plane, nor was the flying wing the only way to build such a plane. To make an airplane stealthy, Stealth expert Bill Sweetman explains, you can take an aerodynamic shape and make it stealthy, or take a stealthy shape and make it aerodynamic. Northrop's Stealth bomber took the first approach: the F-117A Stealth fighter developed at Lockheed by Kelly Johnson's Skunkworks, and displayed in public for the first time in 1989, the year the Blackbird was retired. The F-117A looked completely different from the B-2. It would emerge as a hero of the Gulf War. It was angular instead of smooth and suggested a series of flattened pyramids slapped together. It was a kind of flying wing, too, only rendered as a faceted series of surfaces, like some incomplete computer-aided design diagram of the sort called a "wire model"—all straight lines, a composition of faces designed to bounce radar waves off at oblique angles, away from the receiving antenna.

The shape of the aircraft became only as aerodynamic as it had to be to fly adequately. Pilots nicknamed the F-117A "the Wobblin' Goblin." As a result, Sweetman says, "it doesn't look right. The boxy and angular shape . . . is very difficult for most people to conceive. It doesn't fit their concept of a 'modern design.' "

Now, a new functional requirement—presenting a series of surfaces that divert radar—was laid over that of aerodynamics. That was the irony to the flying-wing shape of the Stealth. In its design the functional shape of aerodynamics was secondary, compromised. The Stealth planes demolished the simple idea of aerodynamic functionalism, of romantic streamlining. They showed that shapes are always compromises.

☆   ☆   ☆

Northrop B-2 Stealth bomber and Northrop YB-49 flying-wing bomber.
(Northrop Corporation)

Lockheed F-117A Stealth fighter. (Lockheed Corporation)

IN OTHER AREAS of flying, too, the mystique of the flying wing continued, its aesthetic appeal masked under all sorts of supposed practical advantages. A letter to the editor of the *New York Times* not long ago advocated that modern airliners be built in the shape of an early flying-wing concept called the "Burnelli lifting body," an eccentric vision aimed at providing passenger safety in crashes developed half a century ago.

And around the time Northrop was testing his flying wings in the desert northeast of Los Angeles, another engineer who aspired to the pure wing was making his own invention.

On the sands of Kitty Hawk, North Carolina, where motels and pizza parlors edge up to the monument to the Wright brothers, the most colorful and prosperous businesses are those that sell hang gliders. These flexible surfboards of the air are the creation of Francis Rogallo.

Rogallo lives nearby. He is retired from the government agency to which he proposed the Rogallo wing in the late forties. He worked for the National Advisory Council on Aviation in those days, assigned to the Langley Research Center in Virginia. Those were the years when the country dreamed of cheap, individual aircraft in every driveway—or on top of every flat-roofed garage—the flying flivver. *Popular Mechanics* was featuring autogiros and car-planes that would soon be parked in every suburban driveway or on every roof. And through tail fins in back and prop spinners in front, the American automobile was attempting to transform itself in the yearning for a flying machine.

Rogallo was one of those who wanted to create a cheap personal airplane—a Model T of the air. His first test model was created in 1948 from a square of chintz kitchen curtain his wife trimmed, tested in the blast of a large fan in the same kitchen, and patented the same year. Half kite, half parachute, with two arcs of material filling the two angles formed by three struts joined at a point, it was a wonderfully simple and flexible design, with all sorts of possibilities. But no one at NACA, his employer, was interested. So, in 1950, he licensed a version called the Flexikite, marketed by the man who promoted Silly Putty. But the Flexikite did not become the next Silly Putty or Hula Hoop.

For a while, the Army seemed interested in the Rogallo wing. In 1961, it experimented with a Rogallo wing hooked up with an engine and called the Flexwing. A later, larger version was to carry seven men—it was called the Flexible Wing Aerial Utility Vehicle, from which the acronym Fleep, for flying Jeep, was somehow derived. But the "Fleep" never made it, either.

NASA considered using the wing for recovering booster rockets and landing space capsules on land. I remember seeing an artist's conception in *National Geographic* of how a Rogallo wing could land a Gemini space capsule, equipped with stubby little skids, on dry land. It was the first Rogallo I had seen, and I tried to duplicate it with a trio of balsa sticks and an expanse of plastic. From our second-story back porch, it worked fairly well.

A lot of people noticed the Rogallo wing in this way, probably, because by the end of the sixties there were those in California who had built versions of the wing large enough to carry a person—a person foolish enough to hurl him- or herself beneath it from a precipice or a tall dune like those at Kitty Hawk. Soon hang gliding—a sort of aerial version of surfing—had caught on, and Rogallo's wings filled the air at places like Kitty Hawk. Hooked up with a little chainsaw engine, a Rogallo wing mutated into an "ultralight" single-person-powered airplane. Technology that was not high enough for NASA was low enough to become a staple of leisure engineering—and the closest you could get to a pure flying wing.

CHAPTER ELEVEN

Streamlines

IN 1911, VISITING a display of machinery, Marcel Duchamp and his friend Constantin Brancusi contemplated an airplane propeller. "Art is dead," Duchamp commented. "How can painting compete with that?" Within a few years, Duchamp would "renounce art" and declare himself "an engineer," displaying his famed "readymades," functional objects taken out of context and put on display.

At the Museum of Modern Art's Machine Art show in 1934, you could see more airplane propellers, and boat propellers and fans, themselves readymades, presented as products of "pure" engineering. Machine art gave American culture a modernist European validation: the objects in the Machine Art show were admired the same way that grain elevators were admired by European modernist architects. It was a cool, intellectual approach; the catalogue quoted Plato and St. Thomas Aquinas on the subject of ideal shapes.

The propeller was the joke of the 1934 MOMA show—the airplane propeller, the boat propeller, the electric fan blade—a kind of vision of the dynamo made portable, or at least the turbine that drove it. It was of a piece with the lines of the DC-3; the same laws of aerodynamics gave it shape.

The Machine Art show also represented part of a new look at American design: the search for a usable material past to match the search for a usable literary and intellectual past advocated by Van Wyck Brooks and carried out by Lewis Mumford in his literary reappraisals of Melville and Whitman and his architectural ones of, for instance, H. H. Richardson (to whom MOMA devoted one of its early shows). Just as Wilde had found American beauties in the functional and unintentional, curators Alfred Barr and Philip Johnson found them in the industrial object, the "undesigned" and anonymous object. The usable past, gleaned with a European trained eye, was to be discovered amid the industrial detritus of the age; the usable design tradition extracted from beneath the great distracting catalogue of the stereotyped, lithographed, engraved, and bric-à-brac nineteenth century, with its frozen Empire styles and classic cast-iron motifs. That past would be found in the "vernacular" tradition, not the "cultivated" one, John Kouwenhoven argued in *Made in America* in 1948.

But just as MOMA and the American celebrators of barns and grain elevators, dynamos and bridges, were finally discovering in design the "usable past" Van Wyck Brooks called for, just as European thinking was again admiring the plain, unadorned, and unencased American object, Americans were deciding to put streamlined shells around them—shells inspired not only by American airplanes but by European automobiles and the wind-tunnel research of former Zeppelin builders. And just as the works of anonymous designers were being celebrated, the anonymity of designers began to vanish.

Now, too, the anonymous creators of mass-produced products were no longer so anonymous. In 1927, the advertising pioneer Earnest Elmo Calkins made an argument for product design as a marketing tool, an adjunct of advertising, in an influential article in the *Atlantic*, "Beauty: The New Business Tool." How could ads work, he argued, if the styling of the product did not express the qualities claimed for it? In 1929, he established the first styling and design department at his agency, Holden & Calkins, and appointed Egmont Arens director.

Soon a new group, the industrial designers, were emerging as heroes to the public and to corporate executives, heroes who could turn losing products into winners and, it was strongly implied, help pull the country out of the Depression. Raymond Loewy, Walter Dorwin Teague, Harold Van Doren, and Henry Dreyfuss were the best known of those who would create a style of shaping products that turned the lines of the propeller into a model for designing virtually anything. Like the wash from the prop itself, streamlining swept across all sorts of products, from vacuum cleaners to pitchers.

The word "styling" became established in the business vocabulary. Earnest Elmo Calkins, who was deaf, hailed design as the "silent salesman."

The popular term "streamlining" meant more: it was a strategy for making the functional into the decorative and the expressive. The industrial designers took streamlines as a particularly American source of ornament—borrowing from promise of the future the way Europe borrowed from the achievement of the classical past. With varying degrees of skill, they replaced older decorative patterns to create a new fashion. Siegfried Giedion shrewdly argued in 1948 that streamlining "perpetuates the showiness of nineteenth century taste."

Neither anonymous, functionalist, nor quite vernacular, the industrial designers were anathema to the Museum of Modern Art. Streamlined versions of the most beloved American machine, the automobile, were in general disdained; they still looked affectionately at the Model T. One of the few exceptions was the Lincoln Zephyr, which was displayed at the Museum of Modern Art. Zephyr—the very name summed up the romantic, the mythic, and the slightly silly sides of streamlining. *Zephyr*, too, was the name the Burlington Railroad gave its streamlined train, whose maiden run carried it in record time from Denver to the 1934 Century of Progress Exposition in Chicago.

High modernists and the celebrants of functionalism found the creations of industrial designers especially offensive. Raymond Loewy's streamlined pencil sharpener, looking like the Pitot pod from a bomber, became notorious; refrigerators, ranges, and toasters were other commonly attacked villains. Alfred Barr, the Museum's founding tastemaker, wrote that "their designs are aimed at sales appeal of the most superficial kind." Siegfried Giedion disdained streamlining as a passing fad.

Some objects lent themselves with particular ease to streamlining—most of all, of course, those that moved. Irons could be sleek as automobiles, fans were like small engine nacelles, vacuum cleaners could suggest car designs by Norman Bel Geddes. To streamline these objects was to boast of, to advertise their speed and efficiency. But to reshape things this way did something else: it made them metaphors for their own qualities, and in the process, both dematerialized and mythified them. A car, Calkins had written, should not just look like a car, but like speed.

Kem Weber's "Zephyr" clock, 1934. (Musée des Arts décoratifs de Montréal. Liliane and David M. Stewart Collection, Gift of David Hanks in memory of David M. Stewart. Photographed by Richard P. Goodbody.)

Because of this attitude, streamlining had an impact on the new electrical and mechanical products with no essential external shapes, simple collections of components—tubes or pipes or gears—in a box. Streamlining gave these things, from radios to refrigerators, a shell that unified their parts and symbolized their power and at the same time offered a style symbolizing modernity and efficiency. It took the functional language of aerodynamics and turned it into the rhetorical, metaphorical language of "styling," of curves and parallel "speed lines." If that language spoke most often in sales pitches, it was nonetheless eloquent and inspirational. It discovered another sense of "function"—the commercial function of dramatizing the product to sell it. In the process, often with a sort of complicity between the seller and the buyer, it gave identity, personality—style—to things.

Before the industrial designer, many mechanical and electrical devices still vainly sought to define what today are called their "typeforms," the basic shape one imagines when one hears, say, the word "toaster," the one Disney cartooned in the animated short film *The Brave Little Toaster.* And in the mind's eye, the toaster became the rounded, bulging shape lent it by American industrial design and the processes of streamlined styling, and not the long low black or white box of the Braun model that MOMA would put in its vitrine. Streamlining also lent personality. In kitchen appliances such as the Rival Juice-O-Mat, which appeared in several forms beginning in the 1930s, streamlining turned the mechanism into a cartoon character, as rounded and abbreviated as its trademark.

It was perhaps in implicit reference to this effect of streamlining that John Kouwenhoven admitted in a 1986 interview that, in looking at the best in "plain" American design in 1948, he had underestimated the importance of industrial design. In fact, styling and streamlining have exerted a major influence on the shape of American products right up to the present.

Streamlining was, of course, a form of packaging—and theater. The classic American industrial designers came out of theater, where they designed stage sets, or department store window displays, where objects were the actors. In the thirties, as powerful consultants to corporate America, they became designers for a larger theater, dressers of the shop windows of capitalism, Busby Berkeleys of the department store. Streamlining had more to do with penetrating the market than penetrating the air, more to do with overcoming customer sales resistance than wind resistance.

Streamlines arrived in the home as part of the marketing line. The great inspiration was not only the success of aeronautical engineering but the success of design in selling automobiles. It was a short step from taking the automobile maker's line of different models as a marketing strategy to taking the automobile's lines as a stylistic model. The idea of the product line co-existed easily with that of the streamline. Streamlining shouted speed, excitement,

Kem Weber's Airline chair, 1934–35. (Musée des Arts décoratifs de Montréal. Liliane and David M. Stewart Collection, Gift of Geoffrey N. Bradfield. Photographed by Richard P. Goodbody.)

Rival Juice-O-Mat
juicer, 1940s.

modern efficiency, promising to deliver in immobile objects the
qualities of mobile ones. Streamlined irons surely would make iron-
ing faster and easier, streamlined toasters or ranges take drudgery
out of the kitchen. Fans and mixers, closely related in shape and
movement to propeller engines, adapted themselves with special
ease to this treatment.

"Streamlining" entered the language in an abstract sense—we
began to speak of businesses or agencies as being "streamlined." On
the desks of bureaucrats in organizations with such aspirations, it
was surely appropriate to find the streamlined Hotchkiss stapler of
1936, or Raymond Loewy's infamous teardrop pencil sharpener, or
even the cardfile wheel, for fast access to addresses and telephone
numbers, that began life sharing the name of a streamlined
automobile—Zephyr—before it was renamed the Rolodex.

And what made streamlining endure—from the first parallel
speedlines to the jet wing and boomerang shapes of the fifties, to
rounded telephones and CD players in the nineties—was its ability
to speak of the performance, convenience, and efficiency of machines whose physical workings were either invisible or unintel-
ligible.

The cries of dismay at the non-functional use of streamlining
ignored its real rationale: placing shapes derived from function not
just in the service of ornament but of metaphor. That these uses

were also good marketing, "silent salesmanship," did not detract from their sheer visual and tactile attractiveness.

Taking the American fetish for the straightforwardly functional and marrying it to the American fetish for the elaborately ornamental created something new—and something even more nationally characteristic. Streamline style and its architectural kin, the Moderne, were at once wonderfully perverse in their subversion of function and wonderfully exciting in the sense of movement they gave to objects and buildings. They were in their way the most truly American sort of ornamentation ever created. That is why they persisted, in varying forms, through the fifties, why they managed to resist European geometrical functionalism, and why they remain the objects of affection during periodic revivals of public and collector interest.

There was an irony in the fact that streamlining was an ideal of design well before the DC-3, and even came relatively late to aviation. The railroads, sensing the future competition of the airplane, began to streamline their locomotives and cars for symbolic as much as functional reasons, hiring Loewy, Dreyfuss, and others to produce locomotives such as the iconic *Zephyr* and *Twentieth Century Limited*. Even buses took on smooth silver metal skins. Raymond Loewy signed on with Greyhound and slimmed down and slicked up its buses exactly as he did the company's logo, replacing the previous "fat mongrel" with a thin "thoroughbred" silhouetted out of an official American Kennel Club photo.

Streamlining entered the home, sheathing appliances, at virtually the same time it took to the skies. The great virtue of aerodynamics was that it supplied a source of form that was "natural," "functional," "inevitable" without being geometric. By contrast with "cold," rational geometry, it was also romantic and spiritual, smooth and tactile. But with the model once developed, it became a kit of shells that could be applied—without regard to "function," of course—to everything from toasters to buildings. The typeforms of electric irons, toasters, refrigerators, and other items reached maturity in the streamlining period. It was an ideal series of shapes for packaging such equipment as vacuum cleaners, irons, and radios. Airplanes *had* to be streamlined, and their packaging was functional as well as beautiful. They were the metaphor of the age, and from them, like streamlines themselves, radiated an aesthetic that continued to hold sway for decades afterwards.

The shapes of streamlining, blended with the parallel lines of Art Deco, also fit in neatly with the requirements of new materials of mass production: the bulges and corrugations of stamped metal, the reinforced edges required of Bakelite and other plastics, which took over the bodies of radios at the beginning of the thirties.

The aircraft designers were the epitome of the creators of unintentional beauty and the ultimate argument for the inherent beauty of function: the lathe of the wind tunnel for shaping structures. The shapes of airplanes appealed to practically every major designer of the modern movement. In 1935, Le Corbusier published *Aircraft*, a celebration of the shapes of the airplane, whose design virtues he had also praised as a model in *Towards a New Architecture*. Located just half a mile from the Bauhaus in Dessau was the Hugo Junkers aircraft factory, where the Junkers JU-52—the first all-metal cantilever monoplane—was produced in 1932, even before the more streamlined and far more successful Boeing and Douglas equivalents.

Symbolically, too, streamlining reflected the surge into modernity and the shapes of the future. It was as if streamlined objects were smoothed and faired to penetrate the winds of change and the resistance to technical innovation instead of any physical atmosphere. Streamlining made technology acceptable and desirable because it romanticized it.

The proponents of streamlining argued, not unconvincingly, that the shapes and proportions generated by wind tunnel were inherently beautiful ones. American industrial designers, who were soon being hired to treat the interiors of the DC-3 and other airliners, took the exteriors of those planes and applied their lines to other objects. Walter Dorwin Teague's paean to the form of the DC-3, with its assertion of the universal appeal of aeronautically derived shape, was the most cogent defense of streamlining, although he specifically avoided the term. "This line," he wrote, "composed of a short parabolic curve and a long sweep, straight or almost straight, expresses force and grace in whatever form it defines. Its nervous tension gives an especially dramatic quality to the vertical fin of the tail of a Douglas plane. . . . There surely is no more exciting form in modern design. But the line is significant wherever it occurs, and in forms that have nothing to do with flight or with 'streamlining.' "

The parabolic line Teague saw as the basis of aircraft shapes was applicable to anything. It was an ideal shape, he said, identical to that possessed by the Doric capital, the profile of Greek vases, ornaments, statues, and the curves of the Gothic style as well. That it also looked good on toasters, barstools, easy chairs, and office machines, he did not say, but in his own work he attempted to prove it.

Just as manufacturers of other products had adopted the methods of the armories, marketing men for other products adopted the methods of the automobile industry. If the "armory system" represented the characteristic American innovation in making things, the application of streamlining to automobiles represented the characteristic American innovation in packaging them. The aerodynamic figured, in the minds of the industrial designers, not just as a found set of functional forms, but as a set of proportions and principles, like the Golden Section, which could be applied to any number of applications. It achieved the ideal of the Moderne: the functional made decorative. Suggestions of speed, modernity, and efficiency all adhered to the chromed streamlining of toaster or radio—even if the only trajectory the object described was that of fashion, the only wind resistance it met was that to change.

The automobile manufacturers, resisting the "pure torpedo," turned streamlining into a vocabulary for defining differences in brand, in year, in position on the ladder from stripped-down to luxury model. To the high modernists, this was mere packaging, decoration, advertising. Of all the objects that looked to airplanes for their shapes and cars for their marketing model, none was more offensive than the refrigerator, which Barr singled out for its rounded curves and top.

☆   ☆   ☆

"WHAT ARE REFRIGERATORS," asked Billy Durant of General Motors, explaining why he had bought the Frigidaire Company in 1918, "but boxes with motors?" And, he noted, GM was already in the business of making boxes with motors.

Refrigerators were in fact to become appliances of status and universal appeal like cars: centerpiece possessions. The marketing and design techniques first applied to cars—along with streamlined bodies that changed annually—were applied to refrigerators, as they were to be to radios, toasters, and other appliances.

Appliances had always enjoyed a special prominence in American life, not unlike that of the automobile. Before the refrigerator became a symbol of the up-to-date house, the stove similarly promised qualities that went well beyond its simple utility. In their adoption of streamlining, refrigerators came to seem as representative of their time as iron stoves had been of the mid-nineteenth century, with its iron horses and iron buildings. Stoves had once been items of prestige, too, their differing types carefully graded in expense, class, and decoration.

Conceived around 1742, the famous Franklin stove was a cast-iron box which reduced heat lost up the chimney. Like so many appliances to follow, the stove was cleverly sold by its inventor with an appeal to personal vanity: Ben Franklin claimed in a promotional brochure that it would suppress the aging of the female face produced by harsh hot air. He added that it was also healthy; drafts, of course, were said to be the source of most physical ailments. Franklin concluded with a poem in praise of his stove. He did not neglect to advertise in the *Pennsylvania Gazette*, and he had the stove cast with the image of the sun, on which he claimed it to be an improvement. Like the long rifle and the Conestoga wagon, the Franklin stove was manufactured around Lancaster, Pennsylvania—and despite its common name, Franklin himself shrewdly capitalized on localism by calling it the Pennsylvania stove.

The iron age brought a wealth of new stove designs and the establishment of the stove as social center, replacing the fireplace. Americans had fuel to waste in their stoves as in their locomotives and steamboats. Efforts were made to create more efficient models, such as Philo Stewart's Oberlin stove, introduced in 1834, but the typical American room of the nineteenth century was often as overheated as the twentieth century one was to be overcooled. By the 1840s, Charles Dickens and other travelers repeatedly commented on the omnipresence of the pot belly stove, searing the faces of those closest to it, as a center of social gathering not only in homes but aboard all means of transportation. Dickens described one—a coal stove—set in the center of the railroad carriage he rode in 1842 as "for the most part red-hot. It is insufferably close; and you see the hot air fluttering between yourself and any other object you may happen to look at, like the ghost of smoke."

But the stove was a key item in the home, especially after its cooking and heating functions were separated, and specialized models for kitchen and other rooms were developed. It was a symbol of the iron age, and its patterns, like the classical orders rendered on the facades of iron buildings, showed how simply the ornament of the Old World could be borrowed and literally recast by the New.

With the arrival of central heating, the stove lost its prominence and prestige in the living room; and with the arrival of the refrigerator, it ceased to become the center of the kitchen as well. Smooth, white-painted steel replaced ornate black iron as the refrigerator took pride of place from the stove.

Billy Durant was right: the refrigerator was like an automobile in many ways. Its body was stamped from metal. It worked with piston and cylinder; it had an electrical and a radiant cooling system. And the refrigerator, to many, seemed an ideal candidate for the sort of expansion into a mass market that the automobile had already enjoyed, and by the same methods. "With the glittering example of the automobile industry before them," *Fortune* noted in 1940, refrigerator makers cut costs and made their product a necessity. In the thirties, "the industry zoomed across the shadow line separating the specialty product from the mass-produced, mass-consumed commodity." This happened with dramatic speed.

Iceboxes, regularly restocked by the iceman, were common by World War I and huge centralized icemaking plants supplemented the "harvesting" of ice. For years, the technical problems of replacing ice with mechanical refrigeration had intrigued engineers and scientists—including Albert Einstein, who toyed with new ideas for a refrigerator pump at intervals during his work on the theory of relativity. But it was not until the twenties that major breakthroughs in pumps and coolants made the refrigerator possible. In 1921, only 5,000 refrigerators were manufactured in the United States. In 1937, there were 3 million and market penetration had reached 56 percent.

The first refrigerator makers came out of Detroit. Kelvinator, whose founders had a background in automobile manufacturing, sold its first unit in 1918. Guardian, the brainchild of inventor Alfred Mellowes, was sold to William Durant of General Motors in 1919 and promptly renamed Frigidaire. The Guardian machines, of which Mellowes had sold fewer than three dozen, utilized a brine tank and water-cooled compressor. They seemed to work only with his constant personal repair visits to customers.

In 1921, Frigidaire was still losing $2.5 million a year. But by 1926, with the application of GM mass-marketing techniques and technical improvements, it owned 45 percent of the market and was GM's third largest division in unit volume after Chevrolet and Buick.

Frigidaire's breakthrough was the development of a non-toxic, non-inflammable refrigerant to replace sulfur dioxide and other gases used previously. It benefited from GM's links to Du Pont, which held a major interest in GM. In 1930 Thomas Midgley, Jr., a Du Pont chemist, developed the refrigerant Freon 12 (or dichloro-difluoromethane), a chemical later to be identified as harming the ozone layer. To make the entire market grow and to standardize its own system, Frigidaire immediately licensed the new chemical to the rest of the industry.

The big push began in the late twenties. The largest firms—GM and GE—started to gear up for the mass market around 1926. The 1927 GE Monitor Top surged to take a third of the market on the strength of the "sealed compressor" perched on its top like a bee-hive, or, as *Fortune* called it, "a gravid crown." Frigidaire and others fought back with the flat or shelf top, placing the compressor out of sight on the bottom. The importance of such cosmetic differences showed that the machines themselves were becoming similar and reliable (most offered five-year warranties).

By 1931, refrigerator manufacturers were spending $20 million on ads, or almost $10 per machine. Now that the machines were in fact safe and clean, it was time to call in the industrial designers to make them look that way. White, the color of cleanliness, had already become dominant. An all-porcelain model arrived in 1926. The GE Monitor Top, the Model T of the industry, represented the shape of the machine before streamlining (so well respected was this shape that when architect Raymond Hood designed an office and showroom for a New York GE appliance dealer, he included a mock Monitor Top on the roof).

In the early thirties, after dallying with Norman Bel Geddes, General Electric hired Henry Dreyfuss to update the Monitor Top. Crosley brought in Walter Dorwin Teague to give shape to its Shelvador model of 1935. Raymond Loewy did the Coldspot for Sears from 1935 to 1939, then switched to Frigidaire. With one basic shape inflected by these styling accents, by the 1930s the image of the refrigerator as what the industrial designer Harold Van Doren called a "sleek, sanitary monolith" was firmly established.

The changes in the refrigerator's shape from now on would closely follow those of the other box with a motor—the automobile. Cabriole legs would vanish under skirts, like tires under wheel pants and mudguards. Motors and heat exchangers would cease to be separately exposed units and retire beneath shells of stamped steel, with rounded corners and raised speedlines. Throughout the thirties refrigerators, like cars, became fatter, with swollen curves or "radii." And like many household appliances—stoves, toasters, and vacuum cleaners, for instance—the imprint of the industrial design look of the thirties remained dominant, with only minor cosmetic changes, in the fifties.

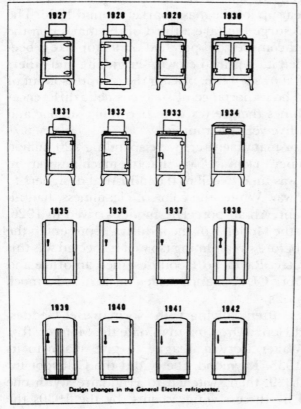

Design changes in the General Electric refrigerator.

Evolution of General Electric refrigerator. (General Electric)

In 1931, Sears had introduced its house brand Coldspot, made in an old washing-machine factory and lacking the "sealed compressor" that had become the quality standard since the GE Monitor Top. At first, the Coldspot was a disaster. Production could not meet demand and quality control was nonexistent. Then, in 1932, Sears hired Raymond Loewy to redesign the Coldspot, whose new shelf pattern was taken from the radiator of the Hupmobile. Sears would introduce model changes that were literally annual throughout the rest of the decade—until, in 1939, Loewy deserted to Frigidaire. And like the auto makers, the refrigerator was offered in a series of models, like GM's, ascending in price from the "nude" box to the Gold Seal or top of the line.

In 1934, Sears doubled its sales of the previous years, with 59,000 models. For 1936, it sold 216,000, and 290,000 in 1939. "Super Six," they called one model, auto-slogan style.

In 1934, total refrigerator sales numbered 1,388,000 units; by 1941, they had increased to 3,783,000. Prices, which in 1920 had been around $600, comparable to those of the Model T, sank as low as $90 by 1940. This price, however, was for the loss leader, the "nude" model, on which the companies by the end of the thirties were losing money. The idea was to move the customer up to the next grade—just as the car salesman did—once he or she had been lured inside the store.

The difference between the nude, low-end models and the higher-grade boxes, *Fortune* claimed in 1940, was dramatic: "The difference is precisely the difference between the bucktoothed schoolmarm at Punkin Crick and the females who pose for the corset advertisements. As for marvels like Polarex meat savers, automatic reset defrosters, stackable Foodex drawers, fly shelves, and finger-tip releases, their gross effect is to strain your craving to the breaking point." A lettuce crisper or a Westinghouse streamlined accessory water pitcher—an item that has today become collectible—could make the difference in a decision.

If the automobile was a time machine as well as a transportation machine, shortening the duration of trips and the psychological distance between places, the refrigerator was a clime machine as well as a transportation machine: leveling out the seasons with frozen and cooled foods, making the Georgia peach last longer in Massachusetts, the California orange in Minnesota. But the

Streamlined refrigerator
pitcher, J. P. Thorley,
designer, circa 1940.
(Hall China for
Westinghouse)

refrigerator really began to enjoy new status after the arrival of
frozen foods, in small force before World War II and in mass
numbers after it. The refrigerator now supplanted the stove as the
center of the kitchen: here notes and pictures and grocery lists were
posted, and inside the wealth of the garden—the irrigated fields of
California—and the bounty of the ranch (and the packaging genius
of Madison Avenue) stood on display.

The refrigerator stood as a powerful icon of Western prosperity
to those in the Third World. Citicorp CEO Walter Wriston reported
a conversation he had with Rajiv Gandhi in which the Indian prime
minister called the refrigerator a revolution. If spices to preserve
food had drawn the West to the East, and in the process led to the
incidental discovery of America, as cartoon history has it, then the
discovery of the refrigerator drew the East toward the West, the
South toward the North, and the Third World toward the First.
Almost as much as the automobile, it stood for the benefits of
industrialization.

☆   ☆   ☆

WHILE THE REFRIGERATOR and other motor-driven machinery
looked to the automobile industry for the shapes of its boxes, the

new technology of electronics was even more separated from "functional" expression.

Radios, the science fiction pioneer Hugo Gernsbeck had prophesied, would never be successful until they were simple to use in the home. And, he could have added, looked at home in the home. The first home radios, strictly for hobbyists, were simple "breadboards"—naked tubes and crystals and tuners on a plank. Soon, however, the contents were surrounded by a box, sprouting dials and knobs whose location, lent by the practical design of the circuit board, made their fronts as abstract as a work by Klee or Arp. This was still the era of earphones and of speakers disguised as papier mâché lyres or seashells. A familiar form was the Gothic-cathedral model, the better to tune in to the new radio priests and preachers perhaps, but also echoing the shapes of clocks.

Later radios took on rounded, almost toylike shapes, and were sold under such friendly and familiar names as "Buddy" and "Chum." Radios looked as if they were designed purposely to be made to dance and talk, as they did when they came alive in the cartoons. Radio cases took early advantage of new plastics: phenolic, casein, and others. With their streamlines and vents, their softening of the box with curves, they began to look like radios instead of modified bread boxes and furniture. Their volume and tuning knobs were often made the same size, so they turned into headlights, speaker into grille—like the face of an automobile. Their colors ran to mustard, burgundy, and tomato. Some looked like globes, some like champagne bottles. Some bore the image of Mickey Mouse, or were housed in little copies of Raymond Loewy Coca-Cola coolers. (Designer names even helped. When Loewy designed radios for Emerson, they put him in the ads.)

Now radios were all over the home, in the living room, the bedroom—and, beginning in the early thirties, thanks to a company called Motorola, in the car.

Radios, refrigerators, toasters, automobiles—whether mechanical, electrical, or electronic, devices were more and more being wrapped in stylistic shells. The use of streamlined styling to introduce fashion and eventual obsolescence to products was an easy target of the defenders of high culture. That critique was put in a nutshell by Raymond Chandler, in *The Long Goodbye*.

"You can't have quality with mass production," one of Chandler's more curmudgeonly characters tells private eye Philip Marlowe, ca. 1950. "You don't want it because it lasts too long. So you substitute styling, which is a commercial swindle intended to produce artificial obsolescence. Mass production couldn't sell its goods next year unless it made what it sold this year look unfashionable a year from now. . . . We make the finest packages in the world, Mr. Marlowe. The stuff inside is mostly junk."

It would take the interventions of Pop and camp sensibilities, via the backdoors of irony and humor, to create a wide appreciation of the grace and beauty of this sort of packaging. The stylist's iconography of speed and smoothness, of efficiency and ease, created what critic Reyner Banham was to call the "throwaway esthetic," reflecting popular taste as well as the manipulations of salesmanship that were all the high culturists saw there.

What was in the package was technology that changed as rapidly as that of the steamboat whose builder had explained to de Tocqueville why his vessels were built to last only a few years. But what was also in the package was personal desire and social aspiration.

Boxes with motors, boxes with tubes. A new class of objects had arrived, without the natural form of geared mills or dynamos. The new sorts of boxes were like those that had enclosed clockworks in the past. And even clocks were updated so they resembled radios, which resembled moving vehicles. The name Zephyr was also applied to a clock designed by Kem Weber, creator of the streamlined Airline chair. In the clock, Weber took the curve of a streamlined car or locomotive's front and gave it a twist: one end sweeps down, the other sweeps back, making the object seem almost to float on air. With the Zephyr clock, even time flies.

Not only mechanical servants—toasters, typewriters, phonographs, vacuum cleaners—but electronic servants of all types began to assume the livery of the streamline. When television arrived, prematurely, in the case of the first model in 1939, it was a streamlined mahogany box by the industrial designer John Vassos. Hi-fi, stereo, and computer systems would, at the very least, round their buttons and corners in testimony to the lingering influence of the power that rounded the first radios.

To join the mechanical maid and entertainer, the mechanical salesman soon arrived: Earnest Elmo Calkins's "silent salesman"

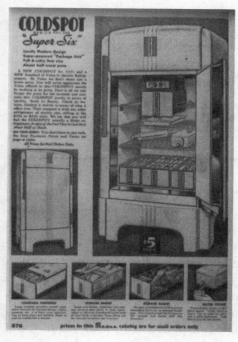

Sears Coldspot
refrigerator, Raymond
Loewy, 1935.

made literal. In the twenties and thirties, vending machines of all
sorts began to figure in the national landscape in such numbers that
they astonished the normally placid writer of the presidential *Report
on Recent Social Trends*, commissioned by Herbert Hoover. Put a coin
changer on a refrigerator, load it up with bottles of Coca-Cola and
some chutes and ladders to dispense, and you had the vending
machine, glowing and humming softly in corners across the country. And their shapes, too, bore the romance and drama of the
streamline. The two-tone red and cream-colored models, rounded
on their tops, looked like a late forties hardtop Buick by Harley
Earl, or a Raymond Loewy Frigidaire or Coldspot. The Coke machine was the ideal streamlined sales machine, in which not only the
case but the process was streamlined; and from which, with a dime
and fingertip, you served yourself.

CHAPTER ☆ TWELVE

Self-Service

EVERYONE KNOWS THE assembly-line scenes in *Modern Times*, Charlie Chaplin's 1936 demolition of the factory system and Taylorism. Far less well remembered is the episode later in the film when Chaplin visits a cafeteria.

The cafeteria is a sort of anti-assembly line, where machines (almost literally) serve men instead of men serving machines. Instead of the line moving past the Tramp, the Tramp moves past the line, taking off dishes instead of tightening nuts or applying parts. He delights in the place; it contrasts to the mechanical "feeding machine" tested on him earlier beside the assembly line, the one mocked in the film's titles as "streamlined," with its "synchromesh" corn feeder: the ear of corn set on a motorized axle, like a typewriter platen. In the cafeteria, the Tramp eats at his own pace, helps himself to a toothpick and idly employs it—the very picture of leisurely consumption.

The cafeteria came along about the same time as the assembly line. It was developed in the 1890s in the Midwest and grew strong in California. (Kansas City is often credited as its birthplace, but there is no clear holder to the title of first cafeteria.) It took lessons of speed from Harvey's, the chain that fed passengers in Western railroad stations during the few minutes of their stop, and added to them the assembly line's economies of scale. It needed volume to succeed; otherwise food cooled or wilted. The cafeteria was for people in a hurry—factory workers, for instance, on a brief break from that assembly line. And like the assembly line, it reduced labor, in the form of waiters and waitresses, and so provided lower prices. But above all it provided a dose of that great American value, convenience.

The cafeteria and its like—the convenience store, the super-market, the vending machine—streamlined retailing the way the assembly line streamlined production and the shapes of the products themselves. And the new retailing created strange new environments. Being inside them was at once like being in a theater and like being inside a vending machine. Americans understood that, like the object itself, the retail environment had to be designed to sell. The theme of the theater of the new retail was individuality—self-service.

Self-service changed with the times, but increasingly seemed central to American life. The ubiquitous soda fountain was a vital center of Wilsonian America, all aglimmer like a new car, with shiny nickel plate and piping.

The cafeteria never possessed the romance of the Automat, although dining from vending machines proved far more influential in legend and more romantic in retrospect than in reality. Despite experiments in other cities, the Automat was never successful outside of Philadelphia, where Horn & Hardart opened the first example in 1906 with a complement of German-made machines, and New York, where the theatricality of the machines and the Deco-inspired exteriors made it a legend—at least in the memoirs of actors and writers who had sipped coffee there before they became famous. But the Automat perfectly summed up the emphasis on mechanical delivery of food for convenience. The Automat was such a piece of received culture that Andy Warhol, that framer of Pop items, wanted to revive the concept as the "Andymat." It was not a practical idea: the Automat survived as an ideal and icon, but fast-food schemes with quicker turnover had beaten it out. Streamlining service depended on more than machines: on the Frederick Taylor–inspired combination of machines and men working like machines that made fast food's assembly line a success. The last Automat, on 42nd Street in Manhattan, was essentially turned into a cafeteria, with the vending machines left in place more for ambiance than for function, before it gave up the ghost for good in 1991.

If the American is a do-it-yourselfer, the parallel American obsession with convenience has attached itself to the object that "does it itself"—from the self-scouring plow to the self-cleaning oven and self-defrosting refrigerator to self-sharpening knives to self-service stores.

In selling, this impetus took the form of "self-service." Perhaps the desire for self-service arose spiritually out of the embarrassment that being served in a society of ostensible equals inspires. Its beginnings could be seen in Jefferson's dumbwaiter and revolving server, two *Popular Mechanics*–style devices that concealed the ugly fact of slavery. But it was also tied to the push to reduce labor costs and dependence on the human vagaries of the employee. And it capitalized on the characteristic impatience of the American customer, who was eager to save minutes or seconds even if he already had time on his hands. Self-service demanded speed, and it was in restaurant preparation that the application of mechanics to food took place—not in feeding, as in the vision of *Modern Times*, but in cooking.

Selling things in America has depended on neat kits of techniques and presentation. In case after case, the pattern has been that of figurehead entrepreneurs—often almost crazy, near-folk figures—who have emerged to magnify their eccentric visions by high levels of organization into sales empires. Along the way they have shared the ideal of the product that will sell itself.

The recurrent theme of the thing "so easy to use, a child can do it" was part of this approach. Just as Kodak showed children using Brownies, and Xerox, fifty years later, showed them able to operate its photo copiers, William Durant of General Motors boasted of an early Chevrolet that "a little child could sell it."

Removing the salesman meant removing the personal factor. Self-service depended on the self-reliant object, the ideal of the thing that would sell itself, the thing as it were suspended in the retail world, liberated from personal relationships, made to float free as in the surreal space of a catalogue or a shop window.

It also focused on the design of the product: Earnest Elmo Calkins saw that advertising began with the look of the product. Corollary to this idea was a recognition that if form followed function, the function of any product was before anything else to sell.

The peddler—with his mobile store—served the hinterlands and frontier with a select offering of objects. He pushed the New England wooden clock, the Hitchcock chair, the eggbeater, apple corer, and later the sewing machine—to say nothing of the wooden nutmeg. He was succeeded by the traveling salesman, a rationalized peddler, equipped with kits of technique—argument joined to psychology—devised by such prophets as National Cash Register's John Patterson.

If Europe—and Imperial France in particular—invented the department store, America added its own innovations to the model: the elevator, the escalator, the paper bag, and, above all, the big catalogue. The fixed price, the physical accessibility of goods, and the public nature of the store space itself, all fit well into the trends of American society. The American department store brought the Parisian one home in a box. Stores like A. T. Stewart were among the early supporters of cast-iron architecture, replicating entire arched bays in their facades, obtaining classical pediments and arcades like their European ancestors, in iron, on the cheap, without hand work.

The department store brought theater and the fixed price to retailing. The department store's innards—the plumbing of elevators and escalators and pneumatic tubing connecting counter with cashier—fascinated early shoppers. They were part of the theater of "the shopping experience," the sights and sounds of the modern bazaar, where advertising replaced haggling. While Stewart, faux-marbleizing his cast-iron store in Manhattan, created a simulacrum of the European department store, and while the catalogues of Sears or Montgomery Ward—along with a rail network and, later, the government parcel post system—distributed it all over the country, F. W. Woolworth downsized it. With his fixed-price system, the mystifications of plate glass and bauble, of red and gilt signage, he created stores like little theaters on Main Street. The price, for Woolworth, was the package. The fixed price was the beginning of the change in retailing just as the fixed form of ax or rifle stock had been the beginning of the change in manufacture.

Woolworth began his career working in a store with a five-cent table. In 1879, he decided to make that table the basis of a whole store. A failure in Utica, New York, was followed by success in Lancaster, Pennsylvania, the country of long rifle and Conestoga. There, people appreciated a bargain—and tended to keep down their outlays on "fancy" items. A nickel or a dime—shiny as the coinage John D. Rockefeller's public relations man had him distribute to worthy urchins—could now both please their imagination and salve their conscience.

By 1886, Woolworth had seven stores; by 1892, he had twenty-eight; and by the turn of the century, fifty-nine, doing $5 million of business a year amid standardized gold and red décor. In 1913 he was ready to build the mother church for this chain of sales chapels, the Woolworth Building in New York, the Cathedral of Commerce, and, to small-town Americans, the greatest advertisement possible for the Woolworth's on their own Main Street.

With the retailer's eye for detail, Woolworth picked everything himself. He nickeled and dimed contractors, suppliers, and even his architect, to whom he proposed a reduction in fee. He picked toilets and urinals—Sanitas brand—and it took him three more months to decide on the handles for the urinals. He chose the buttons on the thirty elevators, which rode like pistons inside cylindrical wells and used air compression instead of ratcheting as a safety measure. He cut a deal with Judge Elbert Gary of U.S. Steel to provide the steel at cost, in return for the publicity of its being used to build the world's tallest building.

By the 1960s, single stores based on the Woolworth model would offer square footage nearly as vast as that of a skyscraper like Woolworth's tower. Kresge, one of Woolworth's competitors, created as an experiment a new kind of store, a huge discount operation that dwarfed the corner five and dime but projected the same idea to gigantic scale. It was called K mart.

☆   ☆   ☆

IT WAS IN the area of selling food, however, that the American tendency toward self-service showed most clearly. The words "consumption" and "consumer" became common in the 1890s, an economic outlook whose origins were clearly alimentary. A 1918 *Scientific American* article on the new phenomenon of self-service groceries, aptly entitled "Cousin of the Cafeteria," included a description of a new kind of store, which it did not name, founded in Memphis by a man who also went unmentioned. In fact, the store was Piggly Wiggly—a brand name too indecorous, perhaps, for a journal of progress.

The origins of the name are unclear. One account suggests the children's toe game ("This little piggy went to market"), and still another turns on the way the new kind of store grabbed hold of the

First Piggly Wiggly store, Memphis, Tenn., 1917. (Memphis Museum System)

customer like a wriggling piglet—an apt reference to the way in which the Piggly Wiggly format channeled customers like stock through a series of fences, turnstiles, and counters. Entering the store, you passed through the turnstile, picked up a basket, walked up and down the aisles selecting items, until at last you arrived at the checkout counter with its Burroughs adding machine.

Piggly Wiggly—the store concept—was patented and the whole system leased for 20 percent of gross to independent entrepreneurs in a sort of proto-franchise. If you look at the patent drawing, filed in 1917 by Saunders, an eccentric Memphis businessman, you see a series of arrows and baffles: a maze, a system for getting customers into, through, and out of the store, and along the way confronting them with every possible item that could inspire their purchasing impulse.

The self-service market is a traffic pattern—a machine for self-service, and, in the best American tradition, for saving labor. Clarence Saunders's patent diagram looks like the coilings of the Kinetoscope film through its box, or like a vending machine's innards. It was a way of getting people in and out, but also of exposing them, like the film in the movie camera, in sequence to all the labels and boxes of the products Piggly Wiggly offered. Piggly

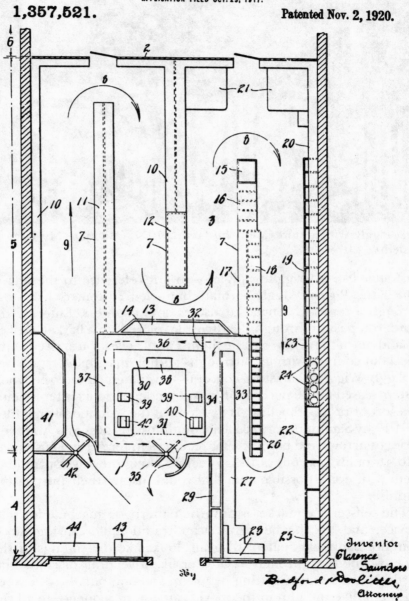

Patent for Piggly Wiggly store system, 1917. (Patent and Trademark Office)

Wiggly also recalled Oliver Evans's automatic mill: customers seem to fall through the store, up and down, up and down its aisle, then finally through its checkout turnstiles, to be processed like the grain in the hoppers and chutes of Evans's mill.

The shelves are straight and sober as in a library, and ranked as high. There are no handwritten price signs or colorful starburst ads for specials, only neatly printed price tags, diamond-shaped, dangling from strings. There is a preponderance of canned food. The brand names and can designs do the selling. The store name echoes the old warning, "Don't buy a pig in a poke." There were no pigs in pokes at Piggly Wiggly. It depended on name brands to sell themselves. There was no haggling—prices were fixed. No credit, very little help from clerks, no delivery. (This last omission was analogous to the cafeteria's omission of the waiter.) At the Piggly Wiggly, the goods tended toward the best entrenched and most vividly packaged trademarks. Only the vegetables and fruits ended up generic, in generic packaging. There was no meat, and only two clerks were required to run the whole store.

Today, Saunders's mansion, "the Pink Palace" and its modern addition, contain the Memphis Museum of History and Science. There, among other exhibits, is the interior of the first Piggly Wiggly, moved lock, stock, and barrel from its original location downtown at 79 Jefferson Avenue, part of a display entitled "America's Food Distribution System."

It is eerie, walking down those aisles, seeing some of the first national brands in wonderful old cans and bags—the Quaker Oats man looking more like William Penn than his generic successors—Pride of Illinois Country Gentlemen Sweet Corn, Kellogg's Krumbles, Arbuckle Ariosa Coffee. On the Tabasco bottle's label, square and diamond quarrel, the result looking as if it could have come out of a Russian Constructivist atelier. There are cans of Heinz vegetables and Campbell's Tomato soup, which bear the same strange relation to their Pop art successors as the old photo of an ancestor does in the presence of his lookalike great-grandchild. You find Aunt Jemima pancake flour priced at 37 cents and Towle Log Cabin Syrup in the cabin-shaped tin. Forty cents would buy you a package of Diamond Clothes Pins (made on the Shaker model, by the same folks who made matches). The established national brands were consumer currency as solid and standard as the nickels and dimes of Woolworth's.

On October 22, 1916, Saunders opened his second store, which he called the Piggly Wiggly Junior. Saunders aimed his advertising at women, significantly, and he seemed to personify or at least animate the Piggly Wiggly in ads in the Memphis *Commercial Appeal*. He spoke of the Piggly Wiggly's "scientific diet"—"100 women an hour wait on themselves at Piggly Wiggly, one every forty-eight seconds."

At its peak, Piggly Wiggly had 2,600 stores that did some $180 million of business; by mid-1918, Piggly Wigglys were to be found in Chicago, Houston, Dallas, Richmond, and Cincinnati. Then Saunders, perhaps impatient with a world he saw himself far ahead of, got greedy. He piggishly tried to create a corner on his own stock. In the backlash, the price fell, and the overextended founder lost control of the Piggly Wiggly in 1923.

It was not until the forties that major chains such as A&P moved to a system similar to that of Piggly Wiggly in its vast network of stores. They were threatened by Piggly Wiggly's successors. Later supermarkets—notably that of Michael Cullen of King Kullen, the first of which opened in 1930, in Queens, New York—would blow the store up to huge dimensions, and with the creation of the island or "gondola" shelf, create a more elaborate set of baffles through which the consumer was pinballed. They would heighten the pitch of the ads to even shriller pitch: "rockbottom prices—world's greatest price wrecker," add the parking lot to the mix, and the idea of shopping by car.

A photo shows the beaming Cullen in the first store, standing among pyramids of mayonnaise jars beneath great trussed girders: a modern bazaar, but one where all the haggling was haranguing, the price setting one-sided.

Saunders helped create the supermarket, understood as a machine for purchasing and for selling, which depends on packaging, national brand advertising, the refrigerator—and, of course, the automobile. Piggly Wiggly, blown up to huge scale, modified to the car, was the supermarket. The supermarket provided something else—parking—and as a concomitant of the automobile, the shopping cart, a kind of lighter for filling the freighter of the automobile. The addition of the carts (whose invention made a miniature folk hero of one Sylvan Goldman, an Oklahoma market owner), of moving belts by the cash register, of laser scanners, all have made the supermarket more like a factory.

The supermarket shopper looked for speed, and the new store exteriors suggested the streamlining of shopping, thanks to self-service. Shopping was not only fast, the buildings seemed to say, but heroic: a Los Angeles supermarket sported a tower borrowed from the model of a Hugh Ferriss skyscraper. Speed brought efficiency, but also impersonality, in contrast to the old corner grocery. The supermarket was not only a clean, well-lighted place; it was an aseptic one. In the film *Double Indemnity*, Fred MacMurray and Barbara Stanwyck meet to hatch their plot in the secure anonymity of the supermarket aisles.

☆   ☆   ☆

SAUNDERS'S SUCCESS INSPIRED others in Memphis, the self-proclaimed "distribution center" of wholesaling and warehousing for the western part of the South. Before Saunders, the Fortune Drug Store, not far from the first Piggly Wiggly, began curb service, catering to automobiles. Honk twice and a soda jerk would come out to take your order. Curbside service, like so many of the innovations of self-service retailing, developed and expanded in cities of what would become the Sunbelt, in Dallas and Houston, and of course quickly in Los Angeles.

To package rooms as neatly as groceries or sandwiches, to standardize the hostelry experience, to generate absolute predictability for the traveler, jukebox magnate Kemmons Wilson chose to begin the Holiday Inn in Memphis, too, where the first opened in 1952.

Years after he had made and then lost millions with Piggly Wiggly, Saunders hatched a new scheme. It was called the "Keydoozle" store, and it took self-service even further, replacing the shelves and clerk with a series of vending machines. The open shelves of the supermarket had been superseded in this new patent with dispensers controlled by personal keys—in essence, vending machines. The model for Keydoozle was obvious: the Automat.

☆   ☆   ☆

THE AUTOMAT EMPHASIZED machines for serving. The development of the fast-food franchise industry depended on the development of labor-saving machines for cooking. The first White Castles even employed a specially designed spatula. Colonel Sanders owed his success far less to the famous blend of herbs and spices than to his discovery that you could fry chicken much faster in a pressure cooker. In his autobiography, *Grinding It Out*—an appropriately mechanical, if inelegant, phrase to sum up a life—Ray Kroc of McDonald's explains his rise from paper cup salesman to multispindle blender salesman to fast-food operator.

The fast-food business can almost be seen as a parody of the factory system. The process was a reversal of "downwards" or "backwards" integration, by which an auto maker, for instance, would move to making his own steel. (At Ford's River Rouge plant, the ideal was for an industrial process that began with piles of iron ore and other raw materials, and ended with Model A's being driven out the factory doors.)

In 1921, in Wichita, Kansas, two men named William "Billy" Ingram and Walter Anderson, the common American duo of tech-

Ice cream stand in shape of ice cream container, Connecticut, 1930s. (Library of Congress)

nician and promoter, opened the first White Castle restaurant. They sold hamburgers and coffee. White Castle summed up the combination of tradition and modernity in which the public craved to serve itself. "The porcelain palace" at once suggested medieval heartiness and modern cleanliness: a sort of Connecticut Yankee reformation of King Arthur's Court. The later buildings for White Castle were movable and made of interchangeable parts. Designed by L. W. Ray, they were eventually manufactured by the firm's own construction company, Porcelain Steel Buildings Company, in Columbus, Ohio. The porcelain-covered steel panels were designed so the restaurant could be assembled without rivets or bolts. The system was related to that of the dime store, with similar prices and similar deployment of a kit of standardized architecture. Other restaurants adopted kit buildings of this sort, which reflected the company's simple mode of operations: selling the burgers jokingly called "sliders" and "belly-bombers" with minimal labor.

The structural packaging of American restaurants was carried through to new packages. White Castle urged the public to "buy 'em by the sack." (Its current offering comes packed in a pasteboard box, patent number 2,435,355, open at the top where the edges bear the silhouette of two vestigial castellations.) And the plastic clam-shell containers for Big Macs and Whoppers echo the shape of the mansarded outlets that dispense them. Today, eco-consciousness has led to blending the buildings into the environment with such "natural" materials as shakes and shingles, and returning to wood-pulp-based burger boxes.

The rise of the hamburger itself as an icon of the American diet corresponded to the drive for self-service. The hamburger—which began as the hamburger sandwich—aspired to be the universal American food. It was beef, the most favored of American foods. It was pre-ground, almost pre-digested (not unlike the "healthful" steak of Dr. J. H. Salisbury). It was portable and required no utensils. And it could be customized with the addition of various toppings and garnishes. It was easy to cook, even by the unskilled.

We have always sought the perfect self-contained universal food. The candy bar and the Life Saver, all these neatly packaged, bite-sized (as it were) foods are as American as the hot dog. Sometimes it was for health, from Dr. Sylvester Graham and his Graham cracker to Kellogg and Post with their cereals, from Gail Borden with his primitive K-ration, the "meat biscuit," to Ray Kroc and McDonald's. The meat biscuit is a sort of Neanderthal ancestor of the hamburger. Perhaps the Deerslayer's pemmican and jerky, adapted from the Indians, should be placed here too, at the very base of the evolutionary diagram.

The hamburger was fun—it was sports food, as there are sports clothes and sports cars. Covered with ketchup and onions, it played out the same cheap melodramatics of salt and sugar, bitter and sweet, that seemed to appeal to the American palate. Pushed at the 1904 Exposition in St. Louis, it caught on much the way Belgian waffles caught on at the New York World's Fairs, first in 1939 and then again in 1964.

James Deetz, in his wonderful little book on the archeology of early American life, *In Small Things Forgotten*, argues that by the end of the eighteenth century, "one pot" meals, as medieval in origins as the post-and-beam house, and which survive in such dishes as the New England boiled dinner, had been replaced by meals of distinct elements. Deetz sees this change as reflecting "the shift from a world view of a corporate nature to one that places great emphasis on the individual. The ideal American meal of meat, potato, and vegetable," he goes on, "is not only tripartite in its structure but very mechanical . . . those paper plates we buy at the supermarket may be divided into three sections because of a very deep tripartite mental structure. Their structural similarity to a Georgian facade may be much more than coincidental." Even baby plates, Deetz notes, employ the three-part structure.

The fast-food schema altered the trinity. It got rid of vegetables, except for the token appearance on some menus of cole slaw. The new trinity was burger, shake, and fries—with cola an acceptable substitute for the shake. Or fried chicken accompanied by biscuit and/or mashed potatoes and cole slaw. But the burger itself, garnished with tomato, onion, and lettuce, and sandwiched between bun halves, replicates the whole meat/starch/vegetable division in a neat assembly.

Beef was the key. A whole history of America could be written through the history of the beef industry: from the open range through the trail drives to the age of the railroad refrigerator car and the corned beef can, and ending in the supermarket steak, lying beneath polyethylene wrapping on a Styrofoam tray.

Some years ago, I was invited to the ceremonial frying of the 50 billionth McDonald's hamburger. The event took place in the great ballroom of a New York hotel, all gilt and draperies. In the middle stood a single, simple grill, like an altar. Onto the grill was deposited the pale pink, round disc—exactly 1.6 ounces in weight, 3.875 inches in diameter—of the official burger. The company president, who began as a frycook, held up the patty for the television cameras in a way (insinuating delicacy and value) that at first reminded me of a round wafer of silicon onto which computer chips are printed and then, a moment later, of the priest's symbolic elevation of the Host. Or perhaps a disc of nuclear fuel.

The exact timing of the event partook of some license, since it could only be estimated, by time, from the production figures of McDonald's outlets around the world. At that minute, determined to be 12:05 EST, the patty was placed on the grill. With a flourish of drumbeats and marching-band music, the slightly reticulated patty was fried and deposited on a bun, to be consumed by none other than Dick McDonald himself, the survivor of the two brothers who created the chain.

"No one takes hamburgers as seriously as we do," Ray Kroc had said, and this was proof. Kroc, as everyone knows by now, had found, not founded, McDonald's. The story of how he did so suggests what it was that made him see the possibilities of packaging— as franchises—the system of self-service the McDonald brothers had developed. Cutting labor, reducing menus, reducing prices, and applying machinery, they had achieved a volume of sales in the late forties that impressed Kroc, then selling Multimixers—six-headed milkshake blenders.

Many other burger joints were operating similar systems; White Castle, White Towers, A&W Root Beer stands, and others had laid the groundwork for a kit system of selling a kit type of food. Kroc had been a radio producer, a pianist, a real estate promoter, and a paper cup salesman. It was this last job that set him on the course to fast food.

Early in the century, public conveyances banned the common drinking cup under pressure from the government health bureaus. The paper cup had been developed shortly after the turn of the century in the face of growing concerns about public drinking facilities. Health officials found the shared tin cups on trains and the silver cups—chained to the base of the ice-water urn—that were common on riverboats dangerous in spreading disease. The answer was the disposable cup.

In 1907, Samuel J. Crumbine, executive director of the Kansas Board of Health, banned the common tin cup on trains. By 1910, a man named Hugh Moore was offering his Healthy Kup on the Lackawanna Railroad. Soon, it was renamed the Dixie Cup. A year later U.S. Envelope, Inc., began selling its Lily pleated cup. Beyond protecting the public health, the paper cup, like the paper plate and disposable utensils that followed, was to sustain new types of restaurants. By the beginning of the twenties, Lily had joined its leading competitor, Tulip, to form the dominant Lily-Tulip Cup Company.

Kroc sold cups to baseball parks in Chicago. At Wrigley Field he met Bill Veeck, soon to become the game's premier showman, and suggested a number of stunts to increase concession—and cup—sales. Veeck was not interested. Kroc was always trying to figure out ways not just to sell more cups, but to create situations in which his customers could use and would therefore need more cups. In the course of this process, he discovered the Multimixer, whose six spindles could mix six shakes at once, and so increase the use of cups. He bought rights to the device and soon began selling it instead of the cups. Promoting the Multimixer, he discovered the operation which was to make him famous: a restaurant in San Bernardino, California, was employing an unprecedented eight Multimixers in a single hamburger stand. The drive-in run by Richard and Maurice—"Dick" and "Mac"—McDonald sold half a million malteds a year.

When Kroc first viewed the McDonald brothers' operation—the two brothers had actually had a previous franchising agent—the whole thing was run by two cooks. A Lazy Susan device toasted two dozen buns at once. "Speedy Service Systems" was the name of the operation, and the concept. (For a while they had considered calling it "The Dimer," an echo of the dime store.) An advertisement in the trade press touted all that the system did away with: all the bother of curb service—waiters and trays hung on car doors:

> Imagine—no carhops—no waitresses—no waiters—no dish-washers—no bus boys. The McDonald system is self-service. No more

Early McDonald's service window, mid-fifties. (McDonald's)

glassware—no more dishes—no more silverware. The McDonald System eliminates all of this: In our model drive-in here in San Bernardino, California, we sell more than a million hamburgers a year.

The model drive-in in the picture looked like a cross between a giant hamburger and a flying saucer, glowing in its airbrushed tones atop glass serving windows with a little semicircular bit taken out of them at the bottom, like ticket booths. "Self-service" was written over the center of three windows. McDonald's symbolized their system with the cartoon figure of Speedee—later dropped due to an unfortunate similarity to Alka-Seltzer's mascot.

After Kroc contracted to franchise McDonald's nationally, the often repeated success story turns—as much as on franchising and real estate—on the development of specialized equipment.

McDonald's' development of special technology became even more intensely scientific—right down to the metal scoop invented by one Ralph Weimer that allows filling the french fry bag in a single pass. James Schindler, a former designer of submarine kitchens, was put in charge of engineering and equipment in 1958. Special air conditioning was developed by the wonderfully named Mammoth Furnace Company to change the air in a stand completely every three minutes.

Ultimately, the aim was quite simple: to cut labor to the bare minimum through the application of machinery, the classic American mechanization story. And the result would be that the customer came as close as possible to serving himself.

☆   ☆   ☆

As SUPERMARKETS GREW bigger, and Ma and Pa groceries suffered, room opened up in the market for a new kind of store—the convenience store. As the service station sold service, the convenience store was selling, of course, convenience. It was also selling coffee concentrated by hours on the burner, microwaveable burritos, and magazines about four-wheel-drive vehicles and body-building women.

Such stores aspired to the vending machine, the ultimate self-service device. An executive at Best Products—the discount catalogue store known for its innovative architecture—once compared its system to that of a vending machine. Samples of goods were displayed in front, in a three-dimensional blowup of the store's catalogue. Customers ordered items which were then delivered from a warehouse in the back by a system of chutes and belts. For such stores, the inner workings remained the same, but the basic and repeated unit of the stores—like a New England clock—could be encased in any number of ways, offering the opportunity for architectural patronage in the exteriors.

Best Products could hire James Wines and his firm SITE to create conceptual works—mock ruins, moving walls, gardens that seemed to penetrate the building—for the shell, without changing the function of the store. Best Products made its founders, Sydney and Frances Lewis, quite rich, although Best would later falter in the face of competition from all kinds of small discounters and expanding chains such as Wal-Mart. They became art patrons: among their collections, displayed in the postmodernist headquarters building they had constructed in Richmond, Virginia, were jukeboxes of the classic era: the vending machine at its most heroic.

☆　☆　☆

SIMILARLY, THE FAST-FOOD outlets encase their basic model in shells drawn from a kit of possibilities. Their signs have shrunk, and their buildings have grown more generic as the task of advertising is increasingly assigned to television. The structures themselves have merely to live up to the appearances on the screen.

Gas stations—designed to feed our cars fast—were also given universal brand looks in the thirties, when the big oil companies hired industrial designers. The best known result was Walter Dorwin Teague's treatment for Texaco, done chiefly by Robert Harper of his office, with his series of porcelain-paneled stations offering a bay for each type of service. Today, self-service stations have replaced service stations. The original service stations, as we all know, never quite lived up to their architecture or to the models in the ads. As stations did less repair and maintenance work, the attendant began virtually to disappear. Self-service gas pumping became dominant, abetted by the energy crises of the seventies, when high prices made discounts for pumping your own more attractive.

Contemporary gas stations, still created by industrial designers, appear virtually overnight, Suprematist sculptures for the roadside, assemblages of kit parts—floating roof slabs, upright simple solids for pumping equipment and only slightly larger for the glass cockpit of the operator, like the cab of a piece of heavy equipment or the toll taker's office. They are barely attended vending machines, environments for self-service.

Gas stations were among the last small-scale retailers to turn to self-service, and the change radically altered their design. But they were among the first to present themselves as a consistent package and model. Early on in their lives, gas stations began to look like temples, miniature courthouses, castles, pagodas, lighthouses, windmills, and Renaissance villas. But, at the same time, they also began to look like giant oil cans, huge boots, ships—and airplanes.

Given that driving was flying, it was no wonder that stations looked like planes or terminals, suggesting both a kind of landing place—almost like an airport, where one put feet on the ground for a brief interval in the course of the long, mobile flight of a car journey—and a vehicle of flight itself, and that brand names like "Skychief" and "Super flite" arose. Mobil's Pegasus summed up the idea: mundane, earthbound transportation transformed into mythological flight.

The gas station, like the highway, had to be invented. From the can of gas tipped into the tank at the livery stable to the pump invented by one Sylvanus Bowser (his name is still the generic title for a portable gas pump), the station evolved from a storefront on Main or Depot Street to a covered shed out at the edge of town.

In the twenties, gas station architecture cast about wildly for stature, precedent, and dignity. Its designers turned, in the case of Texaco, to adobe precedents. In Cambridge, Massachusetts, the Gulf station on the edge of Harvard Square assumed the neo-Georgian look that became the official style of the university in the twenties. Along the parkways in the Northeast, the station was simple, practically a stone hut, solid as the houses of the toll takers on the old National Road through Ohio. The rage for Colonial stations—often based on the New England "saltbox" house, like the later homes in Levittown—became national. In part this was because they could be easily fabricated of standard pre-cut frame parts, clapboards, and shutters. Various companies assumed distinctive architectural styles. Pure Oil specialized in a steep-roofed, Tudorlike station.

At the same time, independent operators were offered a series of off-the-shelf, standardized, metal-and-glass stations that would allow them, at relatively low cost, to look as up to date and standardized as the big boys. One of these—a prefabricated station for Standard Oil (Esso)—was included in the Museum of Modern Art's 1932 "International Style" show, a key event in the propagandizing for modern architecture.

By the thirties, the big oil companies had recognized the key fact about the gas station: that it was a package for their products. The product was not only gas but tires, batteries, and accessories—known in the trade simply as TBA. But above all it was service. Thus the humble "gas station" became the service station.

Service is an abstract and elusive concept, especially for the architect or designer trying to give it physical representation. In the thirties they did so by emphasizing simple, clean, white, curved—

"fast and efficient"—shapes and images. For the TBA goods, they created large windows and stands behind and upon which cans of oil were piled in pyramids and towers. Everything bore the same logo as the rotating sign out front, the pump and even the free road map and the credit card; gas stations were among the first to offer credit cards, handy for the wary traveler, as early as the teens. The sheer architectural possibilities of the gas station, with its requirements of openness, mobility, and technology, drew modern architects. Frank Lloyd Wright designed stations, as did Richard Neutra and Rudolph Schindler.

The history of Texaco's stations, particularly well documented by the company, traverses the entire distance between storefront and contemporary station. Texaco—which began as the Texas Oil Company—first sold gasoline from a storefront. By 1909, it had registered its trademark red star with green T and beneath that sign was selling gasoline from round tile-roofed stations and miniature mission-style ones. The year 1916 saw the arrival of its Oaklawn Filling Station, which was designed by chief engineer E. H. Catlin of the Pipe Line Department in the old mission style, with stucco walls, a roof of Spanish tile, and trademarks of art-glass mosaic in the Alamolike gables. By 1928, Texaco went to a Colonial-style station, and before long to a Mediterranean or neo-Renaissance style.

By the thirties, Texaco was looking for a national image, up to date and non-historical. To this end, it hired the industrial design firm headed by Walter Dorwin Teague. Teague's design is probably the most important gas station design of all time. It was cool and corporate—"icebox" architecture, they called it, after the resemblance of the porcelain-enameled panels to household appliances—and for that reason it was reassuring. The accompanying advertising dared for the first time to speak of rest rooms: the cleanliness advertised by the white architecture of the stations was to be enforced by a traveling patrol of Texaco rest-room inspectors. And soon came the ads with "the man who wears the star," neat and friendly as the buildings themselves.

Not only were new stations built to fit the new image but old ones were updated. Thus the grand facelifts of Texaco and other brands happened by degrees; an older station might have new pumps, a Colonial could add a streamline portico, a Tudor job might take on

eco-sensitive mansarding. The result was a sort of cumulative archeology of gas station design, an incomplete version of the uniform, just like the one commonly encountered on the man who sold
the gas himself—never quite as spiffy as the one in the ad, with his
star, bow tie, and officer's hat.

So influential was the Teague design that by the 1950s most oil
companies deployed stations roughly based on it, and only slightly
updated. At the same time, the new superhighways were making
national image more important than local. To catch the eye of
motorists on the Interstates, stations just off the exit ramps erected
huge tower signs with "modular" letters, like giant Scrabble pieces.

During the sixties, the gas station came under fire for its garish-
ness and disharmony with neighboring buildings. The mixing of
successive architectures and the effort to stand out amid the strip-
side crowd grew too much. In 1966, Lady Bird Johnson, as part of
her drive to beautify America with trees, shrubs, and wildflowers,
summoned the heads of major oil companies to the White House

Texaco Type "A" service station prototype, Walter Dorwin Teague
Associates, designers, 1936. (Texaco)

for a dressing down. The result was a major shift in station design. Colonial stations returned, and some stations began to look like the ranchhouses near which they were located. Older stations were fitted out with little mansard roofs of shingle or shake (or at least plastic replicas of those materials). The hard-sell colors of orange and red gave way to "earth tones" of brown and beige. For Texaco, the sixties brought an unconvincing modernist prototype called the Matewan, with U-shaped concrete roof. The "banjo" sign gave way to a flattened hexagonal that deemphasized the Texaco T. That the stations were not entirely enthused about the change was reflected in the result: wood-grained plastic shingles might end up in electric blue or orange, the old colors in an uneasy compromise with the textured demands of "naturalism."

Mobil reacted differently. It hired Eliot Noyes (in tandem, significantly, with the creators of Mobil's graphics, Chermayeff & Geismar) in the mid-sixties, and the result is still a feature of the landscape, with the near timelessness of the best modern architecture. Mobil softened its modernism with brick and stone, and used indirect lighting to lighten standardized metal-and-glass surfaces. Noyes, who as IBM's house designer created the classic Selectric typewriter, represented the change from the flamboyant industrial design of the thirties and forties to an American corporate version of the International style. For Mobil, he took the familiar Pegasus down off the tall sign (often found offensive by the sixties beautifiers) and applied it to a glowing white disc on the building itself. And at the pumps Noyes outdid himself: they became floating metal-and-glass cylinders, neat as Bauhaus lamps, stationed beneath round, umbrellalike roofs.

The stations of the eighties reflected the rise in self-service, from 20 percent of the market to about 70 percent. First emerging in the independent "gaseterias" of the immediate postwar years, self-service was fought hard by the big brands. But rising costs in the seventies brought self-service back to stay. The energy crises reduced the number of stations dramatically. In 1972, there were 226,500 stations in the United States, the most ever; a tumultuous decade later, the number had fallen to 140,000. It was the toughest time the gas station had ever faced. Those days still give the petroleum folks the shudders. "The events of the early seventies" is the closest that one recent Texaco publication can bring itself to acknowledging them outright.

In addition, the rise of such service chains as Midas or Jiffy-lube, and the need for less lubrication and brake work on modern cars, reduced the service side of the service station; grease pits with their lifts began to sit unused, and new stations were often built without them. This trend and the energy shocks were followed by stricter environmental restrictions—vapor-capture devices and leakproofing of tanks are now mandated in many localities, and the financial burden of installing them has driven many small stations out of business.

Today's stations are sculptural compositions of miscellaneous solids, rendered in luminescent plastic and metal panel. The new generation of stations, with their corporately responsible Helvetica lettering and straightforward geometries, shows the extent to which the gas station is an expansion on, an essay about, the embodiment of the corporate logo.

The axes of Exxon's drive-in lanes cut beneath the rectangle of the roof like the slashes in the X's of the logo itself. The new stations are assemblages of boxes—convenience store, car wash, rest rooms, sales booth—arranged like the pieces of Frank Gehry's houses.

In fact, of course, they are designed by corporate industrial design firms, with heavy emphasis on logo and image. Created on a napkin during a Palm Springs vacation by Raymond Loewy, the Exxon logo is known worldwide. It is also, claims the cultural historian O. H. Hardison, the second most famous "concrete poem" (after Robert Indiana's LOVE logotype) in the English language. It is a word designed for its looks, with little history and dubious meaning.

At the beginning of the eighties, Texaco went to another industrial design firm to create its new-generation, mostly self-service black and red stations. Kenneth Love of Anspach, Grossman, Portugal—the corporate image and design firm that did Texaco's "Concept 2000"—says that the key idea was "the station is a package, but one that's too big to put on a shelf. The shelf in this case is a street or highway."

The packages were prefabricated "modules," produced by the Beaman Corporation in different shapes and sizes to make up what Texaco executives liked to call a "family-of-buildings configuration": a gas sales-office box with pump islands beneath a "canopy," and a separate rest-room box, might be accompanied by a convenience store at one location, and a service portion or car wash at another. All of them were black metal boxes with standardized sans-serif lettering.

Like the Wheaties box Love invoked, the Texaco station was a package that involved an existing image, a tradition, a brand recognition not to be easily discarded. Thus, after consumer surveys showed that the star was still the image the public associated most consistently with the company, Love "extracted" the Texaco star from the "hex logo" introduced in the sixties and went back to images as old as the century. The T was put back inside the star, and the simplified trademark dramatically rendered in red on black. All this, Love admitted, was in part "subliminal, much as . . . the way you dress."

If the employee inside the station rarely wore the star anymore, the building itself would be a star, the dramatic costume of the brand. "The Star of the American Highway" runs the slogan banded across one black beam of the station. The stations were a theater of boxes carefully blocked on the stage, vending machines dancing in place.

CHAPTER ☆ THIRTEEN

Rural Electrification, or Eclectic Guitar

The Mississippi Delta was shining like a National guitar.

—PAUL SIMON, "GRACELAND"

*Electric guitar is copied, the
copy sounds better.
Call this law and justice, call
this freedom and liberty.*

—DAVID BYRNE, "ELECTRIC GUITAR"

THE BEST-SELLING RECORD and the nearest thing to a hit recorded by Robert Johnson, the greatest of the Delta blues singers, was called "Terraplane Blues." It celebrated a powerful, fast, and not too expensive car. When the song was released, in 1934, the most streamlined thing about the Terraplane was its name; but within a few years it had been redesigned to rival the Chrysler Airflow and the Lincoln Zephyr.

In Johnson's music one can see mechanical imagery, always personalized and sexualized, invading and cohabiting with "folk music" and images that go back to hoodoo and African tales of the devil figure Legba. Johnson's songs include references to Elgin watches, Greyhound buses, and the phonograph itself, in addition to the universal blues vocabulary of railroad and riverboat, crossroad and open highway. "Terraplane Blues" and "Phonograph Blues" are tokens of the way that American popular music, the area of some of the country's most vital cultural achievements, was recorded and distributed—and shaped and changed. The intersection of folk and an ancient vocabulary of images, characters, and tales with modern technology produced a new kind of music, eclectic in its images and styles, swirling with influences and bright with innovations.

"Don't shoot the pianist; he is doing the best he can," ran the famous sign Oscar Wilde noted in a Colorado mining-camp saloon. It was, he dryly commented, the most complete and succinct statement of artistic criticism he had ever heard. It was also a call for the appreciation of the vernacular, the amateur, the popular in music. But if democracy was the backbeat of the piece, then the burden of the melody was the amplification of individual passions to public dimensions—not just to audiences of auditorium size but to the scale of those for records and radios. By mechanical imitation and expansion, amateur music became commercial music. The expansion of the audience was to be the driving force behind the development of characteristically American instruments, from the banjo to the electric guitar, and the means of recording and distributing their music.

But Wilde's encounter with the piano player in Leadville could not have prepared him for the player piano—the mechanization of music that stands for the mechanization of many arts in America—nor for that matter for the phonograph. Would he have approved a sign that read "Don't shoot the player piano; it is doing the best it can"? Shortly after Wilde's visit to Leadville, the player piano, directed by rolls of paper punched with holes like a continuous IBM card, had become a feature of American life, a mechanization of art that was vital in spreading such new types of music as ragtime. The first Nickelodeons were coin-operated player pianos, before records and the jukebox.

The jukebox is a machine that sells all by itself, a radiant vendor of experience. It was the quintessential self-service machine, selling the service itself, not even a product, like a can of Coke to drink, but a set of sensations: a beat to dance to, or to drink to.

The arrival of the jukebox stands for the process by which popular music became mechanical, then electric, then electronic, the way a handmade thing changed when made by machine. Robert Johnson got his start, won his fame, and sang his "Terraplane Blues" in the juke joints of Mississippi, but his records ended up in jukeboxes. He achieved virtuosity on the acoustic guitar, played with a bottleneck or metal slide, but was rumored, shortly before his death, to have been experimenting with an electric guitar—the chosen instrument of his heirs, Muddy Waters, Elmore James, and T-Bone Walker, and of his rock-'n'-roll admirers.

The procession from the juke joint to the jukebox was a characteristically American process of turning popular art—a virtual folk art—into the commercial product, with both causing mutual changes in the other. Think of it as cultural backwash, or interference pattern, the simple and sophisticated intersecting in a moiré of wave patterns. In the thirties, the electric guitar and the jukebox (sometimes then called the "piccolo") together arrived in the juke joint. With these new devices and new objects, the center of American popular music changed.

It was no accident that before he created the type of the standardized motel room, the no-surprise, self-service product called Holiday Inn, Kemmons Wilson was the Memphis distributor for Wurlitzer. His career began in the early thirties with a popcorn machine for theater lobbies and five pinball machines, before he moved into jukeboxes. Wilson combined what he learned in these businesses with techniques learned from another, tract housing, to invent Holiday Inns, understanding the motel as a kind of vending machine. From a good motel, the traveler wanted a product as predictable as a bottle of Coca-Cola or a Baby Ruth or a Sinatra record, nationally uniform and at a low fixed price. It was more than legend that Ray Charles, the blind singer, preferred to stay in Holiday Inns while on tour because their standardized layout made his life easier. (Wilson founded the first Inn in Memphis, then sold his first franchise in Clarksdale, Mississippi, the heart of blues country. Today, his new line of Wilson World Inns includes one across the street from Elvis Presley's mansion, Graceland.)

To purists, the jukebox was also the most notorious example of gaudy, functionally insincere commercial design, worse than the streamlined toaster or Deco radio. It was like a model of the little movie theater on Main Street, circa 1935. In style, the jukebox evolved; but for all its lights and buttons, it tended to look back to early Art Deco and even Art Nouveau. "Jukeboxes on wheels" was Raymond Loewy's 1955 epithet for the chrome-laden American cars of the time. But jukeboxes themselves mimicked auto styling; their fronts sported twin red lights and chromed wedges clearly derivative of tail fins. They were Cadillacs without wheels.

The evolution was based on one of the little recorded or celebrated breakthroughs of the century: the development of the electrically driven paper speaker around 1920. Without it, the modern radio, the jukebox, the electric guitar would have been impossible. Before it, speakers had been strictly mechanical, like ear trumpets for the nineteenth-century hard of hearing. After it, the jukebox, like the radio, could take on all sorts of shapes. And the music, exposed to wider influences and larger audiences, could take on new shapes as well.

The ancestors of the jukebox, the Nickelodeon and other public phonographs, preceded the private phonograph. It took nearly twenty years after Edison's development of a tin-foil phonograph, in 1878, for the machine to be commercialized. Edison's method was challenged by the disc system of Emile Berliner and others, and he resorted to the wax cylinder of his "perfected phonograph" by 1888. By 1912, Edison was forced to offer the disc mode of machine (his "Diamond Disc" line) side by side with the original cylinder one, which lasted until the end of the twenties.

The public Nickelodeon had arrived by the nineties. These first coin music boxes, like the first radios, used eartubes like musical hookah pipes. In Brooklyn by the turn of the century there were some 2,000 such nickel machines.

World War I saw the development of cheap, portable mechanical phonographs, wind-up models driven, like clocks, by springs; by 1920, they were common even in the poorest homes. But then came radio, with its free music, whose arrival had reduced sales from 100 million records in 1921 to only 2 million in 1933. To create new markets for records, and especially to capitalize on areas and social segments without radio, the record industry began to scour the hinterlands for music to sell to blacks and poor whites, with their mechanical phonographs, and, soon, to stock jukeboxes. The result was a mad rush to record "true folk" music of all types. At the same time that folk art was winning the affection of art lovers, the record scouts were looking for folk, race, and ethnic music for labels such as

Okeh and Black Swan. The "black belt," music scouts noted, constituted an audience of 100,000 people—the equivalent of a metropolitan market, and one still largely without electricity or radio. The
record industry's commercialism was resolutely eclectic. Vocalion
recorded Zeke Williams and the Rambling Cowboys shortly after
Robert Johnson's later session, in Dallas, above a Buick showroom.
Both were aimed for jukeboxes.

Already the combined effects of the jukebox, the phonograph,
and the radio were changing much of the old music, melting it down
in the vacuum tube, flattening it in the acetate press. For the singers
of such music, mountain balladeers and blues singers alike, simply
to hear themselves on a record was a profoundly disturbing and
exciting experience; to hear others on record was almost as much
so. Many of the singers were not even literate, and in rural areas the
phonograph was still surrounded by superstitions. Even the most
primitive Mississippi Delta blues singers such as Charley Patton
listened to Mamie Smith and Bessie Smith records—and to their
own, which fascinated them and made their styles self-conscious.
Now they had to sound as good as their records.

A key moment of symbolism here was the legendary record
producer John Hammond's vain search for Robert Johnson. Hammond, who had produced Billie Holiday and would discover Bob
Dylan, had already heard Johnson on record: Vocalion engineers
had brought Johnson to the Blue Bonnet Hotel in San Antonio for
a session, and later to Dallas, with an eye to creating records for
the Texas jukebox market. John Hammond happened to own a
Terraplane, which, in the spring of 1938, he drove south from
New York, scouting talent to appear at what became the famous
"Spirituals to Swing" concert of December 1938. In North Carolina he dug up such folk blues masters as Blind Boy Fuller and
Sonny Terry. And from the Terraplane, he also sought the creator
of "Terraplane Blues," Robert Johnson, finally dispatching scouts
to Mississippi.

Hammond's scouts scoured the backroads Delta for Johnson.
They arrived only in time to hear of his death in August 1938—his
whiskey poisoned in a juke joint by a man whose woman Johnson
had seduced. Two of the records made by the creator of "Phonograph Blues" were played on stage at Carnegie Hall that December
in his stead, in a cruel variant on the old "How do you get to

Carnegie Hall?" joke. Not finding Johnson, the scouts brought back Big Bill Broonzy. Soon another group of musical prospectors, from the Library of Congress, latched on to McKinley Morganfield, no mere cotton picker but a tractor driver who was to become Muddy Waters, the creator of records played over and over again by Keith Richards and Eric Clapton and Duane Allman—the future popular guitar virtuosos. Morganfield, who idolized Johnson, also ran a juke joint, with a jukebox.

It was not until amplification and paper speakers, at the end of the twenties, and the demise of Prohibition that the jukebox came into its own. It sold booze: music meant dancing, and dancing meant thirst, which meant drinking. By 1939 there were 225,000 of them—enough to save the record industry from radio. Jukeboxes created a tremendous demand for records, to fill them and to satisfy the home sales they inspired. To feed the boxes, record sales boomed; they rose back to 33 million in 1938.

It was in the thirties (Wurlitzer was founded in 1933) that the jukebox took on its familiar form and name—from the black "juke joint." (The word "juke" goes back to the African *Wolof* for "unsavory." This was appropriate; the original juke joint was a locale for music, moonshine, dancing, and prostitution.) The jukebox had to live up to the romance of the music it played. In the film *Out of the Past*, Robert Mitchum stands in front of a jukebox in a small-town luncheonette. It is one of those classic Paul Fuller–designed Wurlitzers, a 1946 model 1015, which also shows up in *On the Waterfront*, framed in an arch of bubble tubes. The bubbles themselves seem to drift up from the past—mysterious, intriguing, and faithless as the gangster's moll who used to be his girlfriend. This is likely the reason jukeboxes have become so fervently collected in recent years. For many collectors, they represent the physical equivalent of the old songs they play. The Wurlitzer 1015 today sells for around $15,000 a copy; a replica model, made in Germany, was priced in 1990 at $4,000.

If mechanization threatened the old forms of popular music, it could also enable them to survive. Just as electrifying the guitar was far less dangerous to the old songs than the ministrations of piano and orchestra rearrangements, recording served to define a widely distributed body of music newly eclectic in its range and influence.

Most blues singers were street musicians who also played in juke joints. Their repertory was necessarily vast, including not only blues but Tin Pan Alley and hymns. Robert Johnson was said to play the songs of a founder of country and western, Jimmie Rodgers, "the singing brakeman," who developed his "blue yodel" and other features of his style in part from black songs learned from railroad workers. Johnson commonly played in the street, and once, after his guitar had been smashed, he sang and played the harmonica on the sidewalk until he raised enough money to buy a new one. That guitar, a metal resonator model, he would take to plantation dances and informal clubs, and play for contributions. And by the end of

Wurlitzer 1015
jukebox, circa
1946, Paul
Fuller, designer.

his short life, not only were his records played in jukeboxes, but the jukebox had begun to take over the juke joint.

☆   ☆   ☆

WHEN BOB DYLAN appeared on stage at the 1965 Newport Folk Festival playing an electric guitar, his anthem "Like a Rolling Stone" following on the heels of a singer called Cousin Emmy and her rendition of "Turkey in the Straw," the folkies felt he had committed an unforgivable sacrilege. Folk was pre-mechanical, pre-electric, in their view. It was a legacy of Elizabethan England, by way of the Dust Bowl, preserved in the minds of its listeners. Folk music was to be collected, found, and recorded, as Alan Lomax had done. Lomax actually got into a wrestling match with Dylan's manager, Albert Grossman—a literal rendition of the struggle.

For Dylan, as for the best American musicians of all types, there was nothing sacred about the folk forms he admired, from the Delta blues of Robert Johnson to the songs of Woody Guthrie. They were just a beginning, models of emotion and effect, providing a kit of parts to build from, not simply collectibles like old weathervanes.

Those demanding the pure and naive in American music would search in vain. At its best, American music was improvisational and eclectic, cross-influenced, and above all shaped by the possibility of each new mechanical means of expanding its audience that came along. For Dylan, an instrument as "modern" as the electric guitar, in tandem with an instrument as simple as the harmonica on the wire holder around his neck, was a way to project his power further. It would expand the audience for his records in the same way that it physically expanded the size of his potential live audience. But it was also a symbol of the movement beyond the notion of a music still "folk," to one that was both popular and commercial.

☆   ☆   ☆

"PURE" FOLK MUSIC in the 1950s and 1960s revived the banjo, and turned it into a sort of cliché for the beatnik and then the hootenanny world. The banjo is a direct descendant of African instruments and their Caribbean offspring—the African banjar and Caribbean bania. An often reproduced painting from the Rockefeller Folk Art Collection shows a banjo in use on a South Carolina plantation. Benjamin Latrobe described seeing the banjo being played by a slave during his travels through the South.

Thomas Jefferson, one of Latrobe's correspondents, had correctly associated the banjo with the African banjar—observing, along the way, that blacks seemed to be especially endowed with the sense of rhythm—but after freedom, the banjo was associated with minstrel shows. Black folk legend—adapted, like the instrument itself, from African culture to that of American slavery—associated the invention of the banjo with the opossum, whose hair was said to have supplied its strings (the possum's skin might literally have been employed). There was a key association here: the possum was the food of the poor, white and black alike. And like the black slave playing subservient, the possum played dead when threatened.

But the banjo also became a cliché of the happy darkie under the Spanish and live oaks. By Reconstruction, few blacks wanted to be seen playing one. It was left to the white performers of the minstrel show—which rigidly banned blacks from the audience as well as the stage—to keep the instrument alive in a parody of the original art and artists.

And, in an ironic reversal, the banjo was to become a white instrument, while blacks began to adopt the guitar and adapt it to make it their own. White mountaineers built their own sorts of banjos from local materials. Eventually, the only people playing banjos were Southern mountaineers and the college-educated bluegrass aficionados who admired the "Elizabethan purity" of the mountain songs and tones—the audiences for Bill Monroe and Doc Watson and Flatt and Scruggs.

☆   ☆   ☆

In Europe, the guitar figured as a Mediterranean instrument, admired and played in the north as a sign of the warmer, classical clime. The guitar in Jefferson's music room was rarely played except by "French people," his slave Isaac reported. But in the United States, the dominant name in the history of American acoustic guitar making was the Martin Guitar Company, founded by C. F. Martin. A Saxon immigrant fleeing, the story goes, the guild restrictions that allowed only violin makers to make guitars, and to cash in on their rising popularity, after a stay in New York, Martin established the factory in Nazareth, Pennsylvania, that continues as the company's location.

In Texas, the guitar flourished as the German community established itself in the eastern part of the state. Around New Braunfels, where religious refugees and refugees from the turmoil of the 1840s had built a thriving community, Germans encountered Spanish influences from the Southwest. Surely no stranger musical combination was ever mixed than the waltz and polka with the romantic twang of Spanish guitar.

The ukulele, a legacy of the 1915 Panama-Pacific International Exposition in San Francisco, became a fashion through the twenties. Hollywood picked up on the ukulele, and also on the Hawaiian style of guitar playing, which, according to the music historian Nick Tosches, is an invention properly attributed to Joseph Kekuku, a student at the Kamehameha Boys School in Oahu, about 1894. Hollywood and radio brought the Hawaiian steel guitar to even the humblest country picker. One of Bob Wills's steel guitarists was first inspired by the music in the Bing Crosby movie *Song of the Islands.*

The rise of the guitar as a black instrument corresponded to the origin of the blues. Homemade instruments compounded of cigar boxes or lard cans were often the first instruments owned by many blues singers. For both rural blacks and whites, the mail-order catalogues quickly became a main source of guitars.

Even cruder was the first instrument played by Elmore James, B. B. King, and other blues greats. This was the "one strand." A descendant of the single-string instruments of West Africa and Brazil, it consisted of a length of metal wire—often borrowed from the binding of a household broom—drawn taut from a board or the front of a house. William Ferris describes these still being made in the Mississippi Delta in the 1960s. One that he saw strung on the front porch of a shotgun house turned the whole house into the resonator for the single steel string. A portable version of a similar primitive instrument was called the bo-diddley—a name that was to be borrowed for the stage by one of the pioneers of rock-'n'-roll.

The one strand—a primitive steel guitar, if you will—offers the high metallic sound that was to become associated with the blues and the bottleneck or slide guitar playing of the blues. For all the research into field hollers and levee, railroad and prison songs, the origin of the blues remains mysterious. One thing seems clear, however: the development of the blues corresponds neatly in time with the post-Reconstruction "redemption," the clamping down of Jim Crow laws on Southern blacks, and the solidification of the sharecropper system into de facto servitude.

By the 1890s, the guitar was being played in a new way. W. C. Handy first recorded hearing the bottleneck, Delta blues guitar in the 1890s—a music, he said, that he overheard in a Tutwiler, Mississippi, railroad station "as if in a dream."

> A lean, loose-jointed Negro had commenced plunking a guitar beside me while I slept. His clothes were rags; his feet peeped out of his shoes. His face had on it some of the sadness of the ages. As he played, he pressed a knife on the strings of the guitar in a manner popularized by Hawaiian guitarists who used steel bars. His song, too, struck me instantly.
> "Goin' where the Southern cross the Dog. . . ."*
> The singer repeated the line three times, accompanying himself on the guitar with the weirdest music I had ever heard. The tune stayed in my mind.

Handy had encountered the Delta folk or country blues, some of whose proponents were soon providing his string bands with competition at local gatherings in the Delta. Later, touring with his band, he watched a local blues group perform music he knew from workers in field and levee and noted that "the dancers went wild." He noticed something else: the "rain of silver dollars" the grateful dancers tossed to the players. "There before the boys lay more money than my nine musicians were being paid for the entire engagement. Then I saw the beauty of primitive music."

Handy was to soften the Delta blues with the addition of elements from the minstrel and stringband tradition and ragtime that constituted the mainstream pop music of the day. Years later, in 1914, also the year of his "St. Louis Blues" and "Memphis Blues," Handy created his own "Yellow Dog Blues"—a popularized version of the original

---

* The Southern Railway crossed the Yazoo Delta, known colloquially as the Yellow Dog railroad in Moorehead, Mississippi.

made "thematically coherent." "Their music wanted polishing," Handy said of the Delta blues singers, "but it contained the essence."

☆   ☆   ☆

THE BLUES ARE like a chair, John Lennon once said, and they live on in rock-'n'-roll. They can be simple, like the chair in a sharecropper's cabin, or upholstered in an electric rock-'n'-roll production, like a big Grand Rapids easy chair. They are a traditional form on whose frame many changes can be rung and much mechanical upholstery hung, whose structure and function remains the same however new the arrangement seems. When the Rolling Stones sang Muddy Waters, the blues were still there, beneath the electric bass.

The key to the development of the blues has been the arrival of the slide or bottleneck guitar. The bottleneck method made the guitar more plastic—and enabled it to handle the mournful and tragic emotions of that most egoistic music, the blues. "To escape from the too-rigid diatonic scale," writes Frederic V. Grunfeld, historian of the instrument, "the blues guitar had first to be converted into something like a bania, capable of keeping the 'blue-notes' (third and seventh of the scale) hovering with twanging microtones between major and minor. This was accomplished with the simplest of devices: a bottle neck, or a metal slide or pocket knife, applied to the neck with the left hand." Often, the sound obtained replaced a word that was deemed obscene or subversive.

The complaint of the slide guitar, its sound hovering with those twanging microtones, gives the blues their blue notes, lending its characteristic tone to both the Delta blues and the "white" blues of early country music. The white equivalent of the blues can be traced to Jimmie Rodgers, who mingled with blacks in the railroad yards where he worked before his "Blue Yodels" became hits. Hank Williams first struck it big with his yodeling "Lovesick Blues." (In the film version of his life, *Your Cheatin' Heart*, the young Williams is supplied with a venerable black "uncle" who conveys the blues tradition to him.)

The blues attracted intellectuals by their strange juxtapositions— the extemporaneous recombination of known lyrics so that the result bordered on surrealism. When Carl Van Vechten presented the blues to the sophisticated readers of *Vanity Fair* in the mid-twenties, it was in the same way that the magazine presented the strange new art of Europe or primitive African masks.

☆   ☆   ☆

THE ELECTRIC GUITAR grew above all from the need to amplify the sound of the guitar so it could hold its own in noisy clubs and, later, against the horns of jazz bands. At the same time it amplified popular and subcultural music, and gave voice to elements of society outside the mainstream—poor blacks and poor whites.

Common belief credits the Belgian Django Reinhardt and the American Les Paul with inventing the electric guitar. But the instrument was not the work of any single individual. The first attempts at amplification were mechanical: adding various types of metal resonators to the instrument. Los Angeles, of all places, turned out to be the most fertile ground for guitar innovation. John Dopyera, a Czech immigrant, moved there in 1908, when he was eighteen. His brothers Rudy and Ed later joined him and in best melting-pot fashion, they responded to jazz and country-and-western customers with the invention of the mechanical amplified guitar. In 1929, the Dopyera brothers patented the acoustically amplified or "ampliphonic" instrument they called the Dobro—for Dopyera brothers. The instrument offered a spun aluminum resonator, placed over the soundpiece, that looked like a pie plate cut out in Pennsylvania Dutch patterns. Before long, apparently with the technical assistance of an employee called Rickenbacker, the Dopyeras had turned out a full-scale electrical guitar—that is, one with an electrical mode of picking up the vibrations and conveying them to an amplifier.

The guitar, more than the banjo, lent itself to electrical pickup and amplification and its primary role in country-and-western music, in rhythm and blues, and in jazz. Recording, too, helped make the guitar more popular, especially after the introduction of the electrical process of recording in the mid-twenties. The electrical process, so the blues scholar David Evans reports, was "much more favorable for recording the guitar. Formerly the only kind of guitar music to be recorded extensively was Hawaiian guitar playing, which used a special technique for achieving a penetrating tone that could be picked up cleanly by the recording horn. Finally, the electrical process enabled the companies to use multiple microphones and portable equipment that could be easily transported to temporary studios in the South."

The arrival of the phonograph, the radio, and the electrified instrument all worked to the same end. The mechanization of music set popular music on the same footing as high music and created an atmosphere for the one to feed off the other. In West Virginia mining company towns or North Carolina textile mill towns, you would pass the little shotgun houses and hear the blues records from side-wound phonographs. Zora Neale Hurston talked about the intrusion of "the mechanical, nickel phonograph" into the turpentine, logging, and phosphate camps of rural Florida. Rural blues singers at first imitated the records they heard on the jukebox, then resented the music as it replaced them in the juke joints.

The music changed the guitar, but the guitar changed the music too, of course. The arrival before 1920 of the first blues records—the first hit among them is generally agreed to be Mamie Smith's "Crazy Blues," which sold 75,000 copies at a dollar each in the first month after its release—in effect froze in place the folk and vernacular sides of the music. From now on, the musician was influenced not only by other human musicians but by their mechanically distributed music. Local singers did not hesitate to imitate or interpret the songs they heard on the jukebox. Eventually, the machine—cheaper, more flexible, more reliable—would replace the live musicians.

The guitar transcended the differing branches of American popular music. It served as a blending ground for music of different nationalities and styles. But the electric guitar finally found itself most at home with music conceived less for live performance than for recording. It was the cousin of the phonograph, which in turn began as democratic device, do it yourself, cranked by your own hand. The principle remained the same, even as the phonograph was successively superseded by the technological layers of high-fi, stereo, and laser disc.

The key date in the creation of the modern electric may be April 10, 1956, when Clarence L. (Leo) Fender received U.S. patent 2,741,146 for his "tremolo device for stringed instruments." This was the heart of Fender's Stratocaster, the most famous electric guitar of them all. Equally significant was Fender's electric bass, which was to power the backbeat of rockabilly, rhythm and blues, rock-'n'-roll, and soul.

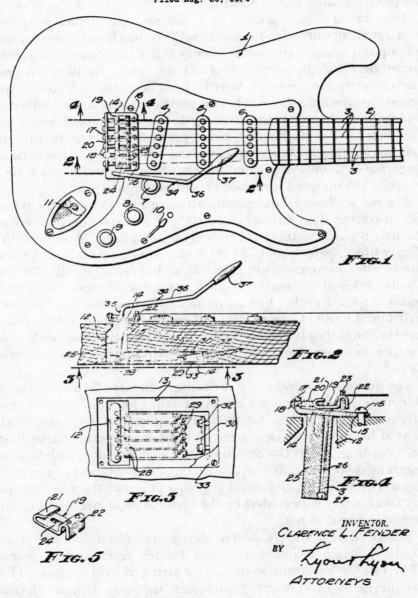

April 10, 1956 C. L. FENDER 2,741,146

TREMOLO DEVICE FOR STRINGED INSTRUMENTS

Filed Aug. 30, 1954

FIG.1

FIG.2

FIG.3

FIG.4

FIG.5

INVENTOR.
CLARENCE L. FENDER

BY

Lyon & Lyon

ATTORNEYS

Fender electric guitar tremolo device patent, 1956. (Patent and Trademark Office)

In country-and-western music, the whining, high-tension pedal steel guitar, a vital part of the commercialized sound of the late forties on, joined the traditional guitar, but abandoned its shape. Beginning with the influence of Buddy Emmons, steel player for Webb Pierce, the Nashville star whose guitar-shaped swimming pool is a Nashville landmark, pedal steel became a vital part of the characteristic sound of country-and-western music after World War II. Emmons and Bud Isaacs developed their Sho-Bud guitar into the most popular model. This highly mechanized guitar, now set as flat as a piano or Hammond organ, corresponded to the mechanization of the old country ways.

An electric guitar no longer needed to be shaped like a guitar. Les Paul's first "solid body" was called "the log"—a solid wooden model of a guitar. Soon all sorts of other shapes appeared. A few of these originated in "organic" or ergonomic needs. Guitar backs were hollowed out to fit the player's rib cage. But most were aimed at giving a futuristic, "space-music" look to the instrument.

The shapes and appearance that the electric guitar took on in the fifties were like those of customized cars. Freed of its function as a

Fender flat steel guitar. (Collection of Country Music Hall of Fame and Museum)

resonator, the body could take the most theatrical of forms. The original guitar shape served as something like the stock bodies the car buffs customized. These guitars were sprayed with Duco paint, sanded and sprayed again, then glitter was added. The candy-apple look of a fifties guitar was like that of a customized '49 Mercury. One guitar of the era sported what was described as "a flush mounted, teardrop body scoop that looked like it came off a Buick." The look of the guitars was perfectly appropriate to their hot-rod sounds—the reverberation and feedbacks. Before long, there was no "natural" look to a guitar at all. Like a radio, its shape had been completely separated from its function, which was now electronic.

Despite their mechanical nature, electric guitars quickly became as much objects of their owners' affections as traditional ones. Folk singers had attributed an almost mystic power to their instruments. "This machine kills fascists," was the inscription Woody Guthrie added to his acoustic during the forties. Rock-'n'-rollers emphasized the power of their instruments with the synonym "ax." Blues singer B. B. King, heir to the Delta blues tradition of Robert Johnson and Charley Patton, named his guitar "Lucille," much as Daniel Boone had named his rifle "Old Tickflicker." In its projections, the guitar writ large the personalities of the players and turned them into stars. Against the democratic backbeat, the theme was of exaggerated individualism, the popular hook the attraction of glory, wealth, and women.

Guitars became part of the stage pantomime. Elvis Presley could barely play his guitar, his biographer Albert Goldman reports, but he nonetheless kept it as a necessary prop. Willie Nelson's acoustic Martin was electrified with the addition, he said, of a 1951 Baldwin electric pickup and just enough spilled beer. Eroded by the action of a pick, it became a relic, covered with the autographs of performers Nelson had admired and worked with.

The instrument seemed to lend its wielder a special heroism. One of electric guitar's first heroes—its Keats, perhaps—was Charlie Christian, Dallas born, signed on by Benny Goodman, dead of tuberculosis like a Romantic poet before age twenty-five. He used the electric guitar to lift the instrument to a level where it could compete with jazz horns. Christian's ambition, he said, was to sound like a saxophone.

With the idolized virtuosity of a Jimi Hendrix, who sometimes played the electric guitar with his tongue, the instrument's sexual overtones were made explicit—although electrification had long since removed the feminine shape, as celebrated in European art from the Dutch, who employed it in the iconology of the panderer, to Picasso, who exploded it in cubism. Eric Clapton, the great guitarist who melds early American blues (numerous versions of Johnson's "Crossroads") with hard rock inspired by the English working-class world, describes how once, under the influence of LSD in San Francisco in the sixties, he imagined that his guitar had turned to rubber. It was an image that summarized the plasticity of the instrument and the music it played. Subjected to the assaults of the electric guitar, basic blues stretched and returned, expanded and changed shapes.

☆   ☆   ☆

PURPLE RAIN IS a contemporary juke joint on Highway 61, near Duncan, Mississippi. The building is a low, wide, windowless shed, which (except for a door made of the metal of tractor trailer bodies) is painted a dull green—the traditional color of the Delta juke joint. And, inside, its center is the jukebox.

The name is taken, like the names of many juke joints, from the wider musical culture—Disco Lounge, etc. Specifically, it comes from the song and film of Prince, the Minneapolis-based omni-media star, who has long contemplated a film based on the life of Robert Johnson, in which he will himself play the blues legend. The blues gone purple, like prose, became rock-'n'-roll and big time, big screen.

CHAPTER ☆ FOURTEEN

The Dialectics of Dude, or the Mechanical Rodeo

BEFORE THEY WERE jeans, they were overalls. Before they were universal, they were fancy. Before they turned fancy again, they were workaday. Blue jeans showed how even the simplest, most functional everyday item can be turned into a figure of fashion, part of a kit of class, a shifting symbol that bounces between sophistication and nostalgia. The basic structure of the garment remained the same, but the structure of its associations shifted behind it. The evolution of the blue jean, from working garment to fashion object, was remarkably like the development of the blues song into the rhythm-and-blues record.

While the students of 1968 tore up the streets of Paris in blue jeans, inside the Collège de France Claude Lévi-Strauss continued work on his studies of myth and totemism. To track the changing totemism of Levi Strauss & Company would require the efforts of Claude Lévi-Strauss himself.

Jeans are an almost promiscuously associative design, leaping class and status associations, combining *nostalgie de la boue* and designer snobbishness, counterpointing the individuality of shrinkage and wear with the uniformity of standard issue. Everyone thinks he knows why we wear jeans: they are comfortable, functional, uniform. They shrink to fit, they wear to familiarity, they are noncommittal, off the shelf—and all-American. Certainly they have their practical virtues. Margaret Bourke-White donned jeans when she took pictures in steel mills in the late twenties. Martin Luther

LEVI STRAUSS COPPER RIVETED OVERALLS

LEVI'S ELECTRIC RODEO — THE TALK OF TREASURE ISLAND

Promotional postcard, Levi Strauss Electric Rodeo, Golden Gate International Exposition, 1939. "A 100% mechanical rodeo. It moves. It talks. Its figures are all hand-carved likenesses of famous rodeo people. And they're all dressed in authentic Western togs . . . miniature replicas of garments made in California since 1853 by Levi Strauss & Company."

King, Jr., put them on anticipating his incarceration in Birmingham jail.

But the story, wrapped in advertising myths, is more complicated than that. This most American of garments bears an Italian name (*jean* comes from Genoa), is made of cloth named for a town in France, and owes its origins to a Latvian immigrant's patent, sold to a German merchant who made it world-famous.

Levi's promotion began with the famous stunt of demonstrating the strength of the jeans by attaching a horse to each leg. That demonstration has been reenacted, using (depending on how much credulity you are willing to grant) locomotives and/or Jeeps. One story features workers at a logging railroad, confronted with a broken juncture pin, tying a locomotive to the log cars with a pair of the pants. There are more recent claims of twin Volkswagen Beetles playing the role of the horses. And of course the jeans never give way. It was one of those Southwestern tall tales, always growing and changing, always one step ahead of verification, turned into a trademark, and the forerunner of all those Samsonite and Bulova abuse advertisements.

In the 1930s, Levi Strauss & Company promoted its blue jeans with a mechanical rodeo in which clockwork figures went through the motions of cowpunching and bareback riding. It was an appropriate way to present the clothes: like the rodeo, which was cowboy work turned into sport, Levi's were work clothes turned into sports clothes. And they were as stylized as the calf roping and bull riding of the rodeo. That is, they bore the same relation to the actual work that the rodeo events did. Levi's' mechanical rodeo was also an appropriate promotion device for pants whose innovation was mechanical (not the denim, but the rivets), which made them "metal pants."

How many characteristically American items of clothing have been discovered on the range, in the shop, at the fort? John Stetson, an Easterner who traveled west for his health, made the cowboy hat into a product even Easterners would buy—an item for dudes. (In one version of the song "Stagerlee," Stagerlee makes a deal with the devil for a hat specifically identified as a Stetson.) The T-shirt as fashion was "found" in the movies: in *It Happened One Night*, Clark Gable set the T-shirt on its course from under- to outerwear. Sweat clothes emerged from the gym to become everyday wear; the loafer and the Topsider boat shoe, both inspired by the Indian moccasin; the basketball shoe, itself heir of the rubber boot, and all its off-spring; the trench coat, removed from the trench to Chicago, proving as useful against the wind as the Thompson submachine gun, brought from the same location, did against law and order.

Blue jeans should be in the Museum of Modern Art design collection, the critic Joseph Masheck argued in *Artforum*. They are "quite as interesting as, say, the workers' clothing designed by Rodchenko in 1920."

Levi Strauss jeans are at least in the Smithsonian's collection. But the blue of blue jeans was incidental; the earliest pair in the Smithsonian are yellow-brown, with suspender buttons, much fuller cut than today's styles, and with a cinch in back. No belt loops—they did not come until the thirties; workingmen did not own belts.

☆   ☆   ☆

THE STORY OF Levi Strauss is familiar—at least in the carefully nurtured corporate version. Levi Strauss arrives in San Francisco, via New York, from Bavaria in 1853 to join his brother-in-law David

Stern in his dry-goods business, bringing with him canvas intended for tents and wagon covers for the prospectors. "Shoulda brought pants to sell," he was advised. "Pants don't wear worth a hoot in the diggin's." Strauss took the canvas to a local tailor and had it made into pants. The pants—Levi Strauss never called them "jeans," which at that time meant cheap pants, but "waist-high overalls"— became a profitable addition to the family dry-goods line. Soon an indigo-dyed *serge de Nîmes*, "denim," version was added to the original duck.

In fact, the key innovation came not from Levi Strauss but from a tailor in Carson City, Nevada, home of the Nevada silver fields, the Comstock Lode, the country Mark Twain described in *Roughing It*. Jacob Davis, a Jew born in Riga, took a joke seriously when he put rivets in pants he made for a silver prospector named—almost generically—"Alkali Ike." Ike, it seems, kept busting out his pockets by packing them full of samples of likely looking ore. Other miners got interested. Davis retailed goods bought wholesale from Levi Strauss, including the cloth from which he had made Alkali Ike's pants.

The rivet had another function. Before his overalls, Davis had made horse blankets using the rivets to join layers of wool to canvas. To keep old Ike warm, he riveted a layer of blanket into the pants as well. No fool, Davis had no doubt seen enough claims jumped to understand he needed a business partner. He wrote to Levi Strauss, his wholesaler, back in San Francisco.

"The secratt of them Pents is the Rivits that I put in those Pockets," runs Davis's letter of July 8, 1872, offering to split the patent rights with Levi Strauss, and enclosing two pairs of the riveted pants, one blue and one of white duck. "My nabors are getting yealouse of these success and unless I secure it by Patent Papers it will soon become a general thing." Davis did not even have the $68 he estimated the patent application would require.

The patent, when it was issued, significantly did not mention ore, but covered the use of rivets to prevent seams "from ripping or starting from frequent pressure . . . by the placing of the hands in the pockets. . . ." It is apt that this most American of clothing should be associated with this most American of postures, the casual hands-in-pockets stance. The innovation was a mechanical gadget to preserve a human gesture.

Davis moved to San Francisco to supervise production, accomplished by the "putting-out" system. Previously, Levi Strauss's finished goods had been produced by New York factories. Now, Davis cut the cloth, delivered it in the morning along with buttons to seamstresses around San Francisco, and picked up the finished pants in the evening. Soon the company estimated that it held 10 percent of the market for workingmen's clothing in the city. Production had grown so much that the putting-out system was replaced with a factory.

The pants were sold not only in San Francisco and the diggings, but at such places as water stops on the Central Pacific Railroad. Miners, not cowboys, first made jeans a success—miners like those whose artless but aesthetic dress Oscar Wilde would praise in Leadville, Colorado. The patent date—May 20, 1873—was emblazoned on the leather patch the company began to apply to the pants in 1886. "Famous patent riveted clothing" was the sales slogan. "The best in use for farmers, mechanics and miners," ran one early line.

Comfort and ease were emphasized along with strength. The jeans came, of course, in standard and necessarily loose sizes—still a far from universal practice before the Civil War. Among the earliest buyers of loose-sized clothing were sailors, who patronized "slop shops"—purveyors of ready-to-wear clothing—in the short interval between hiring, or impressment, and sailing. One of the slop shops was a company that was to become Brooks Brothers. Off-the-rack clothing had pioneered a primitive sort of ergonomics when, in the 1850s, basic sizes were established as the Civil War's demands for clothing made off the rack the center of the clothing business. The sewing machine created the market for sized clothing patterns—Ebenezer Butterick was the innovator here—that furthered the process. Jeans had a natural ergonomics to them: they shrank to fit, as did buckskins before them. But companies such as Sweet-Orr of New York pioneered standard sizing of overalls.*

Sweet-Orr had also used the tug-of-war in its advertising and even sponsored tugs-of-war. Sweet-Orr for years promoted its standing tug-of-war offer to replace any pair of overalls which

---

* When Levi Strauss introduced its children's wear Koveralls in 1912, it quickly found out that no sizing work had been done on children. In 1915, a Levi Strauss executive named Milton Grunbaum walked down the street to the local orphanage bearing a bag of candy and persuaded the children to pose for the measurements necessary to create a proper sizing scale.

failed to hold up under the strainings of as many as six full-grown men. Founded in 1871, it exhibited its pantaloons, overalls, and engineer jackets at the 1876 Centennial in Philadelphia boasting that "the peculiar cut of these overalls must be noticed, by which such seat-room is given that the wearer feels perfect ease, in all positions; no binding or straining in any part: their strength is such that they are warranted never to rip." Even without rivets.

Like Levi Strauss, James A. Orr had ventured to California with the news of the Gold Rush. He was a real prospector, though, not a merchant, and when he failed to strike he returned home to Wappingers Falls, New York. It was another twenty years before he went into the overall business, teaming up with Clayton and Clinton Sweet, two other local men, to produce what he touted as the first factory-made overalls. He emphasized the comfort his size grading brought to the pants—as well as their strength.

Sweet-Orr and Levi's offered free pairs of pants to replace any that ripped. Lee's, an 1886 addition, based in Kansas, were boosted as "tough as mule skin"; its pants were called "Unionalls," and traded on working-class solidarity. Durability was the key selling point, to be sure, and so was that great American ideal, ease in all positions—but jeans makers never wanted the customer to forget that these were workers' clothes.

Sweet, Orr overall advertisement, 1880s. (Sweet-Orr Inc.)

Levi Strauss patch.

The fact that Levi Strauss has always kept the buttons on the basic 501 model suggests the conservatism that went hand in hand with the presentation of jeans as working clothes. Slow and inconvenient, the buttons were self-consciously retained for the sake of style alone. The buttons also suggest the virtues of handcraft; even today they are attached not by robots but by women using machines—a team of four workers can sew buttons on 4,037 pairs of button fly 501s a day, *Fortune* reported in 1990. Levi's also bore suspender buttons and a cinch in back, until these features were deemed a waste of material by the government during World War II.

For Levi's, the zipper was too much a symbol of modernity and aviation—its earliest applications had been in flight suits—though the streamlined, fast-flying name seems to have done as much as anything to assure its success. The zipper, when it arrived, was high technology: its practicality was assured only by Gideon Sandback's development of machines that could make its parts to the requisite tolerances. Lee replaced the button fly with a zipper in 1926; Levi's not until 1955.

By the 1930s, the pants commonly bore the name "jeans"—a word Levi Strauss hated and refused to use. The name came from the West, just as the pants came east. There, it acquired a new set of associations, fostered by dude ranches and the rodeo promotions and students on Eastern college campuses, where the term was generally "dungaree." On dude ranches, the pants long known as "Californias" were becoming better known to greenhorns as "blue jeans." Lon Smith, chronicler of the dude ranch, celebrated and recommended the garment in a chapter of his *Dude Ranches and Ponies* (1936) titled "Dudes, Dudeens, and Duds," saying that guests should wear "all close clothing. In this way there is less slipping and chafing and everything is out of the way when action comes along."

Levi Strauss cashed in on the dude market. Jeans were made in cuts aimed specifically at women on the ranches. Then there was that mechanical rodeo, premiered at the 1939 Golden Gate International Exposition in San Francisco, with its hand-carved, foot-high figures of celebrated rodeo stars, mechanical riders shot out of the chute on mechanical branches, and mechanical clowns dancing through the arena. (The mechanical rodeo constitutes a strange foreshadowing of the mechanical bull at Gilley's Place in Houston, a mechanism that was to become the icon of the "urban cowboy" craze of the late 1970s and early 1980s.) The rodeo later toured the country on a truck to publicize the company. Soon orders were coming in from Oklahoma and Iowa and Texas. It was a familiar pattern, best exemplified by Buffalo Bill Cody: the Old West as show biz. From this point forward, the tug-of-war to which blue jeans would be subjected was the one between utility and fantasy.

By 1940, jeans had become popular as a status symbol on Western college campuses, notably the University of California and the University of Oregon. Sophomores are said to have adopted them as official dress and forbidden freshmen to wear them—an edict seconded at the University of Oregon by a gubernatorial proclamation. This trend was consciously fostered by the Levi Strauss publicity department. In the East, one story has jeans adopted by students at Williams College in response to the chinos of Yale and other Ivy Leaguers. (Those chinos were another upclassed product, the term moving from association with Chinese dress to khaki uniforms in the military to apply to basic khaki pants.)

An earlier sign of domestication came in the 1930s, when some of the rivets were covered to avoid scratching school seats and car bodies. Levi Strauss literature explains how the rivet on the crotch was discarded even earlier: company head Walter Haas, crouching in front of the campfire, found the rivet becoming uncomfortably hot. A touch of human interest; but had no miner ever squatted so close to the flames, no cowpoke stirred the chili and discovered this truth?

Jeans on female dude ranchers were the tightest pants ever seen on American women. The sexuality introduced was a critical turning point. By the sixties, jeans had become so tight that they were a major cause of gynecological disorders—the revenge of the corset!

It was no accident that blue jeans became popular on Eastern campuses about the same time that blues records did. The working-class aura was essential; the plug that the jeans were "guaranteed to shrink, guaranteed to fade" came only when they were worn as non-working clothes. The "dramatic low status assertion," as the sociologists term it, of the appreciation of work wear was given its highest literary expression in *Let Us Now Praise Famous Men*, with James Agee's soliloquy on overalls as a sign of peasant dignity. Overalls, pronounced "overhauls," he wrote, were "the standard or classical garment . . . of the southern rural American working man: they are his uniform, the badge and proclamation of his peasantry." He praised the "complex and slanted structures, on the chest, of the pockets shaped for pencils, rulers and watches.

"A new suit of overalls has among its beauties those of a blueprint: and they are a map of a working man, each man's garment wearing the shape and beauty of his induplicable body. They wear like old money." (Confederate bills, reports William Gass in *On Being Blue*, were called "bluebacks.")

Agee's own clothes are described by his friend and colleague, Walker Evans, in a memoir reprinted in later editions of *Let Us Now Praise Famous Men*:

"His clothes were deliberately cheap, not only because he was poor but because he wanted to be able to forget them. He would work a suit into fitting him perfectly by the simple method of not taking it off much. In due time the cloth would mold itself to his frame. It did seem sometimes that wind, rain, work and mockery were his tailors."

Agee's wardrobe was worn like overalls. But the key to Agee's lyric appreciation is given in Evans's summary of this casual dressing: "He fell over into a knowingly comical inverted dandyism." No phrase—not "cowboy chic," not "low status assertion," not "dudism"—better sums up the post-1940 wearing of blue jeans.

☆   ☆   ☆

DURING THE "urban cowboy" fashion of the late seventies, Texas observers noted that real cowmen wore immaculate jeans, and carefully polished flat boots. In the last century it was noted that many of the Western customers of Levi Strauss kept two pairs of the pants: one for the work week, and an identical one, worn only on Sunday as "fancy pants." The true cowboy was always vain about his appearance; he was the dude before the dude ranch.

David Dary, in *Cowboy Culture*, argues that dudism was rampant among cowboys as early as the 1870s. There was conformity but also showmanship in their dress. Fancy boots, hats, and clothing were one of the few things that cowboys spent their money on. "The cowboy must also have a pair of fancy chaparajos, or overalls, made out of calf skin, or stamped leather," Dary quotes one account. By the 1880s in Montana it was noted that cowboys "wore the best clothes that they could buy and took a great pride in their personal appearance." Boots were $25 a pair made-to-order jobs. Fancy chaps and hatbands were common, along with "brilliantly colored silk handkerchiefs" worn around the neck.

The true rancher, in recent years, preserved cowboy dandyism in the face of urban cowboys and country-and-western stars with artfully torn jeans. *The Wall Street Journal* reported in the mid-eighties that real ranchers, as opposed to executives affecting the rancher look, always pressed their jeans, starched their flap pockets, snap-closed their cowboy shirts, and wore practical, low-heeled boots—no doubt to more conveniently accelerate the pickup trucks that had supplanted the quarterhorse.

Levi's came to be an expensive, premium product; lesser jeans at a quarter the price continue to be prevalent to this day. The critical change came when more expensive models appeared: designer jeans.

☆   ☆   ☆

JEANS MARKED THE beginning of the sixties orgy of lower-class identification. Between the mid-sixties and mid-seventies, Levi Strauss sales increased tenfold. Jeans were supposed to look old: "Beware," Thoreau had warned, "of any enterprise that requires new clothes." After jeans became cliché, other working-folk garb became chic—uniforms, gas station shirts, mechanics' overalls, bowling shirts, and so on. If it was possible to imbibe the virtues of a lion by wearing his skin, it was possible to imbibe the virtues of the working class by wearing their clothes.

But decorated with sequins, glittering new rivets, and ostentatiously displayed patches of differing color and materials, hippie jeans were like blues songs covered by the Rolling Stones. In the sixties, Levi Strauss held a contest for decorated denims.

Just as plain jeans reached their height of acceptance—an apogee marked by the Coty Award bestowed on them in 1974, the thesis found its antithesis: the undesigned encountered the designer. It began when Errol Wetson, a fast-food millionaire, came up with the idea of selling American kids jeans just like those worn by Europeans on the Riviera trying but failing to look just like those worn by American kids. Before long, American companies such as Sassoon had followed, and soon nothing came between little Brooke Shields and her Calvins.

In 1957, the Milan designers Achille and Pier Giacomo Castiglioni put a tractor seat atop a sprung stand to make a simple stool they called Mezzadro. It quickly became fashionable in the United States as well as Europe, and was shown at the Museum of Modern Art's Italian design show of 1972. For the European, this seat was a beloved found object, quintessentially American, molded to the fanny and ergonomic, sprung like Concord coaches and Baldwin locomotives. Today the variously cast, trademarked, and ornamented seats of tractors and other agricultural equipment are sold in antique shops as collectibles and displayed on walls as artifacts.

Siegfried Giedion admired such seats. The first use he mentions is in the 1880s, although the images of reapers and other farm machinery he reproduces show them already in use in the 1850s. For the Castiglionis, it became a piece of high design furniture.

European jeans were like this: they took an American object that neatly fitted the rear end and stylized it, then sold it back to the Americans. Jeans were a found object, turned into fashion.

Lee "heavy stonewashed Relaxed Rider" jeans. (Lee Corporation)

Lee Bib Overall for children. (Lee Corporation)

The status of jeans as "universal clothing" was changed by the arrival of designer jeans into a prestige market, as fragmented and niched as General Motors' was. Blue jeans circa 1965, according to John Brooks's telling analysis in *Showing Off in America*, were "characteristically inexpensive, quintessentially convenient . . . adaptable to work . . . comfortable and most prized when dirty, worn and faded." These qualities seemed to make jeans a striking example of clothing that resisted Thorstein Veblen's theory of fashion as conspicuous consumption. But by 1981 things had changed, and Brooks charts five Veblenesque qualities of "competitive display and invidiousness" that jeans had come to possess:

• Must be wastefully expensive
• Must signify invidious comparison to others
• Must be worn in emulation of others
• Must not be associated with physical work
• Must be worn free of dirt or wear

Sassoons or Calvins, too tight for work, status-laden and dry-cleaned, had transformed the meaning of jeans. Like Duchamp's urinal, the simple dungaree had become something very different in a new context. Like Lévi-Strauss's totems, their symbolism changed with a shift in the status structure: blue jeans had become blue-blooded. Now the working class, migrated to the service sector, labored in polyester during the day and returned home to relax in natural cotton jeans.

Younger and younger children were ensconced in jeans like those their parents had grown up with. The manufacturers traded on the cuteness of a toddler in adult clothes. (One early line of overalls for children was called Brownie, like the sprites that advertised Kodak's camera, despite the fact that they were blue.) The firm called Oshkosh was to find its kids' jeans and overalls completely overtaking its adult sales in the baby-boom eighties: they went from 15 to 75 percent of its sales.

But since Brooks published his chart, circa 1981, the wheel has turned once more, the dialectic dialed again, and plain jeans have returned to be melded with designer jeans. Plain jeans returned in the form of imitations by designers, and increased emphasis on history by the basic jeans makers. The line of designer jeans called

"Guess," which made an ironic cipher of the name on the tag (although the name of the manufacturer, a North African émigré called Georges Marciano, was also invoked in advertising), tried to combine fashion jeans and basic jeans by advertising its products in black and white photographs evoking the West of James Dean.

Even in the mid-eighties Levi Strauss discovered how much stronger the market for its traditional jeans was in the West than in the East, and mounted a massive advertising campaign to expand into the designer jeans turf, with blues songs, *cinéma vérité* street dancing, and rap music. Sales of the 501—the ostensibly "original" Levi's—were increased when Bruce Springsteen wore them on the cover of his *Born in the USA* album, with its songs of workingman's New Jersey.

The late eighties have seen the arrival of a new stage in the jeans dialectic: designer plus basic results in pre-washed, acid-washed, stone-washed. Jeans roughened up before sale at designer jeans prices, combining chic and work symbols. They are pre-worn not to make anyone think the owner had really worn that look into it, or did a good deal of physical labor, but as a simple stylistic emblem, implying some roughness and toughness of character to the callow youths enclosed in their wrinkles and creases. Acid washing is a near relative to sixties tie dying, with the wrinkles and twists the garment possessed when put into the washing machine rendered as a sort of permanent marbleized pattern.

Pre-washing jeans turned them once more into a fashion product; the totemism had changed. These jeans were no longer raw but cooked. The rage of the industry in the late eighties was acid washing—actually accomplished with dilute bleach and pumice stones. (The geology department of the University of Texas was consulted on the proper sort of pumice for the purpose, and the leading pumice supplier attempted to have his own trademark added to garments treated with his product.) Pre-worn jeans, pre-shrunk, pre-faded, even, finally, pre-ripped (another recapitulation of the late sixties) abounded. (Everything but pre-owned, the favorite euphemism of the used-car salesman: though pre-owned were available in some stores, at prices not much below those of new designer models.) Only the laboring classes still wore designer jeans; in the late eighties you could see workers repairing streets in Jordaches.

Jeans sales peaked in 1981, at 502 million pairs, and slumped to level off at 416 million pairs in 1985, when basic jeans made a comeback; they have now dropped below 400 million pairs. The decline was seen as the result of an aging population, and it inspired the industry's powers to seek ever more various ways of treating and trimming the utilitarian fabric.

One industry mogul asserts that there is a five-year cycle to the market between plain and high-fashion jeans, a kind of alternating current in the iconology of jeans that has left them suspended between the opposing magnetisms of the rustic and the urbane. The two poles were European and American, fashion and function. When a French designer created the ripped jeans of the sixties, it was by design, by manipulation of what was seen as an American artifact. And when the American jeans companies began to replicate the style, it was as a European fashion, looked at exactly as a Chanel or Armani style might have been.

The recurrence of plain jeans and the "fashion cut"—an opposition worked mostly by immigrant entrepreneurs from the Mediterranean—meant a sort of recession of quotation marks inside quotation marks that eventually, like terms in a mathematical formulation, cancel each other out and reduce the item again to lowest terms. The initial appeal of jeans as a "raw," real, utilitarian garment, "cooked" by use, plunged fatally, like Narcissus, into its own reflection. When Ralph Lauren touted his Polo jeans as genuine thirties "dungarees"—an old-fashioned term from the Northeast—the name was a tipoff. His product was aimed at people pretending to be preppies pretending to be cowboys.

CHAPTER FIFTEEN

Private Eyes and Public Faces, or Look Sharp!

"IN MY POOR, lean, lank face nobody has ever seen that any cabbages were sprouting out," said Abraham Lincoln, but after he was elected president a beard began blooming there. The reason was a suggestion (the tale could have come from Dickens) from a young girl, Grace Bedell of Westfield, New York. "As to the whiskers," Lincoln wrote her in reply, "having never worn any, do you not think people would call it a piece of silly affection [*sic!*] if I were to begin it now?" This was in October 1860, and soon afterward he began to grow the beard.

Lincoln had contrasted his prairie of a face with the "round, jolly, fruitful face" of Stephen Douglas which, he said, promised political patronage "bursting and sprouting out in wonderful exuberance...." Cabbages—an odd choice of vegetable. One thinks of "cabbages and kings," and of the face as the hard-tilled garden of dreams and ambition. Lincoln's talk of his face was strange, self-deprecating but secretly proud, informed by the notion of the self-made man. There are fashions in faces, as in clothing, and some classics that never go out of style.

The fashion for beards reflected the effulgence of the Gilded Age—the masks of respectability worn by the robber barons. Beards perhaps reflected the decoration of the time—the scroll-saw gingerbread ornamentation of cottages in the suburbs or the stone-work frosting of "cottages" in Newport. The return of the beard also coincided with the rise of the photograph.

With Matthew Brady, Lincoln's face was to become an icon. Brady photographed him the day he was introduced to Eastern audiences in his Cooper Union speech of February 1860, and when the image was emblazoned on campaign badges, Brady took credit for Lincoln's election. No other American face has been as studied. Washington's is a stone face, stiff with teeth and wig, a careful composition for Mount Rushmore, a stern riddle, a sphinx, a great seal of the nation. Jefferson's is too ordinary, too open, too disarming to suggest more than a bright ordinary man; he is no icon. Andrew Jackson's—photographed only late in his life, with sunken mouth and melting geometry—was in its prime a caricature of high-cheekboned frontiersmanship, of raw military and democratic anger, to which Lincoln gently alludes, one allusion among many subtle ones.

With Lincoln began the cult of the face. He was ugly—even apelike—by the standards of the time—but the face was real, a found object. New England intellectuals alternately loathed and delighted in its plainness. In its cheekbones sharp as the hips of a starving mule, Southern claimants to an alternative, Southern birthplace and parentage for Lincoln would go so far as to proclaim John C. Calhoun his father on no evidence stronger than sheer physiognomy. Lincoln was a self-made man who is credited with saying—although it was a proverb that existed before he did—that by the age of thirty every man has made his own countenance. If no cabbages grew in his face, he would grow some.

His beard marked the change from rail-splitter to Father Abraham, the transformation from the son of the Midwest to the father of his reborn country. (Almost blasphemously, he said on the way to inauguration that the task before him was greater than that which had faced Washington.) The beard made associations with Jehovah and Jesus equally possible.

Photography made the cult of Lincoln possible. Lincoln was the first photographic president, as Roosevelt was the first radio president and Kennedy the first television president. The complete Lincoln photographs fill a stately volume, unfolding like the career itself from the wiry young lawyer, with greasy and restless cowlick, to the earnest orator of Cooper Union, to the inaugurate, with half-sprouted beard, and concluding with the hollow-eyed sufferer on the eve of martyrdom, the bones flattened into a grainy parchment map of the Civil War.

In Europe, photography had ridden the crest of the fashion for *cartes de visite*: small prints pasted onto calling-card-size pasteboard. In the United States the Daguerreotype was brought to high precision, not least because Americans used machinery to polish far smoother the metal plates that were the ground for the silvery mirror images of the Daguerreotype.

But the important thing was getting the camera outside the studio. When Brady, Alexander Gardner, and others succeeded in that, albeit with bulky equipment and wagons for their chemicals, they produced views of Lincoln and his generals in camp that deflated the stiffness of portraiture. The new type here was Ulysses S. Grant, so much in contrast to the regular line and staff generals, a remarkable number of whom tended to pose with hands inside coats like Napoleon. His relaxed posture also stood in contrast to Lincoln in his stovepipe, ramrod straight and as immobile as a mortar. Grant hitched his arm up against a pine tree. He bent over a pew pulled from the temporary headquarters of a church to consult his map. He left his hands in his pockets and his jacket unbuttoned. Only later could General Grant's looseness—which delighted, among others, Gertrude Stein, in her profile of Grant—be read as a signal of the moral looseness of the Gilded Age presidency Grant presided over.

But in the second half of the nineteenth century, one did not have to be wealthy or great to be able to study one's face in a photograph. And as the camera became more portable, it took on new uses. One of the first types of portable camera was called the "detective camera." The detective was an emerging figure of the age, part union-buster, part anti-terrorist agent, part security man. Allan Pinkerton, Lincoln's head of the Secret Service, took "the never sleeping eye" as his icon. He went on to serve the railroads, providing not only "detection" but what we would today call "rent-a-cop" services. With the detective came the use of photography: the mug shot, the "suspect book" (the conspirators in the Lincoln assassination plot glowered into cameras in one early instance), and by the end of the century the concept of identifying suspects from a kit of types—the Identikit.

The detective camera invoked this figure only obliquely. And it took pictures obliquely, too, with lens disguised or its whole body hidden inside some innocuous object—a watch, say, or binoculars, with the camera lens in its side, so the photographer appeared to be looking in a direction ninety degrees away from the photographic subject. Small and portable, the "detective camera" allowed the photographer, in theory at least, to surreptitiously record the world around him. But it also captured the emerging fascination with the detective as the spy of ordinary life. In truth, the most extraordinary purpose for which most detective cameras were used was probably capturing pretty women or spooning couples. The camera figured in song and story in its early days in this role—as voyeur. It lent photography a sense of furtiveness and a certain lasciviousness that it would never entirely lose.

The first camera patented by George Eastman in 1886 was described as of "the detective type." Eastman, a clerk, appears in his own early photographs as an earnest young man, who sported a full and rather pointed beard at the time he developed an interest in photography and worked on new types of photographic films. But Eastman's real contribution was in camera film. He created a roll or "cartridge"-type film—not the first but the first widely successful one. His "detective" camera was a box roughly three by three by six inches, which by 1888 he had further refined and renamed. He now called it the Kodak—a word he had coined for recognizability and adaptability to all languages. It was a name as portable as Eastman's folding, pocket model: already he was thinking internationally.

The transformation of the camera from detective to Kodak was crucial. Introduced in June 1888, the Kodak camera cost $25, including case, shoulder strap, and a paper-backed roll of film allowing a hundred exposures. When these had all been taken, the whole camera was to be returned to the factory, the film developed, and the reloaded camera and prints sent back to the owner. The camera, of course, was a device to sell the film.

Eastman also wrote the book that came with the camera, called the *Kodak Primer*, and explained his complete system for photography as a sort of division of labor: "The principle of the Kodak system is the separation of the work that any person whomsoever can do in making a photograph, from the work that only an expert can do. . . ."

The technical keys were the use of a relatively small negative, allowing a short focal-length lens, and great depth of field, which allowed the camera always to be in focus, except in extreme close-ups. There was no viewfinder, only a set of aiming arrows, and no

Eastman Kodak advertisement, circa 1900. (Smithsonian Institution)

indication of the number of exposures. In 1889, Eastman and his researchers perfected the use of celluloid to make film in rolls. Previously, the image was rendered in a gelatin that had to be stripped from paper and transferred to glass for printing. Celluloid made the rolling, loading, and unloading of film easier. By 1891, there were Kodaks whose film could be loaded in daylight. And, of course, celluloid film also made the motion picture possible. Thomas Edison called on Eastman and worked out a scheme for the production of sprocketed celluloid film for his new cinema camera and Kinetoscopes. Inside a Kodak, from now on, would be a tiny assembly line with a belt of film rolling from wheel to wheel, each frame exposed in sequence.

Cinema would repay the debt in the 1920s, when its film lent the 35-millimeter format to the Leica and other press cameras. But Kodak for years resisted 35 millimeter. It made the film, and it desultorily made 35-millimeter cameras, but over the years its more popular films remained in other, constantly shrinking formats. In half a century, the basic Kodak amateur negative shrank along with the camera, to about one twentieth of its original area.

February 1900 marked the arrival of the Brownie, the dollar camera. The Brownie did far more than make the camera inexpensive; it served to celebrate the child and the childlike. (The word "Kodak," Eastman had said, was "as meaningless as a child's first goo." And, to the parents, as meaningful.) With the Brownie, every man could be his own do-it-yourself family archivist.

The camera took its name from Frank Brownell, the key Eastman employee involved, and from illustrator Palmer Cox's figures of spritelike brownies, then popular with children. The cabbages and kings of the future were now to be seen sprouting in the countenances of the very young. The family camera coincided with lower child mortality rates, an increased emphasis on the effects of environment, a new attention to education, and the development of the individual. Children were, and are, the dominant subjects of snapshots.

The Brownie was designed to be simple enough for a child to use. It was one of the first devices whose internal workings remained a mystery to the average person but whose controls were designed for a basic common denominator: the child as model of the average person. Understanding its inner workings was unnecessary. Only knowledge of the "button"—and Kodak was one of the creators of the entire modern motif of the push button—was required. Along with the telephone, the camera was the first black box: the container of magic transformation, of technology that could be employed without being understood. The first cameras physically resembled the first telephones, with speaker/loudspeaker and lens figuring similarly as apertures to the mystery inside. But the telephone depended on outside agents—the operator and the company with its network of lines. The camera stood by itself, except for the development and printing of the film. The viewfinder was at first absent; you simply aimed. The next step was nothing more than two pop-up metal frames.

After the first blush of amateur photography inspired a sense of the photographer as private eye, the succeeding establishment of the camera as an almost universal item fixed its role as the explorer of private life. This was the camera as pet, friend of the family. The *Kodak Primer* and other publications urged families to record their own histories—to create a sort of visual escutcheon. In 1894, the company hired Admiral Peary, the polar explorer, to endorse and explain the idea of "exploring with a Kodak." Although early Kodak lore is still full of the ogler snapping bathing women (still, of course, in their bloomer-length suits) at the seashore, in the company's propaganda the Kodak after the Brownie increasingly became depicted as a family appliance. Any child could use it and almost any parent afford it. The implicit assumption was that family moments were important to record; every man's child and picnic and party were as important as his face.

The Brownie and its imitators established the pattern that American cameras would "be democratic." It is a truism that George Eastman's cleverness was not so much in creating new camera technology as in creating cameras that would sell film. How characteristic that the Germans and Swiss should become the leaders in lens technology, with their history of close observation and grinding things fine. How equally characteristic that the American camera should aspire to the condition of universal appliance.

The product of the Kodak, the snapshot, that crisp and impatient concept, differed from the portrait; it was about the anecdote instead of the countenance. The empty space inside the black box was reserved for the individual—for his own identity, his snapshots, his mug shots. There, individuality had its space, the numerator of the common denominator. The camera was a device for packaging moments, but also the chance that recorded just this turn of profile, just that split second of gesture, just that sweep of breeze-tossed, lens-blurred foliage in the background. Here was Duchamp's canned chance.

"You push the button, we do the rest," "Anyone can use it," "As easy to use as a pencil," and all the rest of the slogans did not disguise one fact: no one ever seemed to get much better at taking snapshots. A few people might be very good from the beginning, but rarely was there improvement. So much was limited by the camera, the crudity of direct flash, the relatively slow exposure times. The product of the Kodak was in significant part the product of accident.

Thirty-four successive models of Brownie showed that the box could be decorated and eventually its shape modulated in all sorts of ways. Not long ago I bought a Brownie Junior, with a black-on-silver, vaguely Moderne, pattern for $7 at a garage sale. It is a simple rectangular solid, the only protrusions a looping strap and the film-advance key, shaped like a musical clef. Set into it like two windows are dual viewfinders—one for horizontal, one for vertical framing—through which one sees an opalescent, dreamy world, as if in a time dissolve to the year of the camera's making.

These cameras, with their two small ports for the viewfinder set above the larger one of the lens, acquired vestigial faces (eyes and mouth or ears and head, like a crude Mickey Mouse) whose more specific and licensed countenance decorated a number of them. My specimen is fronted with a vaguely Deco pattern of abstract lines, but there were all sorts of variations. Commercial and commemorative versions of the basic camera box, bearing Hopalong Cassidy or the Trylon and Perisphere of the 1939 World's Fair, sell for far more. I saw a virtually identical camera, except for its blue and black decorations, in Los Angeles in a Melrose Avenue antique shop for $150. The difference was that it had issued from the design studio of Walter Dorwin Teague, who became Kodak's key designer in 1926. This was a 1936 design—created at the same time Teague had produced the Bantam Special, a 35-millimeter camera with a rangefinder, with nine horizontal brightwork stripes and a door that popped open to expose the lens. Such variants, including licensing cartoon characters and the application of designer colors to appeal to the women's market, marked Kodak design at its height.

My parents' camera represented the zenith of the Brownie's evolution. It was a Brownie Hawkeye—the black box updated and modeled in plastic by Teague for the fifties. This Kodak was to the first ones as the '48 Ford was to the Model T or Model A. It took on a near-cubical shape now—no longer the rectangular solid of the traditional one. A band of Deco-inspired, rolled ridges distinguished the lens facia. The cube possessed rounded corners and horizontal ribs. You grasped it by a plastic handle that looped over its top. Its shutter release was no longer the metal flipper of earlier models but a gray pearlescent push button ribbed for grip, set into the body of the camera. You advanced the film not with the metal key, like that of a wind-up toy, but with a wheel. A metal escutcheon rolled down from the top of the body to frame the lens like an automobile grille. The viewfinder was the magic: a powdery bright viewplate, in which the world seemed magic-lantern sharp and luminous. In the back, a small porthole of colored plastic displayed the smokey orange red of the numbers of exposures, like a window on the dark aquarium and mystery inside the box.

When the Brownie was finally supplanted in the early sixties, a fundamental truth about the company was revealed. The new camera, the Instamatic, housed the roll of film in a plastic cartridge that could simply be dropped into the open camera. With the Instamatic, the rate of film consumption per camera was almost doubled, and since the cartridges for the Instamatic were twice as expensive as the simple roll film, Kodak's profits increased. Beneath its appeals to ease of use and personal archiving, Kodak was a company whose business was selling film—and only incidentally selling the cameras to consume it. The plastic Instamatic was a flattened box, without much geometrical conviction, touched with phony textures and tones. Pseudo-chrome ruled the world of Kodachrome.

Further advances in the Instamatic mold saw the cartridges and cameras grow smaller, as did the negatives. The ultimate step was the disc system, tiny negatives dropped into a circle designed for ease of processing. The images were little better than they had been a decade ago. By the mid-eighties, the push to sell film overwhelmed the quality of the print completely. While serious single-lens reflex cameras were dropping in price and improving in

Kodak Brownie
Hawkeye, 1950, and
Baby Brownie,
1934. (Walter
Dorwin Teague
Associates)

Deco-patterned
Kodak Brownies,
1933.

quality—as well as design—the world of the Instamatic was fading
as rapidly as its feeble flashes.

It was left for the Japanese, in the mid-seventies, to turn out small,
rounded, "idiot-proof" cameras, sleek and efficient as Toyotas or
Hondas. The Japanese had imitated and improved on the original
German professional cameras. Now they added fun and economy
and high performance per dollar. With stylish colors and fast, fun
curves, they were nonetheless still real cameras, not toys. They
seemed to have been listening to some of the American industrial

designers of the thirties, who argued that the box was the wrong shape. "The hand," wrote Harold Van Doren, "unconsciously . . . formed a dislike of the box shape." He wanted the camera rounded, to "make it snuggle in the palm." Canon, Olympus, and others had achieved this snuggly sort of design by the beginning of the 1980s. Save for a few models designed by Hartmut Esslinger's industrial design group, frogdesign, most Kodaks still seemed cameras for children. Once that had been their appeal, but no longer. But Kodak benefited from the boom in 35-millimeter film sales, and the Japanese were unable to do more than marginally increase the sales of their own film.

Kodak abandoned the disc and pushed its own series of small SLRs. By 1987, small 35-millimeter cameras represented half of camera sales and three quarters of total value. In the same year, 35-millimeter film accounted for two in three pictures taken in the world. In 1975, it had been only one in ten.

Also by 1987, the disposable camera arrived to join the disposable pen and razor. Its plastic body encased in yellow cardboard like the familiar film box, Kodak's disposable was named Fling—a word absolutely indicative of the simultaneous American appetites for fun and waste. By the early nineties, some 20 million disposables were being sold annually, and caught short by new environmental awareness, Kodak and other companies began belated efforts to recycle the units. Kodak's slogan for its disposable, "the film that is also a camera," suggested that it had carried the dominance of film sales to camera sales close to ultimate realization.

☆   ☆   ☆

THE POLAROID SEEMED to further the idea of a camera for every man and every face. It aspired to an ideal of a camera you could carry anywhere in your pocket, a sort of visual notetaker—that you could, even more than the Kodak Brownie, use as easily as a pencil, the status Kodak had first aspired to back in the 1890s with its folding pocket Kodaks, the famed FPK's. It, of course, also grew out of the child's interest. "Why can't I see the picture right now?" Edwin Land's daughter had asked him, so the famous story goes. A burst of inspiration later, Land had his basic system.

But Land's real vision was one of giving the photographer imme-

diate results to improve his pictures, providing feedback. It was a
vision limited by technical difficulties—primarily in manufacturing. It came the closest to realization in the SX-70, introduced in
1972, but it was never the great success Land hoped for. The SX-70
was modeled to fold into a shape no bigger than a piece of wood
that Land carried around in his jacket pocket—a stub of one-by-
four, I suspect. The shape of its folding viewfinder was taken from
the Hasselblad, the Mercedes of cameras, the NASA choice for
moon photography. With its real leather panels and brushed-metal
frame, the SX-70 was the last product designer Henry Dreyfuss
worked on.

But Land had higher aspirations. He wanted to go beyond the
crude snapshot; he believed that his camera would release what he
called "atavistic competences" in the user, making him a better
picture taker because he was receiving immediate proof of his
results. Polaroid's appeal was in the great tradition of the self-
operating machine. And the self-developing picture offered its own
special aesthetic appeal.

The practical uniqueness of Polaroid pictures—you could have
them reproduced, but very few users did—restored some of the
precious, jewel-like quality of the Daguerreotype. More than conventional prints, they possessed the status of objects rather than
simply images. They could be manipulated, as Lucas Samaras and
others did, by the application of chemicals, making the image a
starting point rather than end product. And they regained some of
the lasciviousness of early photography because nude—or more
shocking—images would not be seen by the corner druggist or even
the technician in Rochester. Polaroids of this sort figured in mystery
novels. Here was the privacy of eye and image reasserted. The
SX-70 folded into your pocket as neatly as a lepidopterist's pocket
frame.

The company gave free film to leading art and commercial photographers as a promotion. One of the photographers was Walker
Evans, then in his sixties, and he was enthusiastic. The SX-70, he
said, made photography fun again. "A practical photographer has
an entirely new extension in that camera. You photograph things
that you would not think of photographing before. I don't yet know
why, but I feel that I'm quite rejuvenated by it.

"I want to try that camera with children and see what they do with it. It's the first time, I think, that you can put a machine in an artist's hand and have him then rely entirely on his vision and his taste and his mind." Here was the old ideal of simplicity: a camera as easy to use as a pencil, a camera even a child could use, a camera that made you look more sharply at the world around you.

Polaroid matched Kodak in the value of its camera sales by 1960—in just eleven years. But there were distinct limits to the sharpness of the image that provided an ultimate barrier to further growth. Faster film speeds, finer grains, and speedier development methods for conventional cameras cut into the appeal of instant photography.

In the area that mattered—film sales—Polaroid's challenge failed. In the decade 1978–88, Polaroid sales fell by half. The remaining users of the cameras tended to have practical needs for it: they were insurance adjusters documenting smashed fenders, real estate agents snapping houses, and so on. Even after the on-slaught of Japanese cameras and Fuji film, Kodak still commands 80 percent of the film market. Not until electronic photography arrives—and both Kodak and the Japanese, who have introduced "video still cameras," are working on that—will this lead vanish.

☆   ☆   ☆

KING GILLETTE'S RAZOR arrived at almost the same time the Brownie did. "Look sharp!", his slogan, was the cry of the new era. The barbershop shave was relegated to the background, to a spe-cialized luxury, just as the professionally photographed portrait was—and just as the beard retreated in the face of the clean-cut ideal, associated with honesty and progress. To reduce whiskers to a mustache or no facial hair at all was, by 1890, a sign of modernity and efficiency. As the face became an individual's trademark, it also became used more and more often as a commercial one; and no countenance became better known than that of King Gillette, re-produced on every blade wrapper.

Eastman and Gillette made their fortunes by conceiving systems that required the use of large quantities of disposable "ammuni-tion": film and blades. Their products also related to the individual human face and identity. They were little consumption machines, which, once dispersed, would provide a steady stream of business.

Kodak made its money on film, Gillette on blades. Gillette could give away the razor and make money on the blades, the joke went—but it was no joke. His product, like George Eastman's, like so many afterward, was a product designed to consume another product.

The business battle in the world of the razor had to do with patents and blade compatibilities—the careful design of posts and slots so that competitors' blades would not fit the Gillette handle. It was suggestive of how much popular photography and shaving yourself were linked that one man invented major features for both businesses.

In 1914, Henry Gaisman developed a way to write on the images of Kodak film, so a date and caption could be applied to pictures while they were still in the camera. The company paid him $300,000 for the rights to his system. Kodak's "autographic" cameras were part of its ideal of making "a camera as easy to use as a pencil." By 1926, Gaisman was involved with the Autostrop razor company, but he went to the Gillette company with another improvement on a familiar invention: he had developed a new improved razor blade with an aperture in the shape of an H, which was held more securely in the razor, with less danger of breakage.

In 1895, King Gillette, a salesman for William Painter's Crown Cork Company, had followed the advice of his boss, who invented the crimped metal-and-cork bottletop that became standard in the soft-drink industry, to find a product that was disposable and that people would therefore always need. One day, while shaving, the inspiration came to him—there is no getting behind such a neat tale. As Painter had created the lid for the Coke bottle, of which thousands were thrown away and bought daily, Gillette would create a razor blade to be sold and disposed of in similar numbers.

After the unfortunately named William Nickerson devised the means for making the blades, the company began sales in 1903. They moved only 51 razors and 168 blades in that first year, but by 1904 they were selling 91,000 razors and 123,000 blades. World War I was the great boon to Gillette, as the safety razor went to war. Between the wars, Gillette quickly caught on to the techniques of associating itself with spectator sports, and to the emerging advertising media of magazine, radio, and television. Its razors were also widely distributed as premiums for other products. One maker of overalls, for instance, gave away Gillette razors.

The Gillette idea was simple: to make a blade cheap enough to throw away when it became dull, and thereby to make every man his own barber. Gillette's razor brought self-service to the bathroom. "Hundreds of thousands," he intoned, "are now able to shave themselves who could not before, and do so without involving the question of skill in keeping their razor in condition. . . . Gillette has reduced the art of shaving to such a cheap and simple process."

The change in fashion reflected the change in the ideal of individualism between the Civil War and World War I. Now you got the face you deserved—by your own hand. A 1913 Gillette advertisement, a little fuzzy on its classics but true to pitch, ran: "The message of Solon to the Greeks of old was 'Man Know Thyself.' The message of Gillette to all the world is "Man Shave Thyself.' . . ."

And the rest of the story is that Gillette got his face on every wrapper, and became world-famous—but not in the way he hoped. Before he developed his razor, King Gillette was a political theorist, a utopian and something of a crackpot.

Gillette's passion, even before his razor, was writing utopian tracts read by the audiences for George Bernard Shaw's Fabian tracts and Edward Bellamy's *Looking Backward*. This was more than just an eccentric hobby. Gillette's tracts were far from peripheral to his main work. They showed a belief in the perfectibility of man—even if by strange mechanical drawings that rearranged society as if it were a machine. If the individual would eventually get the face he deserved, so would society eventually get the face it deserved. For Gillette, the perfectibility of the face by mechanical means corresponded to the perfectibility of society by business means: "promoters are the true socialists of this generation," he declared in 1894 in *Human Drift*, which envisaged competition supplanted by a benevolent single giant monopolistic corporation that would build glass-domed towers lit by electricity from Niagara Falls.

Gillette went on to incorporate his World Corporation in Arizona, and produced more books, by now branded with the famous name: *Gillette's Social Redemption* and *Gillette's Industrial Solution*, which included a scheme by which, it was argued, the entire human population could live in Texas.

Part of the Gillette vision foresaw the end of advertising—the very means to his success, what he once described as "the Gatling gun" of enterprise, the medium that had made his face world-famous. He envisioned chain stores leading to centralization by taking over their supplying industries. He met with Henry Ford to discuss his ideas, although few sparks were struck, and hired Upton Sinclair to shape up the manuscript of *The People's Corporation*, which appeared in 1924. The way to Utopia was as simple as his slogan—Look Sharp! Feel Sharp! Be Sharp!—with its neat progression from appearance to sensation to reality. Capitalism would improve itself right out of existence. Through the Occam's razor of competition, only the single, rational, economic and social solution would survive.

☆   ☆   ☆

JUST AS GILLETTE and Eastman made their fortunes by the tremendous multiplication of the profit earned on each blade, each picture, the two most famous business product companies that followed them would make their fortunes by the copy, in the case of Xerox, and by the card, in the case of IBM. They devised systems for organizing, measuring, and controlling individual identities, and their mode of business was the same. They leased the razor, as it were, and made money by the razor blade.

Years after he had convinced the men who ran the railroad to use his card-fed tabulating machines, inventor Herman Hollerith recalled his inspirations. One was the card that drove the Jacquard power loom. Developed in France, perfected by Germans, the Jacquard system arrived in the United States around 1830.* Its shuttle was driven by a series of linked punch cards, which in effect digitally represented the pattern of the final weaving. Charles Babbage, the English mathematician and computer pioneer, had proposed the Jacquard system early in the nineteenth century as a means of

---

* Before the card system was used in weaving, the products of hand weaving themselves looked almost computerized. At first the Jacquard loom was used to replicate the handmade product—the overshot coverlet, for instance, with its complex geometries—and it did so very well. Then, as they drove the hand weavers' trade irrevocably toward extinction, the machine weavers gradually added patterns of presidents, landmarks, and so on, images that suggested the standard cuts of newspaper advertising, circulars, or posters—the generic steamboat for the steamboat service advertisements, for instance. The irony was that this sort of imagery looked far less mechanical than the intersecting lines and varying widths of the handmade coverlets, whose patterns fairly buzzed with complexity. The handmade product looked more machine-made than the machine product ever would again. It also looked very much like the patterns later to be etched on computer chips, or printed from computer dot matrix patterns with a sort of simple virtuosity, like an imaginative session musician who could fill in a jazz or rock-'n'-roll recording track, but had few ideas for main themes of his own.

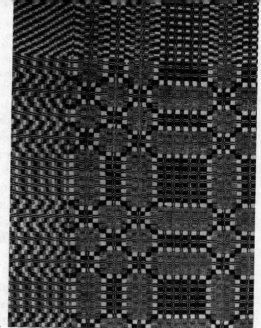

Jacquard coverlet.

"programming" his "Analytical Engine." The dotlike patterns of an original Jacquard card, their faces like a sort of collapsed domino face, were to be echoed in rectilinear form in the first "IBM cards." But while the holes in the original Jacquard system had activated mechanical teeth, as in a player piano, in Hollerith's system they allowed the completion of an electrical circuit that activated counters. For Hollerith, who worked in the Patent Office, the card would compel the great tapestry of the republic to yield up its patterns like an automatic loom in reverse.

There was another, less obvious but perhaps more important inspiration for the punch card. At some point, most likely in the 1870s, Hollerith observed a railroad conductor punching tickets. "One summer," he was to write, "I was traveling in the West and I had a ticket with what I think was called a punch photograph . . . the conductor . . . punched out a description of the individual, as light hair, dark eyes, large nose, etc. So you see, I only made a punch photograph of each person." Among Hollerith's effects was a small conductor's punch used to mark the tickets he described. (Individual conductors, incidentally, carried punches bearing distinct and recognizable shapes.) Each "punch photograph" verified that the passenger occupying the seat was in fact the same who had originally presented the ticket.

Hollerith was himself an amateur photographer—while courting his wife, he had taken hazy, romantic shots of their boating trips and picnics—and so the phrase "punch photograph" appealed to him.

But the vital leap was the idea that this was like a photograph of several simple kinds of information, a "digital" photograph. The form ticket held spaces for eye color, hair color, and so on to form a "digital" description of the ticketholder.

In the late 1880s, years after encountering the punched ticket, Hollerith was working on a related problem when he recalled the episode. Each decade's census was taking longer to compile than the previous one, so much longer, in fact, that one could not be finished before the next was to begin. Inspired, Hollerith proposed using punch cards such as those he saw on the train to record information about individuals, and, by counting them with an electrical device, to compile the general statistics of the census. Counters with clocklike faces toted the numbers read from the punched cards, in the process assembling a numerical profile of the entire population. From these cards evolved the punch card for tabulating machines and eventually for the computer.

Hollerith took the "punch photograph" idea one step further: he compared the census itself to a photograph. The digital compilation of "punch photographs" of individuals produced a portrait of the population and the country. The raw statistics—the punch cards—were analogous to exposing the photographic plate, and their manipulation in his machines he likened to developing: "As the development is continued, a multitude of detail appears . . . giving, finally, a picture full of life and vigor."

This high ideal was not exactly the image that Hollerith's card would later acquire. Instead, the punch card became a symbol of social control. The census was to suggest photography in another way. The Polaroid on the driver's license or other identity card, the punched card warning the user not to fold, spindle, or mutilate, both stood for the reduction of individuality to the lowest common denominators, the Identikit traits of wanted poster and physical description.

Hollerith's punch cards, reducing the individual to a group of constituent numbers and qualities, were to become a prime symbol of depersonalization—not unlike the mug shot or the identity-card photo. There was an echo here of Kierkegaard's famous warning that since everyone could now be photographed, no longer only the privileged, one photograph would serve as a portrait for anyone: we would all begin to look alike.

Hollerith's card was thus a symbol of social as well as technical change. The system that the conductor's punch had inspired cut in half the time it took to compile the 1890 Census. This was the census that to the historian Frederick Jackson Turner showed the end of the frontier expansion to the West which the railroads had helped bring about—and which now threatened the ideal of individualism.

Soon, big businesses and governments around the world were using Hollerith's machines. The railroads were the first: the New York Central signed Hollerith up in 1896; the Marshall Field department store in 1903. Prudential Insurance retained him, and the Russian government hired him to conduct its census in 1896.

By 1911, Hollerith had sold his Tabulating Machine Company. It became part of a new combine called the Computing-Tabulating-Recording Company, which also made time clocks and butcher machinery. Later, it was to become part of International Business Machines. IBM would prosper through the growth of government, obtaining many contracts for New Deal programs. In 1940, IBM was hired by the Pennsylvania Turnpike to work out a system to mark and sort the toll cards for the new superhighway, solving the complex problems of charging by distance and class of vehicle.

The punch card also helped set the pattern of the computing machine as one that served big organizations and abbreviated the identity—and eventually the power—of the individual. The "pantograph punch," an echo of Jefferson's polygraph, levered the power of the individual operator on Hollerith's cards; but in the process it threatened the privacy and identity of the individual customer or citizen.

☆   ☆   ☆

LIKE KODAK, Xerox was a made-up name, whose letters worked as a virtual incantation, promising miracles of new technology. Early demonstrations of Xerox machines featured the use of a paper bag as copying paper—to emphasize that the special paper of photostat machines was not required. And they showed—as the sewing machine makers had showed a century before and Kodak a half century before—a little girl using the machine.

But as a combination of electricity with photography, Xerox was

the machine Jefferson was after in his long search for the polygraph. This mechanical instrument for making two copies simultaneously Jefferson transported in a second version of his portable
desk. He saw it as a vital way of preserving and distributing the
documents of government—perhaps because of loss to fire and
seizure, as had happened to him as governor of Virginia during the
Revolutionary War.

The operation of one government department—the Patent
Office—inspired the invention. The story of the machine's invention is a classic American one, fit for Hollywood. Chester Floyd
"Chet" Carlson (even the name could come from a screenplay
writer's absentminded inventor) created "xerography" in the back
room of a beauty parlor owned by his mother-in-law in Astoria,
Queens. (Nearby, appropriately, lived Joseph Cornell, the master of
the Surrealist box of juxtapositions.) Inspired by the difficulties in
copying he had seen as a clerk in the Patent Office, Carlson conceived the idea of creating a photographic process in which electricity would replace chemistry.

On October 22, 1938, Carlson made what he considered the first
decent copy by the process, which one can see displayed like the
Shroud of Turin or St. Cecelia's handkerchief in Xerox's headquarters in Stamford, Connecticut. This first example was a date finger-
drawn in what was to be known as toner powder. Today, it suggests
some crude piece of art, a combination of Man Ray and Ed
Ruscha—a Ruschagram, perhaps.

Government was to become the largest customer for photocopying, and for the bureaucrat and the clerk, this "machine able to
produce words and drawings not photographically but from electrical signals" (in other words, an electric camera) had all the power
of the miraculous invoked by the company's famous "It's a miracle"
series of ads, in which medieval monks are overwhelmed by the
machine.

But personal papers, as well as personal opinions, could be served
by photocopying. The well-known restrictions on the use of Xerox
machines in the Soviet Union, for instance, showed their intol-
erability to absolutist regimes. And in this country the reproduction
of government documents on the sly—most notably the Pentagon
Papers—demonstrated the power of the Xerox as an addition to
freedom of speech and as a potential weapon against centralized

power. Too good a weapon, many felt, subjecting government to the risk of the photocopied leak, and giving the spy a handy machine to replace the Minox camera and the microfilm dot.

Some of these dangers were evident to all; the latest-generation color copiers reproduce with such laser-driven fineness that they can make simulacra of currency with every detail of fine engraving and color thread paper intact. The few thousand of these machines that existed in the late 1980s were restricted to use by defense-related government agencies and businesses, and kept under tight security. But no security is absolute. Eventually, one such machine will serve as the weapon for some dramatic act of heroism or crime.

For archiving personal papers, for throwing documents back at the bureaucracies that generated them, Xerox was a powerfully personal device. It is possible to look at a decade-old photocopy and see one's own hand, complete to each cut and cuticle, clinging to the edges of the hastily placed original. Then, you might remember that the body replaces its cells every seven years, and realize that it is an image not of you, but only of a body on which your present, older self is modeled. At about the same time that streaking was a campus fad, students, presumably punchdrunk around exam time late in libraries, began a fad for Xeroxing their genitalia. Photocopying can bring the past back or assert physicality in a far more eerie way even than a tape recording or a film or videotape.

Join the copier with the computer and you get "desktop publishing." Join the copier with the telephone and you get facsimile, which by the summer of 1986—it could be dated that exactly—became cheaper than international telex, and so inaugurated an era of the transmission of image instead of letter.

You paid for Xeroxing by the copy, just as you paid by the snapshot or by the razor blade or by the IBM card. Most machines were leased, in the familiar pattern of the computer industry. Xerox the company was, of course, a great business story. A man attending the twenty-fifth reunion of his college class filled in the class report blank for "occupation" with the simple line: "Bought Xerox in 1948." A taxi driver who invested $1,000 on a tip saw his stake rise to $1.5 million. Carlson had sold control of the rights to the process to the Battelle Foundation, which after rebuffs by Kodak, General Electric, and IBM finally interested a company called Haloid. Turning in later years to philanthropy, firmly grounded in the 40 percent of his invention he had retained, Chet Carlson died quietly in 1968, while watching a movie at the Festival Theater in Manhattan.

Xerox understood that it was in "the information business," and that computers and other office equipment were the logical area for expansion. In 1962, Xerox bought Max Palevsky's Scientific Data Systems, then one of the hot young computer companies; and in the early seventies, it set up a research center in Palo Alto to serve as its equivalent of Bell Labs. There, a computer engineer developed a more human "interface" for computers, using "icons" like international road signs and a pointing device called a "mouse." It developed all the essentials for desktop publishing and the personal computer and the computer network, it developed laser printers and the Ethernet office network system, but it couldn't figure out how to get the products into the office. For that, it would take a whole new type of thinking about information machines and their relationship to individual identity.

CHAPTER SIXTEEN

For Whom the Bell Tolled

"OOOH, THAT WHITE telephone!" marvels the waitress played by Mia Farrow in the Woody Allen film *Purple Rose of Cairo*. A small-town girl with big-screen dreams, she has never seen anything like it. Of all the things projected on the silver screen—the fine clothes, the cocktails at the Copacabana, the white piano, white curtains, glittering chromed appointments to white furniture in the luxury apartment—that white telephone is the one she most pines for. It is reminiscent of white phones in any number of black and white movies, some romances, some thrillers. In the thirties, when Woody Allen's movie is set, a white telephone was already as strange and rare as the purple rose itself, and as spiritual as a white whale.

That was because the telephone had recently been standardized. Like the Model T, the standard model 300 came in any color you wanted so long as it was black. White phones were for the rich, and the movies. The standard phone's color and shape was determined by Henry Dreyfuss Associates through exactly 2,240.5 hours of model making, testing, and refinement, of grappling with the "receiver-off-the-hook problem," and, always, of attempting to make the device lower and more compact. Dreyfuss had made the French phone—the handset combining transmitter and receiver—into the standard modern phone.

It was another classic case study of industrial design. In 1929, Dreyfuss had been one of a number of designers offered $1,000 to draw up sketches for a new phone. He turned the money down: the job could not simply be done as a piece of sculpture, he argued; it required a whole system of specifications. The company bought his argument, and Dreyfuss began extensive tests involving a thousand individuals. The 1930 model 300 became *the* telephone.

In 1946, Dreyfuss started the job all over again. Introduced in the early fifties, the model 500 had a lower, more rounded body, floating on feet, that marked a progression not unlike that from a '34 Chevrolet to a 1947. But so successful was the model 300 that the company feared some of the changes. The innovation of making the handle square in section, for instance, was considered risky: would it roll around in unskilled hands?

These phones were models of the idea of the telephone as universal appliance, a model of the universality telephone service had finally achieved, a sculptural icon of the possibilities of phoning— the kit of connections. A 1937 Coca-Cola ad paid indirect testimony to the universality of Dreyfuss's design. Showing female hands reaching for the phone on one side, grasping the Coke bottle on the other, it was headed: "Familiar acts that mark a better way of living. You reach for the Phone . . . or tilt this frosty Bottle." Both objects, the copy went on, were never far from where you are, and each "means a lot." Each represented and in a sense contained the whole service and delivery system that lay behind them. The phone was a package for the telephone company.

Everyone's phone was the black phone, but the black phone was not even your phone. It was rented from the telephone company— already, by the forties, "Ma Bell," the mother of us all. That black phone was a furnishing of the weary workaday world, its shape based on the measurements of a thousand heads and hands selected as samples. It had been designed for everyone so no one could have any complaints. The same model for office and home, requiring all those man-hours of time (the exactitude of the calculations corresponded to the exactitude of the process described), the dozens of clay models, such as a car designer might use, the ergonomic or "psychophysical" studies of exact angles at which people held the phone. The existence of a single color and shape of telephone for office and home helped to break down the barriers Victorian convention had sought to erect between the sordid world of business and the genteel world of the home—and between the man's and the woman's world as well.

☆   ☆   ☆

"Oooh, that white telephone!" Retro-styled "Hollywood" telephone, 1980s.

Refitted 1930s telephones. (Jadis Company, Los Angeles)

By 1880, MARK TWAIN was already noticing the change the telephone had made in social habits. Women could now be reached
more directly, via the telephone, instead of by cards passed into the
parlor. But the other side of it was the woman waiting by the
telephone, a role that made her only more aware of her inequality,
the callousness of men, and the one-sidedness of romance: "Why
can't I call him?" Later, the phone figured as a means of checking in
on a woman. In the fifties, the popular Nashville singer Jim Reeves
wooed his love over the phone, overpowering the physical presence
of his rival just as the phone and teenage girls were becoming so
inseparable that AT&T unbent stylistically enough to introduce the
Princess, exactly right for the purpose. Like the sewing machine
and the typewriter, the telephone quickly became associated with
women. They were soon adopted as switchboard operators because
it was thought their voices were more pleasant, they were physically

Coca-Cola advertisement, 1940.

FAMILIAR ACTS THAT MARK A BETTER WAY OF LIVING

You reach
for the Phone

or tilt this
frosty
Bottle

You probably never wonder how
that telephone, practically at your
elbow, got where it is. But it means
a lot to you to have it there. And as
for the ice-cold Coca-Cola you enjoyed today, or will enjoy tomor-

It could never have come about if
Coca-Cola had not been pure, whole-
some, delicious refreshment. For
Coca-Cola was perfected to be just
two things;—"delicious and re-
freshing." It takes skill to produce

And because of these things you
find Coca-Cola, ice-cold, at more
than 100,000 soda fountains and at
over a million other retail outlets.
Because of them you see the trademark "Coca-Cola" on familiar red
coolers, cartons, trucks, bottles and

more adept at the switching process, and they were more depend-able than the boys at first employed in the role.

☆　　☆　　☆

A UNIVERSAL APPLIANCE implied universal service. The telephone network and the telephone company made up one communications kit, with completely interchangeable parts each of which linked you to the whole.

Among the early celebrators of the phone was Herbert Casson, who pointed out that no single phone has any value in itself and, by extension, each additional phone added to the system increases the value of each previous phone. You use not only your phone but the whole system—the reticulated possibilities of switchboards and long lines. For a nickel deposited in a pay phone, Casson enthused, you can rent a multi-million-dollar system.

The idea of the telephone network took a while to crystallize. At first, no one knew quite what to use the phone for. One early idea was to use it for broadcasting—early demonstrations featured music piped in from distant concert halls and faintly projected. It was conceived as a public device. Only later did the pattern change to one of personal conversation.

"One policy, one system, universal service," was the slogan of Theodore Vail, who, beginning around 1900, pushed the telephone into every home and business and in the process established AT&T as a de facto monopoly. Low basic rates and increasingly sophisticated automatic switching systems helped make this possible. Bell himself had foreseen that automating the system would "so reduce the expense that the poorest man cannot afford to be without his telephone." Compare this with Henry Ford, circa 1915, talking about the Model T. "Soon," he said, "there will be no one who cannot afford a car."

In fact, the phone took a while to become anything like universal. In 1880, there were only 60,000 phones. The takeoff did not really begin until the 1890s. From 1896 to 1899, the number doubled; from 1896 to 1906, it increased eight times. In 1910, there were 6 million phones, and by 1920, 10 million, representing one for every ten people in the United States and 70 percent of all the phones in the world.

But even in the thirties, telephone distribution was far from universal: the infamous *Literary Digest* poll of 1936, which forecast Alf Landon as the winner over FDR, whose eventual victory was a landslide, was skewed because it sampled only those people with telephones.

☆   ☆   ☆

THE WHITE TELEPHONE was just a dream for decades. The white phone promised connection to a higher world, connections the normal standard-issue phone and the normal customer couldn't make—to, say, the world of starlets and producers ("Don't call us, we'll call you," says the agent of whom it is said in praise, "He gives good phone"). It was the symbol of the personal phone, the phone of romantic connection—who would call, what invitations?—and adjunct to the movies.

Until 1984, you could not connect any equipment not approved by the system to the system. The fact that your phone was part of the network was emphasized by the simple fact that the network owned it. Leasing phones was remunerative for the company, and few customers added up their monthly payments, but if they did, they realized that they would buy the phone many times over. Leasing was apparently originally the suggestion of a far-sighted Bell Company attorney who had previously worked for the Gordon McKay Shoe Machinery Company, which leased its sewing machines instead of selling them and charged by the number of shoes manufactured. It was McKay's sewing machine, introduced in 1862, that helped shoe the massive armies of the Union. Adding machines and computers were later marketed on the same basis. The integrity of the network was to be preserved at all costs. No alien apparatus, no strange bells, no competitive switchboards. No white phones. This not only gave the company a monopoly on lines and switching facilities but on the sale and rental of equipment. Never mind that simple connectivity standards for telephones had long ago made the black telephone as obsolete as the Model T, and phones as easy to connect as lamps to a socket.

In the twenties, the very shape of that white phone was exotic. The decade had only recently seen the arrival of the "French telephone," the familiar handset arrangement. (Marshall McLuhan makes much of its association of ear and mouth, corresponding to the nature of the French language.) The handset had been around since 1878, when a man named Robert Brown developed it for the operators. They had to walk distances among the plugs on early switchboards and carry their own plug-in sets with them, combining the mouth- and earpiece. Brown's handset was adopted in France by the Société Générale des Téléphones. The Society used women operators, and for them a lighter handset was created by the company's chief engineer, a man named Berthon. Soon it was used in home and business phones as well.

When the handset returned to the United States, there were problems. Sound quality suffered when the granular carbon of the mouthpieces—an Edison invention that improved on Bell's original—was moved around or addressed from an odd angle, especially on long-distance calls in the days before sufficient amplification. And in the United States, the lines tended to be longer. After the handset was introduced in 1927, equipped with a sort of scoop shape around its mouthpiece to ease the microphone problems, it became as fashionable as straw boaters or bobbed hair.

The term "French phone" connoted Continental sophistication and possessed racy associations with "French kiss" or "French letter." It suggested good girls languishing by the phone, as the popular song had it, but also call girls; the black phone reduced the need for the red-light district, McLuhan argued. The Mayflower Madam spent most of her time on the phone, using it to project an established and businesslike air well beyond the realities of her tiny enterprise—a lesson in the entrepreneurial powers of the phone.

☆   ☆   ☆

AROUND THE TURN of the century, there were competing phone companies: the original Bell Company, a Western Union–directed series of companies, and other local competitors. The Bell system pretended to be a sort of federal system. Herbert Casson called it "a federation of self-governing companies united by a central company," and added that Vail, a sometime gentleman-farmer, had pursued this consolidation "for the same reason that he built one big comfortable barn for his Swiss cattle and his Welsh ponies, instead of half a dozen small, uncomfortable sheds." Only by constituting itself in this kind of mock federalism—local companies sharing a common long-distance network—could it escape the attacks of the trustbusters.

Surely, no good American could object to that constitutional organization; it seemed so natural. But elsewhere Casson's enthusiasm causes him to slip up and acknowledge another sort of organization that was closer to the truth. AT&T was in fact organized on a military staff system, such as General Motors would later employ. "Staff and line" at Vail's AT&T, he argued, had done for companies what Von Moltke did for armies. The phone company aspired to be a Prussian information system. In fact, the telephone system itself would eventually undercut the system that ran it—the direct staff and line control, which was better suited to the linearity of telegraph than the omnidirectional net of the phone. Calling around your boss was now possible.

The original vision of the telephone network had been there with Alexander Graham Bell. He saw it as analogous to a system of plumbing. It was to be like "a perfect network of gas-pipes and water-pipes," he wrote in 1876. But at the same time, in his mind and others, there existed the competing vision of the telephone network as a centralized system of delivery of music, news—and, potentially, political propaganda. This last vision was what attracted Europeans, especially the always centralized French. The Europeans regarded the telephone as a public utility that should belong to the state—to the Post Office.

Americans, always uneasy with big business and central power, for a long time tolerated competition among telephone-switching systems. It took skillful public and political relations for the phone company to weed out its competitors. The process was accomplished by Vail's buying up smaller competitors, mostly between 1908 and 1913. But in 1913, when public suspicion of concentrated wealth and business power came to a head, Vail backed off and in the Kingsbury Commitment compromised with the new Democratic administration of Woodrow Wilson, agreeing not to gobble up any more independent telephone companies. He had already gotten most of them, enough to remain dominant while the others slowly withered at the end of the massive AT&T grapevine to which they had to connect for long distance, surviving as curiosities like old country stores until the forties. The arrival of long-distance wires, requiring a heavy capital investment, made duplication of resources prohibitively expensive. If ever there was a natural monopoly—conditioned by the nature of its business, and not by the tooth and claw of laissez-faire capitalism—it was long-distance telephone.

The genius of Theodore Vail lay in presenting the telephone from the mouthpiece end (from the user's viewpoint) as a network of opportunities. The key change was that of viewing the phone primarily as an outgoing instrument, a transmitter rather than a receiver. It was a simple matter of emphasis for marketing purposes. The phone as receiver was dangerous, full of fear—whether from bad news, from a stranger's abusive calls, a salesman's pitch, or a crank call. In *The Long Goodbye* (1954), Raymond Chandler speculates on the telephone's ambiguous attractions: modern man wants to be in touch, but the ringing of the phone also awakens a twinge of instinctual fear in him. "There is something compulsive about a telephone," says the private eye Marlowe. "The gadget-ridden man of our age loves it, loathes it, and is afraid of it. But he always treats it with respect. . . . The telephone is a fetish." The phone as receiver inspired that fear and respect. But the phone as transmitter emphasized its power and potential—it reinforced universality.

☆   ☆   ☆

IT WAS THE personal power of the telephone, requiring no special training, no office, no third-party sender, that distinguished it. This power was the focus of the company's appeal. In his wonderful letter of indignation to Hartford's gas company in 1891, Mark Twain fumed and cursed about an interruption in service, concluding: "Haven't you a telephone?"—so, that is, I can chew you out more directly, make you squirm, and receive immediate satisfaction? Advertisements beginning in the late twenties featured the powerful modern man in his office as a figure to be admired. Buy our insurance, car, or mouthwash, the appeal went, and you will share the power of this captain of industry. Advertising historian Roland Marchand calls this image "the master of all he sees" image. And the key icon of this picture, the man's scepter of power, his mode of transmitting his control out into the industrial landscape he surveys, was the telephone.

There was some truth to this depiction of the phone as the power lever of modern business. The arrival of the telephone, for instance, helped build the stock exchange—and to facilitate stock manipulation. Edward Harriman, the railroad magnate, was famous for having telephones everywhere he went—in all his houses, in his office, in his private railcar. Accused of being slave to the telephone, all he is said to have replied is: "Not at all; I am its master." His multiple phones were his weapons against the Morgans and Hills. Herbert Casson could barely contain himself: "What the brush is to the artist, what the chisel is to the sculptor, the telephone was to Harriman. He built his fortune on it."

Grover Cleveland put the first phone into the White House—and generally answered it himself. McKinley's "front porch" campaign of 1896, in which the candidate seemed to hold himself above the fray, was in fact the first telephone campaign. While the candidate posed on the white porch sipping lemonade with his neighbors in Canton, Ohio, while William Jennings Bryan crisscrossed the country recapitulating his "Cross of Gold" speech, boss Mark Hanna in the backroom was dunning big corporate contributions and organizing the McKinley campaign by phone.

It was not until the 1960s, however, that our presidents, beginning with John F. Kennedy, were commonly shown on the phone in power positions; before that, direct contact was preferred. Talking on an item anyone possessed seemed to diminish the presidential power. But the publicizing of the hotline, the red phone, in sixties movies—such as *Dr. Strangelove* and *Fail-Safe*—enhanced its iconic power. An acknowledgment of the global reach of power, it was the desktop version of *Air Force One*. If the white phone expressed the power of romance, the red phone expressed the romance of power.

☆   ☆   ☆

ONLY AFTER THE black phone had begun to be the universal standard did the white phone take on its special meanings. In 1928, a Bell vice president named Arthur W. Page illustrated the significance of the black phone when he urged the company to expand its market by emphasizing comfort and convenience. The wealthy could afford extra phones, Page argued. One way the comfort and convenience factor would be expressed was by phones of various colors and shapes. This stood in contrast to sheer necessity, basic service, as expressed by the single black phone. "Give the public what they *desire*," he argued. Already, Kodak had begun offering cameras in colors, and the Model T was in the process of being succeeded by the Model A, available not just in black but in Duco paints of red and blue and yellow. Why, Page asked, did AT&T sell only "one black desk set, a hand set, a wall set, and one of those black buttoned inter-communication systems"?

Ford, Page noted, "made one little black instrument, too, and it did just what ours did: when it got started, it went fine, and so did ours. But, you know, Henry has recently come to the point where he realized he had to make a change and I think now that he has made a lady out of Lizzie, we might dress up these children of the Bell System."

Ford had sold basic transportation, but now people wanted more. Page argued that people wanted more than basic telephone, too. His efforts were only partially adopted. AT&T had no General Motors to compete against. Some advertisements were run on Page's suggested themes. They spoke of the convenience of "enough telephones properly placed" to "prevent annoying delays when one is preparing for bridge, travel or the theater." But the color of the phones did not change—until the 1950s. Color came to phones in extension phones, in décor colors, for extra rooms. Bell now was fearful of a saturated market. It began to offer a few colors, and it replaced the stern black with beige as the most common color of standard issue.

By the 1980s, with the breakup of AT&T, customers could own their phones. New phones appeared in virtually every shape—footballs, cartoon cats and mice, white elephants and white whales. A thousand shapes bloomed. There are phones shaped like Porsches, phones shaped like women's shoes, phones shaped like a set of lips (courtesy of the original by Man Ray), phones shaped like piles of Tyco children's blocks. The "Classique," a "Classic French rotary phone in ivory with gold tone-accents," intones a recent catalogue, is "Perfect for the Boudoir!" It is the great white phone, straight from Hollywood.

Today, however, that still may not be good enough. What the customer really aspires to is a phone so powerful it is completely freed from the network, to the incorporation of the power of the network and its transcendence, entirely inside the individual instrument. In another Woody Allen film, *Play It Again Sam*, the joke is on the Hollywood producer who spews out a list of the telephone numbers at which he can be reached. He must always be in touch, reachable, reeling off a series of numbers that describe the sequence of his daily locations.

This was before the car phone, which would surely have solved his problem, or the briefcase cellular phone. The phone booth, dating back to 1889, first fulfilled this function. It developed from a huge kiosk for the wealthy (early booths in Bell offices were used to showcase long-distance service) to common utilities. The booth became office and tepee for the Times Square con men A. J. Liebling called "telephone booth Indians." Today, the phone booth, softly emitting its light across the empty prairie behind the roadside, or offering a bell jar of privacy in the midst of the noise of Grand Central Terminal, is an icon of individuality—and loneliness.

It represented the beginning of the process of disassociating phone from place that culminated in the portable phone, soon to be shrunk to the size of Dick Tracy's wrist radio—a private ear for private investigation. The absolutely unconnected phone was a sign of prestige. The car phone was so much a status symbol that by the mid-1980s pimps had taken to installing them—or putting on dummy phone antennas if they couldn't afford them. The cordless phone was as much an anomaly as the white phone. The mobile phone was another version of Moby phone, the mystic and magic white phone of infinite possibility.

Cordless phones were boosted by the prime-time soap operas, like *Dallas*. J. R. Ewing, sitting at breakfast on the patio of Dallas's South Fork, carries out his connivances in comfort, using a cordless phone. Like the distant tones lent by Speakerphones, this device offers the convenience of power, never far from command, with the tycoon letting his fingers do the walking all over his enemies and competitors.

☆   ☆   ☆

THE TELEPHONE FIGURES as the *deus ex machina* in many plays and movies of the period when it was just becoming universal. The ring offstage is as compelling a reason to move characters in and out as

Ray Milland in Alfred Hitchcock's *Dial M for Murder* (1954): the telephone as accomplice to murder. (Copyright © 1954 by Warner Bros. Inc.)

any backstage alarum. The ring of the phone can be either a welcome or a sinister intrusion—possible harbinger of bad news or threats. (The first telephone exchange had been developed as a sideline by a burglar alarm company.)

In *Sorry, Wrong Number*, the 1948 adaptation of Lucille Fletcher's radio play, the action centers on calls made through the single white telephone of the bedridden heroine. The film opens with a view of switchboard operators at work, over which flashes the premise: "In the tangled network of a great city the telephone is the unseen link between a million lives. It is the servant of our common needs—the confidant of our inmost secrets . . . life and happiness wait upon its ring . . . and horror . . . and loneliness . . . and *death*!!!" The phone is the film's *deus ex machina*.

Dial M for Murder, the play Alfred Hitchcock brought to the big screen, asserts that the phone magnifies the individual to the extent that he can make his impact felt in two places at once—in the theater lobby, secure with his alibi, and, via his human and electrical agent, back home, killing his wife.

The public phone, it once seemed, would leave no trail. Kidnappers called from phone booths; spies set up drop boxes through them. On television crime shows, efforts to "trace that call" always fail. If only the dirty rat had stayed on the line a little longer . . . Then came the wiretap, the bug, and the criminal could be caught in the coils of his own calls. (Telephonic evidence can also work the other way. A newspaper recounted the story of a man cleared of murder on the strength of the victim's telephone bill, whose timed entries proved the victim was still alive and chatting at a moment the accused man was seen elsewhere.)

By the beginning of the 1990s, it was possible for individuals to "trace" calls. When the telephone companies began to offer technology that could identify the number of a calling party, it used the same fears dramatized in the movies to sell the service. Dramatic television ads for "Caller I.D." service showed a lone woman picking up the phone and hearing an anonymous caller ask menacingly:

"Do you know who this is?"

"Is 123-4567 close enough?" she snaps, giving the caller's number displayed in LCD letters beside her phone. He gasps and hangs up.

Caller I.D. is a highly charged concept. Viewed from the point of the callee, it offers power. Viewed from that of the caller, it is a threat to privacy and power; the identification of callers, the possible compiling of lists of callers, offer potential power to businesses and government.

☆　　☆　　☆

THE POWER OF the telephone system as a whole could only be demonstrated by the criminal. The phone hacker and the blue box of the 1970s demonstrated not only that the system was the solution but that all its mysteries had solutions. You could, if you knew what you were doing, break into the great computer that was the phone system and call the Kremlin. A computer was an aid in dealing with a computer; MIT hackers used the school PDP-1 for this in the mid-sixties. So did the Matthew Broderick character in *War Games*: he found he could start World War III with a computer, a modem, and a phone call. He had the power of the red phone in his simple black phone.

"Phone phreaking" was a preview of hackers invading big computers. Their motive, said one phone phreak, quoted in Ron Rosenbaum's 1971 *Esquire* article, was that "the phone company is a System, a computer is a System. . . . The phone company is nothing but a computer." Some managed to steal, but most simply wanted to exercise power by sabotage, infecting other computers with destructive virus programs or destroying stored information. The phone invaders enjoyed simply hearing the clicks that signaled they had switched their way from Dayton to Dakar. One blind phone phreak would literally feel his way through switching systems, courtesy of tours the company allowed him to take. One game involved making connections that circled the globe.

Part of the attraction was power—power as proven by its destructive potential. "Captain Crunch," another of the phone phreaks Rosenbaum interviewed, and among the most famous, claimed he could "busy up" the nation's phone system with three callers. "Captain Crunch" was a man named John Draper, who took his name from the cereal in which he found a toy whistle that exactly replicated the 2,600-cycle frequency critical to hacking the phone system. (AT&T had slipped up, printing the frequencies used in its long-distance switching system in the *Bell System Technical Notes*, a publication avidly read by all true phone phreaks.)

Draper became a member of the famous Homebrew Computer Club and a friend of Steve Jobs and Steve Wozniak, the founders of Apple Computer, who were inspired by Rosenbaum's article and later by Draper himself to make blue boxes they sold to Berkeley students. Wozniak once called the Vatican using such a box, pretending to be Henry Kissinger. He nearly managed to have the Pope himself called to the line. Draper spent time trying to tap into the Defense Department's ARPANET computer connection system. And when Jobs and Wozniak hired him to design a phone interface for their Apple II computer, he created basically a blue box on a computer board. The outlaw skills of "hacking" the phone system formed the basis of computer hacking. Hacking at last made the single phone the instrument of tremendous power the phone company had all those years proclaimed it to be.

CHAPTER SEVENTEEN

"Fantasy Amplifiers," or Living in the Virtual World

"I WANT TO put a ding in the universe," was a favorite slogan of Steve Jobs, the pitchman for the Apple computer put together by his friend Steve Wozniak. What did that mean? It implied spiritual powers—to make the computer, in fact, a *Ding an sich*, a thing in itself, whose sheer potential to do things overrode the importance of anything it could actually do. Its myth really was that it put the universe in a *Ding*. The claims that were made for the personal computer were philosophical claims—claims about mind and imagination as grand as those made about space and wealth by the proponents of the railroad, the automobile, and the airplane. No device more than the personal computer realized the American ideals of combining technology to magnify the power of the individual and to embody his or her fantasies. The computer was no less than a model of the ideal mind.

The personal computer redeemed the image of the computer from that of the huge machine, covered with blinking mysterious lights, and spewing out cards that categorize, depersonalize, dun, cajole, and coerce individuals. In the Spencer Tracy/Katharine Hepburn film *The Desk Set*, based on William Marchant's play, the bosses bring in such a machine (the mainframe as seen circa 1957) to replace individual researchers. Of course, the initiative of the individual wins out over the machine, and the boss becomes humanized in the process. IBM, in introducing its PC in 1981, had to fight against the associations with such big, impersonal computers, as

GRiD Compass portable computer, 1982, Bill Moggridge, Stephen Hobson, I.D. Two, designers. (I.D. Two; photograph by Don Fogg)

symbolized by the unbendable, unfoldable, unspindleable, unmutilatable IBM card—the means traditionally used to program computers.

The personal computer was created by renegades—by the very phone phreaks who tapped into the great networks of AT&T, coming out of the counterculture of the sixties. Theirs was a model for liberation from big-business coercion, and a kit of revolutionary tools for freeing and empowering the individual. A personal computer, said Alan Kay, who did as much as anyone to create the device, "is a fantasy amplifier." More academic observers called it a device for "augmenting intellect." "The power to be your best" was the way the appeal was boiled down in Apple's slogan.

A vital polemic of the personal computer movement was Ted Nelson's 1974 book *Computer Lib/Dream Machines*. On one level, *Computer Lib* was a strange call for freeing computers, as one would free animals from the zoo. On another, it called for freeing the individual from the oppressions of big computers by giving him access to computing power. It was the inspiration for early ideas of community terminals for "time sharing." One of these was named "community memory," and it involved the talents of Lee Felsenstein, who would later design the Osborne 1 portable computer.

It begat grander visions: of "hypertext" and Project Xanadu, an on-line database of all the world's knowledge, a sort of a Borgesian interactive library. Hypertext was to present the greatest network of all, only completely personalized—a universe, but an individual's universe.

The first publicly available personal computers were literally kits. In January 1975, a small company called Altair announced that it was selling kits to electronics hobbyists. The kit was featured on the cover of *Popular Electronics* magazine, and brought in over 15,000 orders. Such electronics kits had been a staple of the basement do-it-yourself world. Companies like Heath had offered radios, stereo equipment, and even televisions for years. But a computer was something different.

What made the kit possible, of course, was the development of the microprocessor—the computer on a chip. Created at Intel by Ted Hoff in 1971, the microprocessor was originally intended for use in a calculator being produced by a client of Intel, a Japanese firm that went broke before the product was ever released. For a while, no one knew quite what to do with it.

☆   ☆   ☆

THE CREATORS OF the personal computer had solid counterculture credentials. Steve Jobs was dedicated to LSD, Zen, and a communal fruit farm before he ventured into computers. To finance the Apple computer, he sold his VW Microbus (the original hippie wagon), and when he produced the Macintosh he had the names of the original "cadre" of its developers molded into the inside wall of its plastic housing. Mitch Kapor, a former meditation instructor, created 1-2-3, the spreadsheet and graphics program that did for the IBM PC what Visicalc had done for the Apple II.

In *Hackers*, Steven Levy traces the beginnings of hacker culture—a utopian view of computers for all, free access, free software, and free information—to the late fifties and MIT's Tech Model Railroad Club. A huge set of complete HO-scale railroads, it was a world in miniature, a railroad system that—unlike those in the real world—ran on time and well. (Ted Hoff, the creator of the microprocessor, or computer on a chip, that was to form the heart of the personal computer, was the son of an employee of the General Railway Signal Company and worked for the company as a teenager. His name appeared on two of the company's patents.) The center of it all, and the connection to computers, was the switching system. The power center of the club was the Signals and Power Subcommittee, which controlled the schematic beneath the schema, the model below the models. According to Levy, the term "hacker" was born here: "to hack" was to make a brilliant connection in this system of switches.

It was an appropriate association, for railroads were the most systematic organizations of the nineteenth century—as Herman Hollerith found when he discovered inspiration for the IBM card in their punched tickets, and then showed them how to use the cards to run their business more efficiently. And the card would be resented as a symbol of big organization squeezing the small man, just as the railroads had been to the Grangers and Populists of the late nineteenth century. "Do not bend fold spindle or mutilate" had become not only a part of everyone's consciousness but a battle cry of the individual against the mass, the institution, and the bureaucracy that makes use of such cards.

The microcomputer was to the mainframe as the car was to the railroad, ran the hacker analogy. The railroad and the mainframe were owned by big, cold companies. They controlled the schedule— you bought a ticket, you bought "time sharing." With an automobile, you owned the machine and you chose the route. It was a comparison that ran through computer advertising. Apple's "test drive" program, which allowed customers to take a Macintosh home to examine, featured a hand wearing a fancy driving glove touching the mouse. The Apple II was compared to the Model T. Adam Osborne, creator of the first portable computer, said he wanted to be Henry Ford. "Get rid of your buggy whips," his ads intoned. If you weren't using a computer, you were still in the horse-and-buggy era.

The model railroad at MIT was controlled by elaborate switches, made up of sophisticated telephone relays and dials donated by Western Electric. In effect, the model railroad was controlled by a local telephone network. Out in the real world, the telephone system was the epitome of the big switching board, especially since the rail system had gone to hell quite a while ago. AT&T had, of course, paid for the creation of the transistor. The product of Bell Labs research had been transplanted, seedlinglike, to the San Francisco Bay area to grow up in a multibranched family tree of electronics "start-ups"—companies founded by engineers who left parent companies—a genealogy of "begats" whose forkings were depicted on a chart distributed by the electronics trade organization.

In the process, AT&T had handed the computer kids the power to invade its network. The power of the transistor, multiplied in the chip, allowed electronics whizzes to build whole networks in a box— a blue box, specifically, that could bypass and manipulate the phone company's switching system and really "send a foot through the line," in Bob Dylan's phrase—straight to the White House, SAC headquarters, or the Vatican. Long before they knew about personal computers, Steve Jobs and Steve Wozniak were obsessed by blue boxes; John Draper, the infamous "Captain Crunch," wrote programs for their early computers. His background unknown to IBM executives, the master of the blue box even wrote the first word-processing program for Big Blue's personal computer.

The pioneers of the microcomputer tended to see themselves as virtuous outlaws like Butch Cassidy or John Wesley Harding, assailing big business the way Western bandits had assailed big railroads and banks. (In the famous television ad that introduced Apple's Macintosh computer in 1984, an athletic figure smashes a giant screen on which appears the face of Big Brother.) They did not hesitate to buy parts on the black market or "borrow" them from companies where they worked. No wonder that, from the beginning, the personal computer was also viewed as a sort of weapon.

☆   ☆   ☆

COMPUTER MANUFACTURERS WERE among the few American industries that continued to take industrial design seriously. Searching for ways to make their products less intimidating and more understandable, they called in industrial designers to turn a technology that was still alien into one that was friendly. (No wonder several of the resulting shapes looked like, and were nicknamed, "ET.")

But not everyone wanted the machine to be so benign and friendly. Designer Stephen Hobson, who with Bill Moggridge of I.D. Two created the flat black GRiD laptop computer, said: "Beige is anonymous. It is associated with lack of power and authority. It fits in rather than stands out. We saw black, on the other hand, as reflecting a sense of precision and elegance, like cameras or other fine mechanisms. We didn't know what computing looked like so what we decided to do was make the box look as if it held *promise*.

"When you close the machine," he went on, "we want it to have the same precise, sensual quality to the fit that you get when you slam home the chamber of a gun."

The high-capacity hard disc that became vital to the business use of the microcomputer was for a long time known as the "Winchester." The surface pretext for the name was the fact that the device had been developed at an IBM laboratory located on Winchester Avenue (which in turn was named after the widow of the gun magnate, who built an eccentric mansion in the area) in Silicon Valley. But the reason the name stuck was that it also invoked the power of the Winchester carbine, especially when set beside the floppy disc drive—the "revolver" of the personal computer. By providing the mass storage that replaced bulky cassette recorders, the Winchester won the West for the personal computer.

An Epson ad for its easy-to-use QX-10 called it "the great equalizer," and sprinkled the page with revolver bullets. Apple named a keyboard for its Macintosh II the "Eastwood," with overtones of the star of westerns and urbans involving pistols from Colt to .44 Magnums. "The power to be your best," ran Apple's slogan, and that summed it up: a vaguely athletic overtone ("personal best") that made you think of running shoes, an item featured in the first Macintosh advertising to show the power of its graphics. In promising power, the computer showed that it aspired to the great American status of universal appliance, too, and lever of individuality, like the gun or car. Independence and self-reliance were also invoked: the tagline of another Apple commercial was "The power to make you independent." That ad showed a man starting his own company and using the computer to project the impression of having a large staff. Computerized "desktop" publishing was even touted as a means to freedom of the press. Over and over, desktop publishing advocates quoted A. J. Liebling: "Freedom of the press belongs to those who own one."

Steve Jobs compared the computer to a bicycle, accomplishing by some miraculous power of intellect the magnification of power the bicycle accomplished for the legs. Adam Osborne, the former computer-book publisher who created a minimal, portable computer, was fond of saying that "the computer is like a spade. You dig with your hands but you dig better with a spade. A computer is a spade for the mind." Oliver Ames would have been touched.

But Osborne's real success was in packaging. He made available the whole system, short of printer, but including programs, literally in one box, at one good price. He compared his deal to that of the razor manufacturer, who sold the blades—the software—and gave away the razor—the computer.

That Osborne's computer was portable was almost incidental to the real sell. Portability was at least as much a symbol of the self-containment of the computer as a source of utility in itself. The first machines were Jeeplike in their ungainliness, and that was part of the appeal. Even the sleeker case applied later by an industrial designer didn't change the basic form. A frequent comparison was to old Army field radios. The keyboard folded up to cover the face of the machine. At 26 pounds, the device was not easy to carry around, and its screen was the size of those on the old Hallicrafters television sets of the early fifties. Anyone who really used the thing quickly bought a larger separate monitor. Yet the Osborne users' magazine featured journalists carrying the computer on trips along with Afghani guerrillas, who carried their Stinger missiles on the backs of Tennessee mules. It was the sort of juxtaposition of primitive and high tech that Americans had always loved.

But just as the Model T had given way to more varied and romantically encased automobiles, the basic personal computer gave way to computers that were streamlined in shape and on screen. They possessed the equivalents of the automobile's self-starter and automatic transmission, and they came packaged as slickly as the Airflow or Zephyr. The first personal computers re-packaged the mainframe-shaped model of computing into individual units with a new culture. The next generation of computers repackaged the office inside the computer in a simulacrum that would supposedly make work easier and more fun.

"Computing environments" and "user interfaces" rebuilt the world of the desk in silicon and software. These user interfaces were also called "shells," covering the nuts and bolts of computer codes like the shells the industrial designers placed over machinery, leaving only the buttons and controls showing, hiding the mystifying and ugly innards. Windows, says the manual for one such software package, "is a graphical environment that introduces new, *more streamlined* ways for you to work with your personal computer."

The Apple Macintosh of 1984 was the first computer that aimed to make the machines easy enough for a novice to use—by packaging computer code in visual "icons." "The computer for the rest of us," was the slogan. It was to be everyman's computer, or, in another concept, "a bicycle for the mind," only without the crudeness of everyman's car.

The Mac introduced the first popular "graphic interface," in which the computer was controlled by moving a device called a mouse that, in turn, moved a pointer on the screen among "icons," shaped like international road signs, which indicated programs and files. When the button on the mouse was clicked, these icons turned into words or pictures. The icons tended to resemble ordinary objects on a desk—the computer, according to the ruling metaphor of the system, was an electronic desktop, and the images of calendar, pencil, Rolodex, calculator, and so forth made the machine familiar. It aimed at making all the programs work similarly. The Mac was the first computer that seriously addressed the key problem with computers: learning how to use them easily.

The term "icon" for visual symbols of the computer screen was dreamed up by David Canfield Smith, who claimed that in 1975 he had adopted it from religious sources: in the Russian Orthodox Church, he would later write, an icon "is more than an image, because it embodies properties of what it represents; a Russian icon of a saint is holy and is to be venerated." Combining these icons, which at the click of the mouse blossomed into programs, with a series of windows "tiled" together like so many Post-it notes on glass, or "cascading" like cards laid down in a game of Solitaire, the computer screen recreated the old world of the office in electrical charges. The icons resembled corporate logotypes, but they also looked like old-fashioned preliterate shop signs—the bootmaker's boot, the tailor's spool of thread in front of his shop—or the *kanban* system of Japanese signs.

One brilliant icon was Apple's apple itself: evoking the fruit of knowledge (the "byte" out of its side), Newton's apple, with its implications of discovery and inspiration, the healthiness and naturalness of the apple—Steve Jobs had picked apples on a collective farm and was a self-described "fruitarian" for a time—Johnny Appleseed and his all-American proselytizing, and even the Beatles record label, whose halved fruit appeared at the center of the records. The Macintosh development team spent $100,000 a year on fruit juice.

Development of the Macintosh was driven by these fuzzy ideas

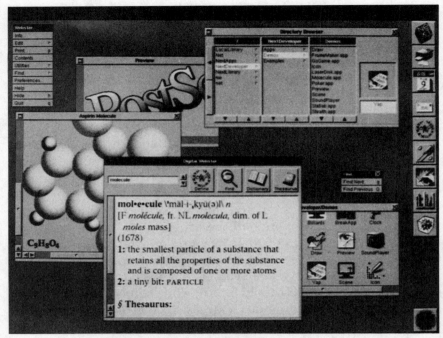

NeXTStep computer interface, 1989. (NeXT Inc.)

about making a universal appliance for "information" workers—a term taken from Peter Drucker, the management analyst. If the Apple II had been the closest thing to a Model T the industry produced, the Mac was aimed at being as universal as the telephone—but a telephone liberated from the telephone company.

Jobs compared the difference between the Macintosh and previous computers to the difference between the telephone and the telegraph. The telegraph was like earlier computers, he argued, in that you had to learn a "whole sequence of obscure codes." With the Mac, by contrast, "we want to make a product like the first telephone. We want to make mass-market appliances. . . . And, beside that, the neatest thing about it to me is that the Macintosh lets you sing the way the telephone did. You don't simply communicate words, you . . . express yourself."

And like the phone, the argument went, the Mac would create its own market. No one, Jobs said, could predict its impact on the world. When Jobs became particularly worked up, he would always call the Macintosh "insanely great."

Jobs also wanted the Macintosh to resemble the Cuisinart. Michael Murray, head of marketing for the Macintosh computer, used to hand out small toy food processors to other Apple staffers. They were a reinforcement of the vision of the Mac as a "desktop information appliance," a "thought processor."

These comparisons lent the Macintosh a personality: partisans called it whimsy; critics called it cuteness. One of the Mac's program categories—for games and so forth—was called "Goodies." The Mac was derided by the more serious types, who preferred IBM PCs: they called the Macintosh "a Yuppie etch a sketch." The Mac was West Coast, open collar, laid back. The PC was East Coast, coat and tie, intense. Television and ad types used Macs; accountants and lawyers used PCs.

Part of the Apple II's success came from its bright colors, part from the shell in which it was encased—a low beige box that gave it the air of serene self-containment and compactness. With the Macintosh, both the look of the machine—upright as a woodchuck, with a face as simple as a smile button—and the look of the screen helped make it become iconic.

The Mac's graphics inspired a whole look in the world of graphic design. Entire careers, such as April Greiman's, were built on its pixel-tweedy black and white look. Computer graphics were to echo the products of the Jacquard loom, which had inspired the machines to begin with.

With the zigzag edges that the pixels or picture elements assumed trying to shape objects and silhouettes, the screen possessed a herringbone, textural look, a swirl of regular geometries like early modernist patterns. The Macintosh suggested Charles Rennie Mackintosh.

Macs, having achieved the status of objects like the Braun coffeemakers and Dreyfuss-designed telephones on which they were in part modeled, were now free to be customized. Just like the telephone after the breakup of AT&T, the basic computer box, too, was decorated, colored, and textured. Special programs allowed you to "customize" your on-screen desktop, and a firm named Aesthetics Technology promised to silkscreen your computer in wood grain, marble, granite, and other patterns. Even Apple chief John Sculley used a marbleized Macintosh.

There was a reason for this sense of the Mac as a toy for adults. The technology for the Mac was a decade old at the time of the machine's appearance. It had been developed at Xerox's Palo Alto Research Center, a sort of Bell Labs of computing, and at the Stanford Research Institute, and used in a computer called the Alto. The work at Xerox PARC had looked to childhood development as its model: the child and his/her learning process as a stripped-down version of the human learning process in general. Jerome Bruner's theories were key to the development of Smalltalk, the language that drove the Alto computer. The computer was intended as a model of the mind, and the child's mind and learning process was a model of the adult's. Key to this thinking was play and fantasy. Alan Kay, reports Frank Rose, told one group of computer executives that they were "in the fantasy-amplification business— fantasy being nothing more than the process of building a convenient microworld, a world inside our heads that's smaller and more manageable than the world that really exists."

☆   ☆   ☆

THE THINKING OF Jean Piaget and Jerome Bruner held that learning begins from play and fantasy. Fantasy is tested experimentally against reality; the child recapitulates the scientific method. Learning is not a passive, receptive process, but an active, generative one.

Kay listened to, among others, MIT's Seymour Papert, who developed the language for children called Logo and a toy/tool called the turtle that he termed "a thing to think with." That is what the mouse was to become: a vehicle, if not necessarily for thinking, for manipulating symbols and programs, for moving about in the "microworld." Kay's work was also in keeping with the theory, confirmed so many times with all sorts of things technological—with the Brownie, with the Xerox machine—that if the kids can't work it, the adults sure won't be able to.

These ideas were linked to Kay's ideal of a computer the size of a book and as easy to use—the Dynabook. In 1972 he made a first sketch of the Dynabook, and showed children using it. Inside the Dynabook was Kay's "microworld."

No wonder the key device in this technology was called a "mouse," with its appropriate overtones of Mickey.* The mouse, so named because it possessed a wire tail and two button eyes, was invented in the mid-sixties by Douglas Engelbart. Properly known as an "X Y pointing device for digital systems," the mouse consisted of a sort of ball bearing mounted between two perpendicular rollers. When the mouse case was moved across the table on the ball, the rollers read its motion electronically and translated it into two coordinates. The analogue—twist of wrist, sweep of arm—was converted into the digital. A button (or two, three, or even five) on the mouse transmitted specific commands. Engelbart was the father of a vision of computers he called "augmentation." Inspired in part by the cybernetics theory of Vannevar Bush, Engelbart saw computers as devices by which to "augment" the individual intellect.

In October 1958, when Engelbart went to work at the Stanford Research Institute, the Soviet Union launched *Sputnik* I, sending American government, education, and business scrambling to close a perceived science and technology gap. One result of the scramble was the creation of the Defense Department's Advanced Research Projects Agency (ARPA), the R&D funding branch of the Pentagon. ARPA is probably the single most important influence in the development of computing in the United States: after Sputnik, it began to pour funding into promising efforts to increase the power of computers—often with only distant military potential. From ARPA research programs emerged such basic computer techniques as time sharing, computer graphics, and artificial intelligence. There was an irony here: the champions of personalizing computers were able to capture new technologies paid for by their arch enemy, the Pentagon.

By 1963, ARPA agreed to support Engelbart's ideas for a whole new series of devices to allow an individual to communicate directly with computers in an "augmentation research center" at Stanford

---

* The mouse—the Model T of biology, the basic lab animal, and the cartoon Mickey— provided a highly versatile set of icons. Claes Oldenburg made the mouse a theme of some of his best art. His mouse echoed Mickey but simplified him even further, to two circles (ears) and a square (head). The graphics of the Disney Channel showed how variable Mickey could be as the company logo. Essentially one large face circle and two smaller circles, he was depicted in those graphics as a TV dinner, or as amoebae seen in a microscope.

European and Japanese designers used Mickey's imagery to soften office equipment, as in Ettore Sottsass's secretary's chair, which he called "a little bit a mickey mouse object," or Toshiyuki Kita's Wink Chair, with its little flexible ears made into a headrest. At Radio Shack, Mickey showed up as a radio, with ears for controls, mouth as speaker.

Research Institute (SRI). Over the next few years he developed computer systems into "a new medium" that, he argued, would change the way we think as much as printing or the introduction of Arabic numerals did. One of the first fruits of his effort was word processing. "Automated external symbol manipulation," he called it in 1963, achieved through four classes of "augmentation means" (or tools): artifacts (physical objects), language, methodology, and training.

"It is likely," Engelbart argued, "that each individual develops a certain repertory of process capabilities from which he selects and adapts those that will compose the processes that he executes. This repertory is like a toolkit. Just as the mechanic must know what his tools can do and how to use them, so the intellectual worker must know the capabilities of his tools and have suitable methods, strategies, and rules of thumb for making use of them."

In the fall of 1968, Engelbart's tool kit of "augmentation means" was ready for demonstration. At the Civic Auditorium in downtown San Francisco, fronting the neoclassical variety of civic center, library, and city hall, Engelbart and his staff set up all sorts of sophisticated audiovisual equipment linked, via microwaves, to mainframes back at SRI.

Standing in front of a huge, crisp display screen, wearing headphones, Engelbart held in one hand a five-button control device and in the other his mouse. For two hours, he flashed windows of information and images on the screen, moving them, opening and closing them with a push of the buttons, tracing in his lecture the development of the lecture itself—from outline to the word processing of the final script. Engelbart appeared as a pilot: earphones and controllers lending the suggestion of movement (flight) through information displayed in the windows on the screen (the cockpit). It was a preview of the personal computers of fifteen years later.

But what was it all for? Engelbart's demonstration, showing a system to produce more systems like itself, offered the sort of recursive answer that was to plague understanding of the technology's potential. Engelbart had an answer of sorts. Picking up on the ideas of Peter Drucker, he noted that "information" or "knowledge" workers were supplanting manufacturing workers as the dominant sector of the economy, just as those manufacturing workers had supplanted agricultural ones early in the century.

Many of Engelbart's staff left SRI for the new research lab in Palo

Alto that Xerox created in the early seventies, where they would give his mechanistic augmentation theories a context in personal psychology. Recognizing that its business was not just making copies but "information processing," Xerox chief Peter McCulough believed the company needed a version of Bell Labs to move toward the office machines of the future—notably computers.

At PARC, Alan Kay, who had been in the audience for Engelbart's introduction of the augmentation system, saw its relevance to his Dynabook ideal. He worked on a new, visually oriented language that adapted many of its features such as the mouse, windows, and sharp graphics. Kay countered and extended Engelbart's "intellectual augmentation" with his notion of the computer as "fantasy amplifier," as if balancing the promise of new power for the mind with an equal appeal to imagination. His vision was of mind expansion—an idea that had much in common with counterculture interests of the time. Called Smalltalk, Kay's language was to be easy enough for a child to use.

At PARC, the window, icon, mouse, and pointer sort of interface (soon to be mocked under the acronym "WIMP") were made vital elements of the Alto desktop computer. Arguably the first personal computer, it arrived in 1974, doing many of the things that would not be generally available in personal computers until the 1984 Apple Macintosh.

☆   ☆   ☆

THE GRAPHICAL INTERFACE used in the Macintosh quietly became something like a standard. By the time it introduced its second line of personal computers, even IBM joined in, ratifying Apple's success with Windows, the program that allowed its computers to use a mouse-driven, graphical interface. The mouse, a device that symbolized the small computer's aims to give power to the weak, its cuteness, its connection with moving objects, became almost universal.

Today's mouse, as one wit put it, looks "like a Porsche 944 or a vibrator." Careful ergonomic testing goes into the creation of mice, but their look possesses fashion, too. Companies such as Microsoft became almost as obsessive in searching for the perfect mouse shape as car stylists. The mouse, after all, was a vehicle—a vehicle for moving through the microworld, for maneuvering through the information landscape. Fifty models and twelve working prototypes, for instance, were tested in the development of the "new" Microsoft mouse of 1988. The works inside remained the same, but the case and controls were redesigned, like a new car. The industrial design firms of Matrix Design and I.D. Two, working together, changed the mouse's "moment of inertia" to allow for "fingertip control" after a two-day test involving sixty mouse users revealed that it is the fingers, rather than palm, that most users employ to move the device. This ergonomic research produced a shape that could have come out of a wind tunnel. Microsoft's competitor, Logitech, proponent of the three-button mouse, offered a stylistic twist with its special "clear case" model, which revived the old gimmick of the transparent plastic shell—applied to radios, to telephones, and to a special "edition" of Polaroid's Spectra camera. It also hired frogdesign—the industrial designers for Apple and NeXT—to create a playful, curved mouse.

The mouse had become more than a simple functional device; like the car or the computer itself, it was a symbol of individual power and potential, whose shape and contours should suggest speed, strength, and some abstract, intangible future.

The mouse corresponded to the new realm of software design. While more and more programs mimicked desktop devices such as the Rolodex, they also attempted in the process to turn them into something more dramatic. The Rolodex was the model for the popular Hypercard program for the Macintosh, in which information was handled on the model of the notecard. It featured a succession of cards, each of which could contain words, images, or even entire programs, but each of which could also have relations to one another.

Hypercard attempted to create a less ambitious version of Ted Nelson's hypertext for the personal computer. This was the computer as cosmic Rolodex, following not the rigidities of the grid— the spreadsheet—or information processing, but the "natural" thought processes of the individual. It was stream of consciousness, "free" association. This was the way people such as those at Apple saw the future: a sort of sci-fi landscape of "information," through which one traveled with a "knowledge navigator." It was a utopian and romantic vision, turning on a comparison with the characteristic American obsession with transportation.

☆   ☆   ☆

THE COMPUTER OFFERED a tool kit for rearranging the world—or at least the world of "information," a word that has taken on far more weight since the arrival of the computer. The computer is an electric desk: Alan Kay's Dynabook or Apple's Knowledge Navigator is smaller than Jefferson's portable desk. They remained ideal products, like Detroit's "concept cars," but more practical laptop computers did catch on, shadows of the Dynabook dream as pale as their blue-on-white screens. Today, businessmen discuss their Toshiba laptops or GRiD laptops with gas plasma display with the pride Babbitt felt in his Buick. "It really increases my productivity," they say of the machine.

Such machines touch the most basic American aspirations: to rearrange the world painlessly from the comfort of a chair and desk, to make a model of the world and manipulate it like a kit. And it may be the last American technology. With the material reduced to the abstract, with "power" and "individuality" so generally projected and so specifically wedded to chips and pixels and icons, it is an appliance that aspires to the most abstract sort of universality.

To express such high-flying aspirations, computer makers, more than any group of manufacturers since the 1930s, embraced the world of the industrial designer. IBM retained Richard Sapper, designer of the Tizio lamp, to direct work on its PS/2 line of computers. Steve Jobs's interest in the field led him to hire frogdesign— which has offices in Germany, Silicon Valley, and Japan, and is the most prestigious organization since Dreyfuss's or Loewy's—first to shape Apple's computers and then to design NeXT, the computer Jobs built after he left Apple.

Esslinger and frogdesign playfully combined the European appreciation for American pop culture with the high-modern traditions of the Bauhaus and Ulm, the latter giving strength to the former and the former giving spirit to the latter. frogdesign was the sort of design office that made streamlining into an international business in the 1930s. Even the group's name reflected its approach: no one designed better mice for computers than frog.

The NeXT machine was a three-dimensional rendition of the logotype Jobs had commissioned from Paul Rand, creator of the classic IBM logo. As such, it objectified Jobs's personal legend as well as the technological power the machine promised. Jobs took the appearance of his machines very seriously. Computers he said, were "the metaphor of our time and should have a higher esthetic standard." He had decided on the shape of the computer long before he knew what it would do. It would be a cube, and Rand's logo suggested a cube as well.

Hartmut Esslinger at frogdesign shaped the machine as the ultimate black box: a basic cube, dark and mysterious as the black stone of Mecca. At trade shows, it was displayed inside gauzy curtains, cloaked in the secrecy of its development and the fascination Jobs and computers held for the public. It took on the mysticism of "computing" in general. Its black monitor was shaped to resemble a proscenium, a stage for the dancing icons displayed there. "Your whole world on a disc" was the slogan with which Jobs touted the optical disc-storage system. Jobs hired movie makers fresh from working on *Roger Rabbit* to turn out a promotional film for his machine. The film is a paean to automation; it follows the machines down the robot assembly line—recapitulating in contemporary terms the magic of the Ford and other assembly lines that fascinated film makers of the 1920s and 1930s. While IBM's ads parodied the assembly line in the manner of Chaplin, to show they were on the side of the little guy, Jobs made a hero of the assembly line at the same time that he made a myth of the product that came off that line.

"Wouldn't it be great," Jobs told the film makers, "if at the end we could have light just pouring out of the box?" The final shot of the film shows just that: a kind of *2001* monolith, a philosopher's stone, a black box from whose vents and apertures pours a mysterious brightness, the light of the universe streaming out of a single abstract thing. It is the computer apotheosized.

☆   ☆   ☆

THE COMPUTER SCREEN, with its dancing icons, marks a beginning and an end. Now that the desktop had become electric, it could sing of an electric world—a world of things reduced to icons, through which, with the best devices available, the individual was forced to make his way.

With the computer, the solidity of the world is deflated to iconology. Completing the process begun by film and television, electronics helped turn objects into mere conduits for sensation—gave them a soul that overshadowed their bodies, software that over-whelmed their hardware. The post-electronic world may be a flat world, but it is a world that has the advantage of being controllable, accessible, maneuverable through. It becomes a landscape.

When pilots of modern fighter airplanes see the world, it is through a display of images—grids, symbols, and gauges—projected onto their cockpits or helmets. All hovering lights and technical cartoons, such systems are called "heads-up displays." They show direction, speed, altitude, and attitude. Through their grids and boxes, lasers guide "smart bombs" to targets. The de-signers of these systems, which make flying the real sky similar to flying in simulated skies, call the information they present a "virtual environment."

The virtual environment is futuristic, iconic—like living in com-puterized simulation. It can be a strange thing to deal with, looking at the world this way, but as one Air Force officer explains: "You just have to get used to the symbology." We will all have to get used to the symbology. More and more the objects around us labor under the burden of their histories, styles, and associations. More and more we are surrounded by the icons of a virtual world.

Imbued with a sense of spiritual mission, the idealists who once created personal computers aspire to nothing less than "virtual reality," a simulacrum of a world created by electronic gloves and helmets, where the user feels sensations of movement, manipulates objects, and sees images—all created by electronics.

Virtual reality is suffused with the mysticism of technology, which has such a long tradition in this country. It promises a little America of its own—a Lockean New World where electricity re-places bodies, an epistemological tabula rasa, settled by electronic means, with the same utopian aspirations and the same potential for corruption. It is a garden of machines, an electronic Eden.

But the snake is already there. Alan Kay had warned against simply using computers to create a computer model of real things that were sufficient and useful unto themselves. "The display is the reality," said one outline for the Xerox Star workstation scheme in the mid-seventies, already far along the path Kay feared, and well on the way to virtual reality. Often the manipulators of these virtual realities suffer. Like seat-of-the-pants pilots suddenly forced to train in electronic simulators, they come down with simulator sickness—the result of the war between calculating mind and the inferring body, between brain and inner ear, pattern and instinct.

☆   ☆   ☆

THE ICONS OF electronics aspire to universality—to instant recognizability beyond language. These symbols are supposed to be more rapidly and more universally recognizable than words. A book by Henry Dreyfuss called *Symbol Sourcebook* attempted to define the basic language. It is a strange production—an encyclopedia of iconology—ranging from electrical contractor's symbols to hobo symbols. Dreyfuss's aim was nothing less than to create a universal visual language, an iconic lingua franca.

Dreyfuss, significantly, was the industrial designer most concerned with ergonomics. He did studies in the field—then called "physical psychology"—in preparing the design for Bell Telephone's basic desk instrument. He created the ideals of "Joe and Josie," ergonomically typical and average Americans. It was in his office that Niels Diffrient began his career shaping airline seats and interiors. From this beginning, Diffrient developed his own more exacting ergonomic models of the average American body that he adapted to chairs and office furniture. Dreyfuss's *Symbol Sourcebook* attempts to do for the eye and mind what ergonomics did for the body: find a universal model derived from the average. It proposes a "symbolics" that is the mental equivalent of ergonomics—and provides a visual and psychological streamlining. The sort of icons Dreyfuss tracked are the sign of international markets and cultures. The little snowflakes and trumpets on automobile dashboards indicate air conditioning and horn button in the United States, Japan, or Europe.

American culture had become a world culture, fed back to Americans by the Europeans and Japanese. There was no such easy definition of the three-dimensional icons—of the change that had taken place in the status of objects. But the more objects become vessels for information, rather than things in themselves, the more they begin to function in consciousness as fleeting icons on a screen. Things have to live up to the reputations of their images. Society had absorbed the changes in the nature of the object created by the advent of mass production; it had yet to absorb those created by the advent of mass marketing.

In response, we hunger for meaning, solidity, texture in objects—the qualities they seem in danger of losing. "High touch" is the phrase of the futurologists and management gurus, a response to "high tech." The word "icon" itself is evidence of the change. It has been devalued, widened in use, the way "tragedy" has been called on to serve in any news account with loss of life. A movie star or entrepreneur or artist is suddenly an icon. It is a process akin to that Daniel Boorstin analyzed by which celebrity replaces fame.

The process has made us more aware of the meanings of objects, the software behind all our hardware, the aura of associations and implications that runs beyond the crisp edges of objects, the networks of history in which they are bound up. Packaging has become as doubtful an idea as progress in a world whose frontiers of consumable resources and landfills are shrinking. And consuming itself has become an adversarial process, like a political or legal process, as iconoclasts from the new Puritans of leftist and greenish inclination deflate the claims of the marketers, and *Consumer Reports* charts counterbalance the blue sky promises of *Popular Mechanics* or *Popular Science*. Products today are put under cross examination.

☆   ☆   ☆

"I THREW THE bottle rack and the urinal in their faces as a challenge," lamented Marcel Duchamp in 1962, recalling the "readymade" works of art he had dreamed up back when the century was in its teens, "and now they admire them as something aesthetically beautiful."

"Aesthetically beautiful"—even his phrases gilded that particular lily. He had tried, he had long maintained, to introduce objects without aesthetic value into art by simple intellectual strategy. But the strategy had not worked. Instead of defeating "retinal art," Duchamp's ploys had helped make people look at the artistic qualities of ordinary objects. Once, Duchamp had seen a world full of things that could become readymades. But those who followed him saw a world already full of readymades.

In 1962, Duchamp was reacting to the new generation of Pop and "neo dadaist" artists—who were to appreciate the plain thing in artistic quotation marks they had learned from Duchamp himself, with cool and poker-faced irony. He was as shocked in his own way at the Pop artists taking ordinary objects seriously as others had been at his urinal.

Joyce wrote of irony as "the spiritual-heroic refrigerating apparatus, invented and patented in all countries by Dante Alighieri." To treat something ordinary this way was to cool it down. Duchamp's refrigeration went further than Dantean irony: it was the predecessor of the cool that now confronted him. Warhol with his soup cans, Jim Dine with his pastel-mystified tools, Oldenburg with his soft toilets and fans, and Jasper Johns with his casts of ale cans cooled down the individual thing, preserved it in the deep freeze of the gallery as in a big white refrigerator. These appreciators of the plain thing rendered into art had begun springing up with *Fountain* as one of their philosophical wellsprings.

And by making us look at ordinary objects in that way, Duchamp changed them. He removed part of their ordinariness, their commonality with the world of the everyday. He unplugged them, and spiritualized them. When Oldenburg changed size, changed materials, changed textures, he liberated the shape of the thing—the Gestalt. His floppy toilets are descendants of Duchamp's urinal, in which the software—shape—has been freed from the hardware—material.

Duchamp's strategy had a result he never anticipated: creating a perspective on the world in which things may be considered as works of art before they are considered as tools or devices, beautiful shapes that also just happen to do what they do: "Design Objects" reads a sign in the Museum of Modern Art store. In the wake of the modernist fast came an almost ravenous appetite for ornament and the periods when ornament grew most lush. *Victoria* magazine became a success by resurrecting the world of lace curtains and doilies. A hunger for new meaning—or new versions of the old meaning in objects—grew up as their images overtook them and they were commoditized.

"People want things that are hard to find. Things that have romance, but a factual romance, about them. . . . People want things that make their lives the way they wish they were," wrote cataloguer J. Peterman at the beginning of his catalogue of things he called classic, functional, but also romantic—things depicted not in photographs but in rough ink sketches, so the customer was required to imagine—and therefore personalize—them.

There was an increasing sense of the danger of things somehow slipping away, of thinning down to mere icons on the screen, simple trademarks—and a resulting thirst for texture, pattern, and patina. Zona, a chic SoHo boutique whose offerings mix Oriental and Southwestern "design" items, displays a sign that reads: "At a time when the ever expanding pressure of electronic tools and high technology is so pervasive, the need to balance our lives with products that celebrate the textural and sensorial [*sic*] becomes essential."

Awash in images and things which threaten to flatten into image, we also crave solidity. We seek more and more personalized icons. The consumer, assembling an identity from the kit of products, is the subject of "targeted marketing"—using more and more specific and individualized demographic and taste profiles that the computer has produced. The old ideals of mass production and the "American system" are increasingly displaced by notions of flexible assembly lines, short-run production, and customized products.

The ad and marketing folks have caught up with this need. They try to use it to lure customers to products offering solidity and warmth. They speak of creating a "bond" between person and thing. They evoke Daniel Boorstin's idea of "consumption communities"—of Tocquevillian association forming not with politics, religion, or hobby in common, but product ownership—the Airstreamer adapted. They also demonstrate quickly that the mixing of kits produces kitsch, the model become mode only.

By contrast, the designs of the thirties that were the subject of so much of Pop's quotation and irony seem newly attractive. Streamlining's contribution of "metaphorical" representation of function is echoed in the desire of contemporary designers to guide use and inspire affection by shapes. This is design by "symbology": the talk of industrial designers today is of metaphor and typeform, of Gestalts rendered in caricature. It is as if Pop art had fed back into the source of its own subject matter: the forties electric fan that Claes Oldenburg modeled in droopy kapok inspires the playful nineties electric fan. Objects should be like the icons on the computer screen—cartoony and easy to recognize and use.

Style now seems not a progression, as the thirties industrial designers saw it, in countless charts depicting the gradual streamlining of things, but a dialectic, turning back on itself every few years, alternating forgetfulness with nostalgia. In the midst of such a dynamic, the warnings of those who decried manipulative design and advertising seem slightly hysterical. Even the most cynical design dictates of artificial obsolescence—that, as the influential early advertising man Earnest Elmo Calkins said, "styles wear out faster than gears"—seem quaint in a time when technology offers its own ready obsolescence. It is no accident that the artifacts of the era inspired by Calkins—the era of streamlining and classic American industrial design—have become among the most prized of "collectibles." Old toasters and irons sell for over $1,000, prewar Harley Earl automobiles and ornate jukeboxes are almost clichés of wealthy collectors.

The word "collectibles" suggests a new function to collecting, a recognition of a world full of objects that are more than objects, that are as much virtual as material things. Collectibles are the furnishings of one's life, placed in the living room to illustrate character, like those deployed by art directors in films.

In an increasingly skeptical market, the old commercial sells that dictated the shapes of things—toasters, etc.—were seen like the old pretensions of the movies, taken with a grain of salt, with late-show mellowness. We gained an appreciation of the streamline shapes of the props of film noir, bathed in light and shadow to best swell the contours of their shells, bubbling with nostalgia like the physically bubbling tubes of Paul Fuller's Wurlitzer jukebox played by Robert Mitchum in *Out of the Past*.

The irony was that objects depicted in two dimensions had more power and appeal than simple, solid, real ones. It said a lot about collectibles that film props were the most prized among them. Steven Spielberg bought the "Rosebud" sled from *Citizen Kane* ("I thought they burned that," someone said). Melrose Avenue in Los Angeles once housed many of the leading sellers of film props. Now the stores there sell antiques—thirties Deco, forties Moderne, fifties modern. Period objects to dress a set. They sell white telephones, just like you see on the screen. Melrosebuds—the objects of childhood, or films seen in childhood—turned into icons of character in the firm conviction that you are what you collect.

No wonder that manufacturers seek "product placements" for their goods in films, almost subliminal ways of marketing that lend contemporary products some of the quality of relic. Those films had always sold American culture not only to Americans but, more important now, to the world, which sold it right back to us.

☆   ☆   ☆

THE PHOTOGRAPHER Walker Evans, who was interested in artless art, argued that taking the picture of something is the same thing as displaying a found object. He was particularly taken with commercial and official signs, which he called "graphic 'found objects.'" After years of collecting, as he put it, images—particularly, in his last years, of painted signs—he at last began to take the signs themselves home.

In the text prepared for a show simultaneously of Evans's photographs of signs—Chesterfield, Nehi, Coke, "No Trespassing"—and of the signs themselves, mounted at Yale University in 1971, he wrote:

> Assuredly, these objects may be felt—experienced—in this gallery, by anyone, just as the photographer felt them; in the field, on location. The direct, instinctive, bemused sensuality of the eye is what is in play—here, there, now, then.
>
> A distinct point, though, is made in the lifting of these objects from their original settings. The point is that this lifting is, in the raw, exactly what the photographer is doing with his machine, the camera, anyway, always. The photographer, the artist, "takes" a picture: symbolically he lifts an object or a combination of objects, and in so doing he makes a claim for that object or that composition, and a claim for his act of seeing in the first place. The claim is that he has rendered his object in some way transcendent, and that in each instance his vision has penetrating validity.

"The object rendered transcendent" is a definition of the icon. And a summary of the expectations not only of the artist but of the ordinary person, for the things he/she uses and owns. The means by which, moving through the world of things and the imagination of icons, one can not only constantly create new worlds but chase the classic American goals: individual definition, self-assurance, empowerment, and individualism amid standardization. It is a means by which the individual can independently pursue happiness.

☆ **Bibliographic Notes** ☆

General Sources

Banham, Reyner. *Theory and Design in the First Machine Age*. New York: Praeger, 1967.

_____. *A Concrete Atlantis*. Cambridge, Mass.: MIT Press, 1986.

Bayley, Stephen. *In Good Shape: Style in Industrial Products, 1900–1960*. New York: Van Nostrand Reinhold, 1979.

_____. *Harley Earl and the Dream Machine*. New York: Knopf, 1983.

_____, ed. *The Conran Directory of Design*. New York: Villard, 1985.

Boorstin, Daniel. *The Americans: The National Adventure*. New York: Random House, 1965.

_____. *The Americans: The Democratic Adventure*. New York: Random House, 1973.

Bush, Donald J. *The Streamlined Decade*. New York: Braziller, 1975.

Burlingame, Roger. *Machines That Built America*. New York: Harcourt, 1953.

Caplan, Ralph. *By Design*. New York: McGraw-Hill, 1982.

Chandler, Alfred D., Jr. *The Visible Hand: The Managerial Revolution in American Business*. Cambridge, Mass.: Harvard University Press, 1977.

Chevalier, Michel. *Society, Manners and Politics in the United States*. Augustus M. Kelley, 1966 (first published 1839).

Christenson, Erwin O. *The Index of American Design*. New York: Macmillan, 1950.

Clark, Victor S. *History of Manufactures in the United States*. New York: McGraw-Hill, 1929.

Doblin, Jay. *One Hundred Great Product Designs*. New York: Van Nostrand Reinhold, 1970.

Drexler, Arthur, and Greta Daniel. *Introduction to Twentieth Century Design*. New York: The Museum of Modern Art, 1959.

Dreyfuss, Henry. *Designing for People*. New York: Simon and Schuster, 1955.

Edwards, Owen, and Betty Cornfeld. *Quintessence*. New York: Chronicle, 1983.

371

Edwards, Owen. *Perfect Solutions*. New York: Chronicle, 1989.

Ewen, Stuart. *All-Consuming Images: The Politics of Style in Contemporary Culture*. New York: Basic Books, 1988.

Ferebee, Ann. *A History of Design from the Victorian Era to the Present Day*. New York: Van Nostrand Reinhold, 1970.

Forty, Adrian. *Objects of Desire: Design and Society from Wedgwood to IBM*. New York: Pantheon, 1987.

Furnas, J. C. *The Americans: A Social History of the United States, 1587–1914*. New York: Putnam, 1969.

Giedion, Siegfried. *Mechanization Takes Command*. Cambridge, Mass.: Harvard University Press, 1948.

Gies, Joseph and Frances. *The Ingenious Yankees: The Men, Ideas and Machines That Transformed a Nation, 1776–1876*. New York: Crowell, 1976.

Greeley, Horace, Leon Case, et al. *The Great Industries of the United States*. J. B. Burr & Hyde, 1872.

Greenough, Horatio. *Form and Function: Remarks on Art, Design and Architecture*. Berkeley: University of California Press, 1947.

Greif, Martin. *Depression Modern: The Thirties Style in America*. New York: Universe Books, 1975.

Hanks, David, et al. *High Styles: Twentieth-Century American Design*. New York: Whitney Museum, 1985.

Hawke, David. *Nuts and Bolts of the Past: A History of American Technology, 1776–1860*. New York: Harper & Row, 1988.

Heide, Robert, and John Gilman. *Dime-Store Dream Parade: Popular Culture, 1925–1955*. New York: Dutton, 1979.

Heskett, John. *Industrial Design*. New York: Oxford University Press, 1980.

Hess, Karl. "The Good Machines," *Quest/80* (February–March 1980).

Hiesinger, Kathryn B., ed. *Design Since 1945*. Philadelphia: Philadelphia Museum of Art, 1983.

Hindle, Brooke, and Steven Lubar. *Engines of Change: The American Industrial Revolution, 1790–1860*. Washington, D.C.: Smithsonian Institution Press, 1986.

Hine, Thomas. *Populuxe*. New York: Knopf, 1986.

Holbrook, Stewart. *Machines of Plenty: Pioneering in American Agriculture*. New York: Macmillan, 1955.

Kasson, John F. *Civilizing the Machine: Technology and Republican Values in America, 1776–1900*. New York: Grossman, 1976.

Kaufman, Edgar, Jr. *Introduction to Modern Design*. New York: The Museum of Modern Art, 1950.

Kiernan, Thomas. *The Road to Colossus: Invention, Technology and the Machining of America*. New York: Morrow, 1985.

Kouwenhoven, John A. *Made in America*. New York: Doubleday, 1948.

Loewy, Raymond. *Never Leave Well Enough Alone*. New York: Simon and Schuster, 1951.

———. *Industrial Design*. Woodstock, N.Y.: Overlook Press, 1979.

Lucie-Smith, Edward. *A History of Industrial Design*. New York: Van Nostrand Reinhold, 1983.

Lynes, Russell. *The Domesticated Americans*. New York: Harper & Row, 1963.

Marx, Leo. *The Machine in the Garden*. New York: Oxford University Press, 1964.

Morison, Elting. *Men, Machines and Modern Times*. Cambridge, Mass.: MIT Press, 1966.

Mumford, Lewis. *Technics and Civilization*. New York: Harcourt, 1934.

Norman, Donald A. *The Psychology of Everyday Things*. New York: Basic Books, 1988.

Oliver, John W. *History of American Technology*. New York: Ronald Press, 1956.

Papanek, Victor. *Design for the Real World*. New York: Van Nostrand Reinhold, 1984.

Pevsner, Nikolaus. *The Sources of Modern Architecture and Design*. New York: Oxford University Press, 1968.

Pulos, Arthur. *American Design Ethic: A History of Industrial Design to 1940*. Cambridge, Mass.: MIT Press, 1983.

———. *The American Design Adventure*. Cambridge, Mass.: MIT Press, 1988.

Rybczynski, Witold. *Taming the Tiger: The Struggle to Control Technology*. New York: Viking, 1983.

Schaefer, Herwin. *Nineteenth Century Modern*. New York: Praeger, 1970.

Sexton, Richard. *American Style: Classic Product Design from Airstream to Zippo*. New York: Chronicle, 1987.

Sheldon, Roy, and Egmont Arens. *Consumer Engineering: A New Technique for Prosperity*. New York: Harper, 1932.

Singer, Charles, ed. *A History of Technology*. Oxford: Clarendon Press, 1954–84.

Sobel, Robert, and David B. Sicilia. *The Entrepreneurs: An American Adventure*. Boston: Houghton Mifflin, 1986.

Sparke, Penny. *An Introduction to Design and Culture in the Twentieth Century*. New York: Harper & Row, 1986.

———. *Design in Context*. New York: Chartwell, 1987.

———. *Electrical Appliances: Twentieth Century Design*. New York: Dutton, 1987.

Stevenson, David. *Sketch of the Civil Engineering of North America*. London, 1838.

Strasser, Susan. *Never Done: A History of American Housework*. New York: Pantheon, 1982.

Susman, Warren. *Culture and History: The Transformation of American Society in the Twentieth Century*. New York: Pantheon, 1984.

Teague, Walter Dorwin. *Design This Day: The Technique of Order in the Machine Age*. New York: Harcourt Brace, 1940.

Van Doren, Harold. *Industrial Design: A Practical Guide*. New York: McGraw-Hill, 1954.

Wallance, Don. *Shaping America's Products*. New York: Reinhold, 1956.

Weingartner, Fannia, ed. *Streamlining America*. Dearborn, Mich.: Henry Ford Museum, 1986.

Wessel, Joan, and Nada Westerman. *American Design Classics*. Design Publications, 1985.

Wilson, Mitchell. *American Science and Invention*. New York: Simon and Schuster, 1954.

York, Neil Longley. *Mechanical Metamorphosis: Technological Change in Revolutionary America*. Westport, Conn.: Greenwood Press, 1985.

1   The World in a Package

On Jefferson's desk: Silvio A. Bedini, *Thomas Jefferson and His Copying Machines*, University of Virginia Press, 1984, and Susan R. Stein, "Thomas Jefferson's Travelling Desks," *Antiques* (May 1988). Also see Jack McLaughlin, *Jefferson and Monticello: The Biography of a Builder*, Henry Holt, 1988; *Memoirs of a Monticello Slave As Dictated to Charles Campbell in the 1840's by Isaac, One of Thomas Jefferson's Slaves*, ed. Rayford W. Logan, University of Virginia Press, 1951; Garry Wills, *Inventing America: Jefferson's Declaration of Independence*, Doubleday, 1978; and Charles Eames, *The World of Washington and Jefferson*, Smithsonian Institution Press, 1976.

Poe's words are taken from his "Philosophy of Furniture," in *Selected Prose, Poetry and Eureka*, ed. W. H. Auden, Holt, Rinehart and Winston, 1950.

On the chuck wagon: J. Evetts Haley, *Charles Goodnight, Cowman and Plainsman*, Houghton Mifflin, 1936.

On the cowcatcher and railroad, see Brooke Hindle and Steven Lubar, *Engines of Change*, Smithsonian Institution Press, 1986.

The term "fantasy amplifier" was coined by Alan Kay.

H. L. Mencken, "The Libido for the Ugly," in *The Vintage Mencken*, ed. Alistair Cooke, Vintage, 1956.

Reyner Banham, "The Great Gizmo," in *Industrial Design* (September 1965).

Julian Harrison Toulouse, *Fruit Jars*, Nelson/Everybodys, 1961.

Oscar Wilde, "Impressions of America," in *The Artist as Critic: Critical Writings of Oscar Wilde*, ed. Richard Ellmann, Random House, 1969. Also see Richard Ellmann, *Oscar Wilde*, Knopf, 1988.

On Duchamp: The ancestry and meanings of the urinal in Duchamp's *Fountain* are ably unfolded in Kirk Varnedoe and Adam Gopnik, *High and Low: Modern Art and Popular Culture*, The Museum of Modern Art, 1990. See also Pierre Cabanne, *Dialogues with Marcel Duchamp*, Da Capo, 1987, and Duchamp's anonymous defense of *Fountain* and the ready-made in the periodical *The Blind Man*, 1917.

On the ergonomics of the urinal: Alexander Kira, *The Bathroom*, Viking, 1966.

On the Museum of Modern Art design collection: *Machine Art*, with essays by Alfred Barr and Philip Johnson, Museum of Modern Art, 1934. Also Sidney Lawrence, "Clean Machines at the Modern," *Art in America* (February 1984), and Joseph Masheck, "Embalmed Objects: Design at the Modern," *Artforum* (February 1975).

2   Blades

Walker Evans, "The Beauties of the Common Tool," *Fortune* (July 1955), reprinted in Lesley K. Baier, *Walker Evans at Fortune, 1945–1965*, Wellesley College Museum, 1977.

On American axes: Hawke and Kouwenhoven. Also *Collins Axe: 100 Years*, The Collins Company, 1926; Eric Sloane, *A Museum of Early American Tools*, Funk & Wagnalls, 1964; Henry J. Kauffman, *American Axes: A Survey of Their Development and Their Makers*, Stephen Green Press, 1972; and James Fenimore Cooper, *Notions of the Americans*, New York, 1829.

For the Ames shovel: Azel Ames, Jr., "A Day with the Shovel Makers," 1870, republished in *History of Easton*, Vol. II, *1886–1974*, Easton Historical Society, and William Chaffin, *History of the Town of Easton, Massachusetts*, John Wilson & Son (Cambridge, Mass.), 1886. H. H. Richardson's role as Ames family architect is recounted in James F. O'Gorman, *H. H. Richardson: Architectural Forms for an American Society*, University of Chicago Press, 1987.

On the plow: Thomas Jefferson, Letters, in *Writings*, Library of America, 1984; Edward C. Kendall, "John Deere's Steel Plow," *Contributions from the Museum of History and Technology*, vol. 218, no. 2, 1959; and Wayne G. Bruehl, Jr., *John Deere's Company*, Doubleday, 1984.

On the Bowie knife: Robert Abels, *Bowie Knives*, Ohio Historical Society, 19th, and Harold L. Peterson, *American Knives*, Scribner's, 1958. On Bowie's fight at the Vidalia sandbar, see R. Hofstadter, ed., *Violence in America*, Vintage, 1968.

3   The Equalizer, or the Confidence Machine

J. Frank Dobie, *Stories of Christmas and the Bowie Knife*, Steck (Austin, Tex.), 1953.

On the long rifle: John G. W. Dillin, *The Kentucky Rifle*, National Rifle Association, 1924; Neil Longley York, *Mechanical Metamorphosis: Technical Change in Revolutionary America*, Greenwood Press, 1985; M. L. Brown, *Firearms in Colonial America*, Smithsonian Institution, 1980; and Louis A. Garavaglia and Charles G. Worman, *Firearms of the American West, 1803–1865*, University of New Mexico Press, 1984. George Shumway, *Longrifles of Note*, self-published, 1953, and Shumway's study *Conestoga Wagon*, self-published, 1968, suggest parallels in the development of the two.

On violence in the American West: Robert Dykstra, *The Cattle Towns*, Knopf, 1968; Gary Wills, *Reagan's America: Innocents at Home*, Doubleday, 1987; Richard Hofstadter, "America as a Gun Culture," *American Heritage* (October 1970); Roger McGrath, *Gunfighters, Highwaymen and Vigilantes: Violence on the Frontier*, University of California Press, 1984; K. D. Kirkland, *America's Premier Gunmaker: Remington*, Exeter Books, 1988; Alden Hatch, *Remington Arms, an American History*, Rinehart, 1956; R. L. Wilson, *Colt: An American Legend*, Abbeville Press, 1985; Charles Haven, *A History of the Colt Revolver*, Morrow, 1940; and Charles W. Sawyer, *Firearms in American History*, Cornhill, 1910.

On the development of the "American system": H. J. Habakkuk, *American and British Technology in the Nineteenth Century*, Cambridge University Press, 1962; and David Hounshell, *From the American System to Mass Production 1800–1832*, and *The Development of Manufacturing Technology in the United States*, Johns Hopkins University Press, 1984.

On the Whitney myth: R. S. Woodbury, "The Legend of Eli Whitney and Interchangeable Parts," *Technology and Culture*, vol 1, no. 3, 1960, reprinted in *Technology and Culture: An Anthology*, eds. Melvin Kranzberg and William H. Davenport, Schocken, 1972. See also Nathan Rosenberg's Introduction to his edition of *The American System of Manufactures: The Report of the Committee on the Machinery of the United States*, 1855, Edinburgh University Press, 1969; Constance Green, *Eli Whitney and the Birth of American Technology*, Little, Brown, 1956; and Merritt Roe Smith, "Eli Whitney and the American System of Manufacturing," in Carroll Pursell, ed., *Technology in America*, MIT Press, 1981.

On the decline of Colt: "Army Dumps Colt," *New York Times*, Oct. 8, 1988, and John Holusha, "Colt to Sell Unit That Won the West," *New York Times*, April 29, 1989.

4 Little Factories

On clocks: David Landes, *Revolution in Time: Clocks and the Making of the Modern World*, Harvard University Press, 1983, and J. R. Dolan, *The Yankee Peddlers of Early America*, Clarkson Potter, 1964. The story of Henry Ford's assessment of the watch business is to be found in *My Life and Work*, Doubleday, 1926, by Henry Ford with Samuel Crowther.

On the sewing machine: Grace Rogers Cooper, *The Sewing Machine: Its Invention and Development*, Smithsonian Institution, 1968; Ruth Brandon, *Singer and the Sewing Machine*, Lippincott, 1977; and J. R. Dolan, *The Yankee Peddlers of Early America*.

On typewriters: Bruce Bliven, *The Wonderful Writing Machine*, Random House, 1954, and Richard N. Current, *The Typewriter and the Men Who Made It*, University of Illinois Press, 1954. The Densmore, Yost & Company advertisement is reprinted in *Asher & Adams Pictorial Album of American Industry* (1878), Knopf, 1976. Also see Cynthia Monaco, "The Difficult Birth of the Typewriter," *American Heritage of Invention & Technology* (Spring/Summer 1988). On the keyboard, see Stephen Jay Gould, "The Panda's Thumb of Technology," *Natural History* (January 1987), and "Whence Qwerty?" *PC Computing* (October 1988). On the IBM Selectric typewriter, see Joan Wessel and Nada Westerman, *American Design Classics*, Design Publications, 1985, and Paul B. Carroll, "Americans' Love of IBM Typewriters Outlasts Company's," *Wall Street Journal*, Aug. 3, 1990.

On the cash register: Isaac F. Marcosson, *Wherever Men Trade: The Romance of the Cash Register*, Arno Press, reprint, 1972, and Gerald Carson, "The Machine That Kept Them Honest," *American Heritage* (August 1966).

On kitchen appliances: Earl Lifshey, *The Housewares Story*, National Housewares Manufacturers Association, 1973. For blender marketing, *New Times*, April 29, 1977. Also see James Beard, *New Recipes for the Cuisinart*, 1978, and Penny Sparke, *Electrical Appliances*, Dutton, 1987.

5 Easy Chairs

David Hanks, *Innovative Furniture in America from 1800 to the Present*, Horizon, 1981.

Siegfried Giedion's extensive treatment of patent furniture and railroad seating in *Mechanization Takes Command* (Harvard University Press, 1948) is famous.

On the Shakers: Edward Deming and Faith Andrews, *Work and Worship: Economic Order of the Shakers*, New York Graphic, 1974; June Sprigg, *Shaker Design*, Whitney Museum of Art/Norton, 1986; and John Kassay,

The Book of Shaker Furniture, University of Massachusetts Press, 1980. Charles Dickens described his visit to a Shaker village in *American Notes* (1842).

For "plain" plantation furniture and the ladderback tradition: John T. Kirk, *American Furniture and the British Tradition to 1830*, Knopf, 1982; *Neat Pieces: The Plain-Style Furniture of 19th Century Georgia*, Atlanta Historical Society, 1983; and Derita Coleman Williams, "Early Tennessee Furniture," *Southern Quarterly* (Fall 1986).

On Hitchcock chairs: John Tarrant Kenney, *The Hitchcock Chair*, Clarkson Potter, 1971, and Dolan, *The Yankee Peddlers of Early America*.

For the easy chair, I have consulted Barcalounger and La-Z-Boy Company promotional materials.

On ergonomics: For Niels Diffrient's explanation of how ergonomics made design acceptable to engineers and managers, see Arthur Pulos, *American Design Adventure*, MIT Press, 1988. Also, Diffrient's *Human Scale* series, and U.S. Department of Defense (Harold P. Van Cott and Robert G. Kinkade, eds.), *Human Engineering Guide to Equipment Design*, 1972. K. F. H. Murrell, *Ergonomics: Man in His Working Environment*, Chapman & Hall, 1965. Also see Stephen T. Pheasant, *Bodyspace: Anthropometry, Ergonomics and Design*, Taylor & Francis, 1986.

6 Kit Homes: Log Cabin and Balloon Frame

On log cabins: Harold R. Shurtleff, *The Log Cabin in America*, Harvard University Press, 1939; reprint, P. Smith, 1967; Roger Fischer, *Tippecanoe and Trinkets Too: The Material Culture of American Presidential Campaigns, 1828–1984*, University of Illinois Press, 1988; Robert Gunderson, *The Log Cabin Campaign*, University of Kentucky Press, 1957; and C. A. Weslager, *The Log Cabin in America from Pioneer Days to the Present*, Rutgers University Press, 1969. On Lincoln's home in Springfield, see Geoffrey Ward, "Lincoln House," *American Heritage* (April 1989). Also telephone interview with Fran Krupka, architect. For contemporary descriptions, see Henry David Thoreau, *Walden* (1854) and *The Maine Woods* (1864), and Frederick Law Olmsted, *A Journey in the Seaboard Slave States in the Years 1853–1854* (1859).

On variations in cabin types: Henry Glassie, *Pattern in the Material Folk Culture of the Eastern United States*, University of Pennsylvania Press, 1969, and Dell Upton and John Michael Vlach, eds., *Common Places: Readings in American Vernacular Architecture*, University of Georgia Press, 1985.

On fences: John R. Stilgoe, *Common Landscape of America, 1580 to 1845*, Yale University Press, 1988.

For the kit tradition in commercial and industrial building: James Bogardus, *Cast Iron Buildings, Their Construction and Advantage*, J. W. Harrison, 1856. On Jefferson's nail factory, see Jack McLaughlin, *Jefferson and Monticello*, Henry Holt, 1988.

On the balloon frame: Carl Condit, *American Building*, University of Chicago Press, 1968, and Walker Field, "Reexamination of Origins of the Balloon Frame House," *Journal of American Society of Architectural Historians* (October 1942). Also Robert M. Cour, *The Plywood Age: A History of the Fir Plywood Industry's First Fifty Years*, Douglas Fir Plywood Association, 1955.

Also see Clifford Edward Clark, Jr., *The American Family Home, 1800–1960*, University of North Carolina Press, 1986; Alan Gowans, *The Comfortable House*, MIT Press, 1986; David Handlin, *The American Home: Architecture and Society, 1815–1915*, Little, Brown, 1979; and Allen G. Noble, *Wood, Brick and Stone: The North American Settlement Landscape*, vol. 1, *Houses*, University of Massachusetts Press, 1984.

7 Wonders of the Modern World, or Bridges and Plumbing

Robert Hale Newton, *Town & Davis Architects: Pioneers in American Revivalist Architecture 1812–1870, Including a Glimpse of Their Times and Their Contemporaries*, Columbia University Press, 1942. See also John Stilgoe, *Common Landscape of America*.

Alan Trachtenberg's splendid *Brooklyn Bridge: Fact and Symbol*, University of Chicago Press, 1965, covers far more than the structure of its title; it is a basic resource for understanding the special meanings of bridges in the United States. See also Elizabeth B. Mock, *The Architecture of Bridges*, The Museum of Modern Art, reprint, 1972; Daniel L. Schodek, *Landmarks in American Civil Engineering*, MIT Press, 1987; and Carl Condit, *American Building*, 2nd ed., University of Chicago Press, 1982.

On the George Washington Bridge: Pamphlets and press releases from the Port Authority of New York and New Jersey. Also Le Corbusier, *When Cathedrals Were White*, McGraw-Hill, 1964 (first published 1947).

On dams: For *Life*'s first issue, with the Fort Peck Dam story, see Loudon Wainright, *The Great American Magazine: An Inside History of Life*, Knopf, 1976. On Hoover Dam, see Joseph E. Stevens, *Hoover Dam: An American Adventure*, University of Oklahoma Press, 1988, and Marc Reisner, *Cadillac Desert*, Viking, 1986.

For imaginative philosophical and phenomenological interpretations of the skyscraper, see Rem Kohlhaas's *Delirious New York*, Oxford University Press, 1978, and Thomas A. P. Van Leeuwen, *The Skyward Trend of Thought: The Metaphysics of the American Skyscraper*, MIT Press, 1988.

On the Woolworth Building: Spencer Klaw, "The World's Tallest Build-ing," *American Heritage* (February 1977). The story of the opening of the Chanin Building is to be found in "Dramatic Mr. Chanin Presents 'A Thrill for All,' in 56 Stories," *New York Times*, Jan. 15, 1929.

The Chrysler tool kit is described in Walter Chrysler with Boyden Sparkes, *The Life of an American Working Man*, Dodd, Mead, 1950.

Hugh Ferriss, *The Metropolis of Tomorrow*, Princeton Architectural Press, reprint, 1986. For the grocery store with Ferriss-derived facade, see David Gebhard and Robert Winter, *Architecture in Los Angeles, A Compleat Guide*, Gibbs M. Smith, 1985. Also see Dickran Tashjian, *Skyscraper Primitives: Dada and the American Avant-Garde, 1910–1925*, Wesleyan University Press, 1975, and Paul Goldberger, *Skyscraper*, Knopf, 1987.

For the Ferris wheel: John Kasson, *Amusing the Millions*, Hill & Wang, 1970, and John Kouwenhoven's essay, "The Eiffel Tower and the Ferris Wheel," in his *Half a Truth Is Better Than None*, University of Chicago Press, 1981. Also see David Braithwaite, *Fairground Architecture: The World of Amusement Parks, Carnivals and Fairs*, Praeger, 1968.

For stadium history: Harold Seymour, *Baseball: The Early Years*, Oxford University Press, 1960; the editors of the *Sporting News, Take Me Out to the Ball Game*, The Sporting News, Inc., 1983; and Noel Hynd, *The Giants of the Polo Grounds*, Doubleday, 1988. Also Phil Patton, "A Close Encounter with the Superdome," *Geo* (September 1981), and Patton, "The Wall and Other Bizarre Afflictions Pertaining to Boston's Crypto-Mythical Red Sox," *Connoisseur* (September 1986).

On grids: Thomas P. Hughes, *Networks of Power*, Johns Hopkins University Press, 1983, and his article, "The Inventive Continuum After the Discov-ery: How the Telephone, the Automobile, and the Electric Grid Grew Up," *Science 84* (November 1984). Also Hughes, *American Genesis: A Century of Invention and Technological Enthusiasm, 1870–1970*, Viking/Penguin, 1989.

On space frames: Daniel A. Cuoco, "Today's Space Frame Structures," *Architectural Record* (June 1982).

8 Ship Shapes and Land Yachts

On lift models and ship design: Carl Cutler's *Greyhounds of the Sea: The Story of the American Clipper Ship*, 3rd edn., United States Naval Institute, 1984; Charles Boswell, *The America: The Story of the World's Most Famous Yacht*, David McKay, 1967; A. B. C. Whipple, *The Challenge*, Morrow, 1987; Howard I. Chapelle, *The History of American Sailing Ships*, Norton, 1935; *The Search for Speed Under Sail 1700–1855*, Bonanza, 1967; Bruce M. and C. Gardner Lane, "New Information on Ships Built by Donald McKay," *The American Neptune*, 1982; and Charles K. Stillman, "The

Development of the Builder's Half-Hull Model in America," *Mystic Seaport Log*, Fall, 1979. John Griffiths's *Treatise on Marine and Naval Architecture* appeared in 1850.

Louis C. Hunter, *Steamboats on the Western Rivers: An Economic and Social History*, Octagon, 1969.

Melville's *The Confidence Man* (1857), Mark Twain's *Life on the Mississippi* (1883), Charles Dickens's *American Notes* (1842), Michel Chevalier, *Society Manners and Politics in the United States* (1839), and Augustus M. Kelley, 1966, give a picture of the technology of steam riverboats and life on board as it emerged in the 1840s and 1850s.

On the development of the railroad passenger car: John H. White, *The American Railroad Passenger Car*, Johns Hopkins University Press, 1978; Lucius Beebe, *Mansions on Rails*, Howell-North, 1959; and *Mr. Pullman's Elegant Palace Car*, Doubleday, 1961. Wolfgang Schivelbusch, *The Railroad Journey: Trains and Travel in the 19th Century*, Urizen, 1979, is a treatment of the subject in the manner of Walter Benjamin. Siegfried Giedion in *Mechanization Takes Command* traces in detail the development of the Pullman car and its variants.

On the town of Pullman, Illinois: Stanley Buder, *Pullman: An Experiment in Industrial Order and Community Planning, 1880–1930*, Oxford University Press, 1967.

On the Airstreamers: Wally Byam, *Trailer Travel Here and Abroad*, David McKay, 1960; John Huey, "This Unusual Group Is Always Traveling, Never Leaves Home," *Wall Street Journal*, Aug. 1, 1985; and Doug Stewart, "To Airstreamers, A Nomad's Life Is the Good Life," *Smithsonian* (December 1985).

9   Model T to Tail Fin

It is amazing how little has been written about automobile design, one of the most characteristic areas of American design. Most material is uncritical and for buffs. Exceptions are Edson C. Armi, *The Art of American Car Design: The Profession and the Personalities—"Not Simple Like Simon,"* Pennsylvania State University Press, 1988; Paul C. Wilson, *Chrome Dreams: Automobile Styling Since 1893*, Chilton, 1976; and Jerry Flint, *The Dream Machine*, Quadrangle/New York Times, 1976.

Vital to understanding the auto industry in general are Brock Yates's classic summary, *The Decline and Fall of the American Automobile Industry*, Vintage, 1984, and David Halberstam's *The Reckoning*, Morrow, 1986. Also see James J. Flink, *The Automobile Age*, MIT Press, 1988.

On car culture: Gerald Silk, et al., *Automobile and Culture*, The Museum of Contemporary Art, Los Angeles/Abrams, 1984, and Stephen Bayley, *Sex, Drink and Fast Cars*, Pantheon, 1986.

On the Model T: Jane Heap in *The Little Review* (Spring 1927), and Jay Doblin, *One Hundred Great Product Designs*, Van Nostrand Reinhold, 1970; Henry Ford with Samuel Crowther, *My Life and Work*; Reynold Wik, *Henry Ford and Grass-Roots America*, University of Michigan Press, 1972; and David L. Lewis, *The Public Image of Henry Ford: An American Folk Hero and His Company*, Wayne State University Press, 1976.

On Henry Leland and the Cadillac: Mrs. Wilfred C. Leland with Minnie Dubbs Millbrook, *Master of Precision: Henry M. Leland*, Wayne State University Press, 1966, and Alfred Sloan, *My Years with General Motors*, Doubleday, 1964.

On the Chrysler Airflow: Howard S. Irwin, "The History of the Airflow Car," *Scientific American* (August 1977), and James J. Flink, "The Path of Least Resistance," *American Heritage Invention and Technology* (Fall 1989).

On General Motors: Harley Earl, "I Dream Automobiles," *Saturday Evening Post*, Aug. 7, 1954.

On Jack Telnack and the aero look: Phil Patton, "The Shape of Ford's Success," *New York Times Magazine*, May 24, 1987, and Earl Powell, "On the Road Again," *ID* (November–December 1987).

On Frank Lloyd Wright's Zephyr: Herbert Muschamp, *Man About Town: Frank Lloyd Wright in New York City*, MIT Press, 1983, and Brendan Gill, *Many Masks: A Life of Frank Lloyd Wright*, Morrow, 1986. Also see Michael Lamm, "The Beginning of Modern Auto Design," *Journal of Decorative and Propaganda Arts* (Winter/Spring 1990), and Charles Willeford, *Sideswipe*, St. Martin's Press, 1987. The striping machine was described in the "New Patents" column, *New York Times*, April 67, 1990.

On the Cadillac: Robert C. Ackerson, *Cadillac: America's Luxury Car*, TAB, 1988, and Ed Cray, *Chrome Colossus: General Motors and Its Times*, McGraw-Hill, 1980. Peter Drucker tells the story of Cadillac's marketing to blacks in *Adventures of a Bystander*, Harper, 1979. Also see Stephen Bayley, *Harley Earl and the Dream Machine*.

On the Jeep: A. Wade Wells, *Hail to the Jeep*, Harper, 1946; Dickson Hartwell, "The Mighty Jeep," *American Heritage* (December 1960); and miscellaneous U.S. Army publications.

10   Airlines

On the Wright brothers: Fred Howard, *Wilbur and Orville: A Biography of the Wright Brothers*, Knopf, 1987, and Lynanne Wescott and Paul Degan, *Wind and Sand: The Story of the Wright Brothers at Kitty Hawk*, Eastern Acorn, 1983.

On the development of the airline industry: Carl Solberg, *Conquest of the Skies: A History of Commercial Aviation in America*, Little, Brown, 1979; Roger Bilstein, *Flight in America, 1900–1983*, Johns Hopkins University

Press, 1984; Ronald Miller and David Sawers, *The Technical Development of Modern Aviation*, Praeger, 1968; Joseph Corn, *The Winged Gospel*, Oxford University Press, 1983; Richard P. Hallion, *Legacy of Flight: The Guggenheim Contribution to American Aviation*, University of Washington Press, 1977; John B. Rae, *Climb to Greatness: The American Aircraft Industry, 1920–1960*, MIT Press, 1968; and Marylin Bender and Selig Altschul, *The Chosen Instrument: Pan Am, Juan Trippe: The Rise and Fall of an American Entrepreneur*, Simon and Schuster, 1982.

On William Stout: Rich Taylor, "The Prophet," *Special Interest Autos*, January–February 1976.

Walter Dorwin Teague's paean to the curves of the DC-3 is found in his *Design This Day*, Harcourt Brace, 1940. The DC-3 story has been turned into legend by a spate of articles, books, and television treatments on occasion of its fiftieth anniversary. See Michael Parfit, "The Only Substitute for an Old DC-3 Is Another Old DC-3," *Smithsonian* (September 1988).

On the flying boats: Kenneth Gaulin, "The Flying Boats: Pioneering Days to South America," *Journal of Decorative and Propaganda Arts* (Winter/Spring 1990).

On the P-38, Constellation, and SR-71: Clarence "Kelly" Johnson with Maggie Smith, *More Than My Share of It All*, Smithsonian Institution, 1985. Also William E. Burrows, *Deep Black: Space Espionage and National Security*, Random House, 1986.

On flying wings and stealth: E. T. Wooldridge, *Winged Wonders: The Story of the Flying Wings*, Smithsonian Institution, 1983, and Theodore von Kármán with Lee Edson, *The Wind and Beyond: Theodore Von Karman, Pioneer in Aviation and Pathfinder in Space*, Little, Brown, 1967.

On the Stealth bomber: Eileen White Read, "They Sneak Around, Learning What They Can About Stealth," *Wall Street Journal*, April 26, 1988. John Northrop's viewing of the Stealth bomber model is recounted by Rick Atkinson, "Stealth: From 18-Inch Model to $70 Billion Muddle," *Washington Post*, Oct. 8, 1989. The world's most involved observer of Stealth aircraft is Bill Sweetman. See his *Stealth Aircraft*, Motorbooks International, 1986, and *Stealth Bomber*, Motorbooks International, 1989.

On the Rogallo wing: Michael Rozek, "Flexible Fliers," *Air and Space* (December 1987/January 1988).

11   Streamlines

For general histories of the early years of industrial design and streamlining in America, see Donald J. Bush, *The Streamlined Decade*, Braziller, 1975; Martin Greif, *Depression Modern: The Thirties Style in America*,

Universe Books, 1975; Jeffrey L. Meikle, *Twentieth Century Limited: Industrial Design in America, 1925–1939*, Temple University Press, 1979; and Fannia Weingartner, ed., *Streamlining America*, Henry Ford Museum, 1986.

One of the classic statements of industrial design turned to the purposes of marketing is Earnest Elmo Calkins, "Beauty: The New Business Tool," *Atlantic Monthly* (August 1927). The classic American industrial designers in the first blush of their success are treated in "Both Fish and Fowl," *Fortune* (February 1934).

For a contemporaneous statement of the aesthetics of Art Deco turning streamline, see Paul Frankl, *New Dimensions: The Decorative Arts of Today in Words and Pictures*, Payson & Clarke, 1928. Each of the big four of thirties American industrial design celebrated his vision in print: Henry Dreyfuss with *Designing for People*, Raymond Loewy in *Never Leave Well Enough Alone*, Walter Dorwin Teague in *Design This Day*, and Norman Bel Geddes in *Horizons*, Little, Brown, 1932.

On appliances and packaging: Susan Strasser, *Satisfaction Guaranteed: The Making of the American Mass Market*, Pantheon, 1989, and Stuart Ewen, *All-Consuming Images*, Basic Books, 1988.

The story of the refrigerator market between the wars is told in detail in "The Nudes Have It," *Fortune* (May 1940); Richard S. Tedlow, *New and Improved: The Story of Mass Marketing in America*, Basic Books, 1990; and Sears's role in Donald Katz, *The Big Store*, Viking, 1987. Raymond Loewy boasts of his role in making the Sears Coldspot a success in *Industrial Design*.

12   Self-Service

On the Automat: J. T. Farrell, "The Last Automat," *New York* magazine, May 14, 1979.

On the proliferation and variety of vending machines: *Report of the President's Research Committee on Recent Social Trends in the United States*, 1933.

The original Piggly Wiggly has been replicated at the Pink Palace Museum, Memphis, Tennessee. It is described in "The Cousin of the Cafeteria," *Scientific American*, Sept. 7, 1917. Kemmons Wilson tells his tale in *Holiday Inn Story*, Holiday Press, 1973. See also "Rapid Rise of the Host with the Most," *Time*, June 12, 1972.

On the dime store: James Brough, *The Woolworths*, McGraw-Hill, 1982, and Simon J. Bronner, "Reading Consumer Culture," in Bronner, ed., *Consuming Visions: Accumulation and Display of Goods in America, 1880–1920*, Winterthur Museum/Norton, 1989.

On Texaco service stations: I consulted Texaco Company publications and Howard Mansfield, "Birth of a Station," *ID* (September–October 1990).

On the cafeteria: Harvey Levenstein, *Revolution at the Table: The Transformation of the American Diet*, Oxford University Press, 1988, and Chester Liebs, *Main Street to Miracle Mile: Roadside Architecture*, New York Graphic Society, 1985. Also Daniel Cohen, "Cold or Hot, Put Your Nickel in the Slot," *Smithsonian*; John F. Love, *Behind the Arches*, Bantam, 1986; and Ray Kroc, *Grinding It Out*, Contemporary, 1977.

On the basic tripartite meal: James Deetz, *In Small Things Forgotten: The Archaeology of Early American Life*, Doubleday/Anchor, 1977.

13   Rural Electrification, or Eclectic Guitar

On the jukebox: Christopher Pearce, *Vintage Jukeboxes*, Chartwell, 1988, and Bill Barol, "The Wurlitzer 1015," *American Heritage* (September–October 1989).

On the phonograph and record industry: Roland Gelatt, *The Fabulous Phonograph*, Macmillan, 1977.

On radio's influence on music mechanization: Philip Collins, *Radios: The Golden Age*, Chronicle, 1987.

The story of Dylan at the Newport Festival is recounted in Robert Shelton, *No Direction Home: The Life and Music of Bob Dylan*, Morrow, 1986.

On country-and-western music: Bill C. Malone, *Country Music, USA*, University of Texas Press, 1968; Nick Tosches, *Country: The Biggest Music in America*, Stein & Day, 1977. I have also consulted a number of photocopied and typescript biographies from the Country Music Hall of Fame Library.

On the one-strand and primitive guitars: William Ferris, *Blues from the Delta*, Doubleday/Anchor, 1978.

On the guitar: Frederic V. Grunfeld, *The Art and Times of the Guitar*, Macmillan, 1974, and Pete Welding, "The Birth of the Blues," in Allan Kozinn, *The Guitar*, 1984.

On the blues: Lawrence W. Levine, *Black Culture and Black Consciousness: Afro-American Folk Thought from Slavery to Freedom*, Oxford University Press, 1977; W. C. Handy, *Father of the Blues: An Autobiography*, Collier, 1970. Also Robert Palmer, *Deep Blues*, Viking, 1981; Peter Guralnick, *Searching for Robert Johnson*, E. P. Dutton/Obelisk, 1989; *Feel Like Going Home: Portraits in Blues and Rock 'n' Roll*, Outerbridge, 1971, and *Lost Highway: Journeys and Arrivals of American Musicians*, Godine, 1969.

For the country music connection: Michel Bane, *White Boy Playing the Blues: The Black Roots of White Rock*, Penguin, 1982; Greil Marcus, *Mystery Train: Images of America in Rock'n'roll Music*, Dutton, 1975; David Evans, *Big*

Road Blues: Tradition and Creativity in the Folk Blues, University of California Press, 1982; Giles Oakley. *The Devil's Music: A History of the Blues*, Taplinger, 1977.

Jimi Hendrix's links to early rock-'n'-roll, rhythm and blues, and the great blues tradition are skillfully sketched in Charles Shaar Murray, *Crosstown Traffic*, St. Martin's Press, 1989.

The inscription, "This machine kills fascists," is reported in Joe Klein, *Woody Guthrie: A Life*, Knopf, 1973.

14   The Dialectics of Dude, or the Mechanical Rodeo

The clothing of average people is one of the great neglected areas of cultural history. For the United States, two contributions are Claudia Kidwell and Margaret C. Christman, *Suiting Everyone: The Democratization of Clothing in America*, Smithsonian Institution, 1974, and Sandra Ley, *Story of Ready-to-Wear, 1870's–1970's*, Scribner's, 1975.

Valerie Carnes, "Icons of Popular Fashion," in *Popular Icons* ed. Ray B. Brown and Marshall Fishwick. Bowling Green State University Press, 1978.

Ed Cray, *Levi's*, Houghton Mifflin, 1978, and various company promotional materials, and Sweet-Orr Company brochure, *Diamond Jubilee, 1871–1946*. See also Lawrence Breese "Lon" Smith, *Dude Ranches and Ponies*, Coward-McCann, 1936, and the 1930s periodical *Dude Ranches*.

Also see David Dary, *Cowboy Culture: A Saga of Five Centuries*, Knopf, 1981.

On Levi's as design objects: Joseph Masheck, "Embalmed Objects: Design at the Modern," *Artforum* (February 1975).

John Brooks's status chart for jeans is included in *Showing Off in America*, Little, Brown, 1981.

For stone-washed jeans: "The Prince of Pumice," *Sportswear International* (Fall 1989).

For ranchers with pressed jeans: Paulette Thomas, "Searching for the Past in Texas, Dick Reavis Is Taking Every Road," *Wall Street Journal*, Oct. 22, 1980.

15   Private Eyes and Public Faces, or Look Sharp!

Lincoln's letter about his beard is included in *The Living Lincoln*, ed. Paul M. Angle and Earl Schenck Miers, Rutgers University Press, 1955.

On the Pinkerton Detective Agency's use of photographs: Frank Morn, *"The Eye That Never Sleeps": A History of the Pinkerton National Detective Agency*, University of Indiana Press, 1982.

On George Eastman and Kodak: Carl W. Ackermann, *George Eastman*,

Houghton Mifflin, 1930, and Richard Coniff, "George Eastman Said 'Kodak,' " *Smithsonian* (June 1988).

On Kodak and Polaroid sales: "Now for Kodak," *The Economist*, July 30, 1988. See also Peter C. Wensberg, *Land's Polaroid*, Houghton Mifflin, 1987. Walker Evans describes his reaction to working with the SX-70 in *Walker Evans at Work*, Harper & Row, 1983.

On colored cameras: Roland Marchand, *Advertising the American Dream: Making Way for Modernity, 1920–1940*, University of California Press, 1985.

The official Gillette history is by Joseph Spang, *Look Sharp! Feel Sharp! Be Sharp! Gillette Safety Razor Company: Fifty Years, 1901–1951*, Newcomen Society, 1951. See also Russell B. Adams, Jr., *King C. Gillette: The Man and His Wonderful Shaving Device*, Little, Brown, 1978.

For King Gillette's political and social theories, see his *World Corporation*, Northeastern News, 1916, and *The People's Corporation*, Boni, 1924.

On punch card development: *Warm and Wonderful: The Jacquard Coverlet*, Hirschl & Adler Folk Art Gallery, 1988.

On Xerox: John H. Dessauer's *My Years with Xerox: The Billions Nobody Wanted*, Doubleday, 1970, and on Xerox's Palo Alto Research Center (PARC), Douglas Smith, *Fumbling the Future*, Morrow, 1988.

On Herman Hollerith: Geoffrey Austrian, *Herman Hollerith*, Columbia University Press, 1982.

On IBM: Robert Sobel, *IBM Colossus in Transition*, Times Books, 1981, and William Rogers, *Think: A Biography of the Watsons and IBM*, Stein & Day, 1969.

16   For Whom the Bell Tolled

On the history of the telephone: Herbert Casson, *History of Telephone*, Books for Libraries Press, reprint, 1971; Ithiel de Sola Pool, ed., *The Social Impact of the Telephone*, MIT Press, 1977; *A History of Engineering and Science in the Bell System: The Early Years (1875–1925)*, Bell Telephone Labs, 1975; James White, "What Has Happened to Pay Telephones; Change, Of Course," *Wall Street Journal*, Sept. 8, 1983; and John Brooks, *Telephone: The First 100 Years*, Harper & Row, 1976.

Robert V. Bruce, *Bell: Alexander Graham Bell and the Conquest of Solitude*, Little, Brown, 1973.

On the formation of the telephone monopoly: Theodore Vail, *Views on Public Questions*, privately published, 1917.

On the telephone as instrument of power: Marchand, *Advertising the American Dream: Making Way for Modernity, 1920–1940*. Ron Rosenbaum's account of Captain Crunch and other phone phreaks is reprinted in *Rebirth of the Salesman: Tales of the Song and Dance Seventies*, Delta, 1979.

17   "Fantasy Amplifiers," or Living in the Virtual World

On the development of the personal computer: Joel Shurkin, *Engines of the Mind*, Washington Square Press, 1985. *Hackers: Heroes of the Computer Revolution*, Doubleday, 1984, is Steven Levy's wonderful history of the hacker ethos, before the word possessed overtones of breaking into systems and personal computing was commercialized.

See also Ted Nelson, *Computer Lib/Dream Machines*, reprint, Tempus, 1987, and Howard Rheingold, *Tools for Thought: The History and Future of Mind-expanding Technology*, Prentice-Hall, 1985, where Douglas Engelbart is discussed and quoted at length.

For biographies of Ted Hoff, Adam Osborne, and other computer innovators, see Robert Slater, *Portraits in Silicon*, MIT Press, 1987.

On the psychology of the computer "microworld": Sherry Turkle, *The Second Self: Computers and the Human Spirit*, Simon and Schuster, 1984.

On the design of the Grid Compass computer: Phil Patton, "The Magic Box," *Connoisseur* (December 1985).

On Xerox PARC: Smith, *Fumbling the Future*.

For the Apple story: Frank Rose, *West of Eden: The End of Innocence at Apple Computer*, Viking, 1989, and Michael Moritz, *The Little Kingdom*, 1984.

On the search for a universal iconology of signs and directions: Henry Dreyfuss, *Symbol Sourcebook*, McGraw-Hill, 1972. David Canfield Smith's claim in the journal *IEEE Spectrum* to have invented the computer meaning of "icon" is quoted in Donald Katz, "So Whose Side Are You On?" *Esquire* (November 1990).

On NeXT: Phil Patton, "Steve Jobs: Out for Revenge," *New York Times Magazine*, Aug. 6, 1989.

On artificial reality: Stewart Brand, *The Media Lab: Inventing the Future at MIT*, Viking/Penguin, 1988; Trish Hall, " 'Virtual Reality' Takes Its Place in the Real World," *New York Times*, July 8, 1990; Steven Levy, "Brave New World," *Rolling Stone*, June 14, 1990; and Doug Stewart, "Artificial Reality: Don't Stay Home Without It," *Smithsonian* (January 1991). Walker Evans's quotation comes from a wall label at a show he assembled of his signs at the Yale Art Gallery in December 1971, quoted in *Walker Evans at Work*, Harper & Row, 1982.

☆ **Acknowledgments** ☆

In the more than five years from its beginnings to its publication, this book has enjoyed the ministrations of five versions of word processing software and three editors. Unlike the software, John Herman, Bill Strachan, and Walter Bode have all conduced to clarity, consistency, and economy, while demonstrating high levels of backward compatibility. It was Walt who finally took the manuscript in hand and skillfully did the hard work of putting it in shape. My thanks also for Ann Adelman's copy editing, Karen Wyatt's photo research, and all of Anne Sikora's help. My agent, Melanie Jackson, was, as always, supportive and efficient.

This project proceeded along with journalistic ones whose paths it repeatedly paralleled and crossed. Anita Leclerc at *Esquire* has for years provided almost daily doses of intelligence, humor, and downright wisdom. I have been fortunate to enjoy the support, ideas, and perspective of fine editors, including Constance Bond, Katherine Bouton, Charles Bricker, Penelope Green, Philip Herrera, Joanna Krotz, Randall Rothenberg, Donna Sapolin, Ben Schiff, Peter Sikowitz, Margaret Simmons, Richard Story, Alex Ward, Michaela Williams, and Sam Young.

Many people have helped, often without knowing it, to supply raw materials of information and ideas. A few of them are Morison Cousins, Peter Drucker, Anne Edgar, Hartmut Esslinger, Chuck Harrington, Robert Hicks, Roy Hoffman, Steven Holt, Alan Kay, David Laskin, Michael Letwin, Ken Mandel, Bruce Porter, John Rheinfrank, Frank Rose, Dick Rutan, Charles Salzberg, Neil Selkirk, J. C. Suares, Alison Thomas, Charles Walden, Victoria Wilson, and William Wilson.

I am also grateful for all sorts of help from the staffs of the Smithsonian Institution, the Museum of Modern Art, the Port Authority of New York and New Jersey, the Detroit Public Library, the Abraham Lincoln House, the Hermitage, the Memphis Pink Palace Museum, the Country Music Hall of Fame and Museum, Mystic Seaport Museum, Peabody Museum, the New York Public Library, and the Montclair, New Jersey, Public Library. For dealing with particular inquiries, I must thank Joan Shedlovsky at Texaco, Arnold Ginsberg of Sweet-Orr, Michael Sullivan at Kodak, Louise Kenneally of Stonehill College Library, Stan Skarzynski of the Springfield Armory National Historic Site, and Art Volpe, TACOM.

To include my wife Joelle Delbourgo here runs the risk of trivializing her place in my life, since her patience with and suggestions for the project represent a tiny part of all the love and support she has given me.

Caroline and Andrew Patton, whose lifespans do not extend beyond this project, demonstrated heroic restraint during its course in sometimes even knocking before flinging open the door to their father's office where, in truth, they are always welcome.

☆ Index ☆

Page numbers in *italics* indicate illustrations.

☆ **About the Author** ☆

Phil Patton, one of the country's leading explicators of design and technology, writes the Living Quarters column for *Esquire*. He is the author of *Open Road: A Celebration of the American Highway*, named by the *New York Times* one of the notable books of the year, and of *Razzle Dazzle*, a book on television and football. He is also the co-author of *Voyager*, the story of the home-built airplane that flew around the world on a single tank of gas.

His essays, blending humor, social criticism, and technological expertise, have explored the designs of things as diverse as American muscle cars, easy chairs, computers, athletic shoes, slave-made furniture, postcards, motels, quilts, razors, and the M-1 tank. His work has appeared in *Art in America, Connoisseur, Metropolitan Home, The New Republic, New York,* the *New York Times Magazine, 7 Days, Smithsonian,* the *Village Voice,* and other publications.

Mr. Patton graduated from Harvard, where he was an editor of the *Crimson*. He has taught at the Columbia Graduate School of Journalism.